Environmental Footprints and Eco-design of Products and Processes

Series Editor

Subramanian Senthilkannan Muthu, Head of Sustainability - SgT Group and API, Hong Kong, Kowloon, Hong Kong

Indexed by Scopus

This series aims to broadly cover all the aspects related to environmental assessment of products, development of environmental and ecological indicators and eco-design of various products and processes. Below are the areas fall under the aims and scope of this series, but not limited to: Environmental Life Cycle Assessment; Social Life Cycle Assessment; Organizational and Product Carbon Footprints; Ecological, Energy and Water Footprints; Life cycle costing; Environmental and sustainable indicators; Environmental impact assessment methods and tools; Eco-design (sustainable design) aspects and tools; Biodegradation studies; Recycling; Solid waste management; Environmental and social audits; Green Purchasing and tools; Product environmental footprints; Environmental management standards and regulations; Eco-labels; Green Claims and green washing; Assessment of sustainability aspects.

More information about this series at https://link.springer.com/bookseries/13340

Elena G. Popkova · Bruno S. Sergi
Editors

Sustainable Agriculture

Circular to Reconstructive, Volume 1

 Springer

Editors
Elena G. Popkova 🅳
MGIMO University
Moscow, Russia

Bruno S. Sergi
University of Messina
Messina, Italy

ISSN 2345-7651 ISSN 2345-766X (electronic)
Environmental Footprints and Eco-design of Products and Processes
ISBN 978-981-16-8733-4 ISBN 978-981-16-8731-0 (eBook)
https://doi.org/10.1007/978-981-16-8731-0

This Springer imprint is published by the registered company Springer Nature Singapore Pte Ltd.
The registered company address is: 152 Beach Road, #21-01/04 Gateway East, Singapore 189721,
Singapore

Introduction: Sustainable Development of Agriculture as a Current Need and a Global Trend

Sustainable development of agriculture implies its compliance with several Sustainable Development Goals (and their corresponding criteria). First, agriculture must comply with SDG 2 in terms of food security. Agriculture can be considered sustainable only if it fully ensures food security—sufficiency of food (zero hunger), financial accessibility of food to the general population, and proper (consistent with national standards) quality and safety.

By this criterion, most of the world's agricultural economies cannot be considered sustainable. The countries of the Global South experience acute and chronic food shortages (i.e., they cannot eliminate hunger). In contrast, the countries of the Global North have a food surplus, which leads to the common practice of disposing (waste) of fresh, quality, and safe food. Therefore, the agricultural economies of the countries of the Global South are not fully sustainable as well.

Second, agriculture must comply with SDG 10 in terms of reducing inequality in the level and pace of development of the agricultural economics and food security of territories. In this case, territories refer to countries and regions within countries. Thus, some territories have favorable conditions for agriculture, which contributes to high productivity and efficiency of agricultural economies, the total satisfaction of domestic needs, and the possibility to increase exports of food. In other territories, the conditions for farming are unfavorable, which results in the chronic dependence of these territories on food imports. International and practical experience in sustainable agriculture needs to be studied and systematized to identify prospects for reducing and overcoming the inequalities of agricultural economies.

Third, agriculture must comply with SDG 9 in terms of high productivity and innovative or high-tech agriculture, SDG 7 in terms of using clean energy and energy conservation in agriculture, SDG 6 in terms of water conservation, and SDG 13 in terms of adaptation of agriculture to climate change. Technologies that ensure the sustainability of agriculture play a key role in achieving the SDGs listed. On the one hand, the Fourth Industrial Revolution has expanded the possibilities for improving agricultural technology and efficiency. On the other hand, it has created high technological barriers in international food markets. Areas with favorable farming conditions started to dictate standards of food quality, productivity, and availability. In the past, other countries could not always meet these standards. The situation radically changed with the spread of climate-smart agricultural technologies, which leveled the playing field and put the territories with less developed agricultural economies in an advantageous position.

In the current digitalized environment, agricultural economies compete not on the basis of natural factors (land as a factor of production) but on the basis of the novelty of technologies and the success of their application. As shown by the experience of recent years, climate-smart technologies can be implemented in different territories and are a promising vector of the development of the global agricultural economy since these technologies allow bringing the quality, productivity, and cost of food to a whole new level.

Fourth, agriculture must comply with SDG 12 and SDG 15 in terms of responsible environmental management, SDG 11 in terms of support for the acceleration of rural development, and SDG 8 in terms of crisis-free development of the agricultural economy and support for unlocking the human potential of the workforce in agriculture. This set of SDGs integrates agricultural business management for sustainable development, encompassing government and corporate management.

Based on the criterion of implementation of the listed SDGs, current agricultural economies also do not achieve sustainability. The existing agricultural practices mostly assume a consumerist attitude toward the environment and an almost unlimited depletion of renewable natural resources (e.g., soil). However, agriculture has a great potential for regenerative land use—environmental improvement—that is not being realized. The very idea of responsible use of natural resources, common in other sectors of the economy, is new to agriculture and challenging to adapt to.

Despite the strategic importance of the agricultural economy for food security, rural areas in most countries significantly lag behind urban areas in terms of the socio-economic situation (quality of life, income, rate of economic growth). In recent decades, the contribution of agriculture to economic growth has steadily declined under the influence of industrialization, post-industrialization, and neo-industrialization (transition to Industry 4.0). Human capital, which plays a crucial role in other sectors of the economy, is underestimated in agriculture; the effectiveness of human resource management (HRM practices) in the agricultural economy is relatively low.

Therefore, the problem of agricultural sustainability is urgent, and its solution has not yet been found. The first volume of this book is devoted to the formation and development of scientific and methodological foundations of agricultural sustainability to solve the indicated problem. It shows that sustainable agriculture is a pressing contemporary need against the backdrop of growing global hunger and other human problems, as well as a global trend—sustainability criteria are being increasingly met by agricultural economies under the influence of the spread and support of the SDGs.

The first volume of the book systematically examines the four identified criteria for agricultural sustainability, each of which is discussed in a corresponding section of the book. The first section defines the contribution of agriculture to sustainable development and food security. The second section focuses on international and practical experiences in sustainable agriculture. The third section reveals the technology of agricultural sustainability. The fourth (final) section explores the management of agricultural entrepreneurship for sustainable development.

This volume is multidisciplinary—it contains research at the intersection of different fields of knowledge. It is of interest to a wide range of sciences, in particular economics (environmental and agricultural economics), management (public and corporate governance), information and communication technology (as applied to the agricultural economy), and environmental sciences.

Contents

Contribution of Agriculture to Sustainable Development and Food Security

Scientific Foundations for the Formation of the Organizational Structure of the Grain Market 3
Roman R. Araslanov, Andrey F. Korolkov,
and Rafail R. Mukhametzyanov

Priorities and Efficiency of Government Support for the Agricultural Sector of Ukraine 13
Leonid D. Tulush, Oksana D. Radchenko, and Maryna I. Lanovaya

Development of Methods of Revision Control of Financial and Economic Activities of Agricultural Consumer Cooperatives 25
Oxana V. Boyko, Tatiana V. Ostapchuk, and Liubov V. Postnikova

Sustainable Agriculture for Food Security: Conceptual Framework and Benefits of Digitalization 35
Alexander A. Krutilin, Aliia M. Bazieva, Tatiana A. Dugina,
and Aydarbek T. Giyazov

Methodological Approach to the Polycriteria Assessment of Agricultural Sustainability: Digitalization, International Experience, Problems, and Challenges for Higher Education in Russia 43
Elena V. Patsyuk, Alexander A. Krutilin, Nadezhda K. Savelyeva,
and Karina A. Chernitsova

Environmental and Energy Efficiency as a Criterion for Sustainable Agriculture 55
Svetlana E. Karpushova, Igor V. Denisov,
A. L.-Muttar Mohammed Yousif Oudah, and Yuliya I. Dubova

Agriculture in Developing Countries: Cultural Differences, Vectors of Sustainable Development, Digitalization, and International Experience 65
Lyudmila M. Lisina, Aydarbek T. Giyazov,
A. L.-Muttar Mohammed Yousif Oudah, and Yuliya I. Dubova

Potential and Opportunities of Organic Agriculture in Russia 75
Kseniya A. Melekhova, Xenia G. Yankovskaya,
and Alevtina G. Demidova

Agflation as a Threat to Food Security: Analysis of Inflationary Factors 83
Oleg P. Chekmarev, Pavel M. Lukichev, and Alexander N. Manilov

Food Subsystem: Innovative Technologies and Prospects for Rural Development 93
Svetlana A. Chernyavskaya, Taisiya N. Sidorenko,
Nadezhda A. Ovcharenko, Elena E. Udovik, and Tatiana E. Glushchenko

Features of Performance and Development of the Vegetable Production Market in the Context of Transition to New Economic Conditions 101
Gulzat K. Kantoroeva, Elena V. Kletskova, and Galina G. Vukovich

International and Practical Experience of Sustainable Agriculture

Ensuring the Competitive Advantage of the Agricultural Sector of Kyrgyzstan (Case Study of Cultivation, Processing, and Marketing of Plum) 113
Saibidin T. Umarov, Kubanych K. Toktorov, and Keneshbek M. Maatov

Problems of Managing Large-Scale Breeding in Russian Cattle Breeding 123
Vladimir I. Chinarov, Sergey A. Shemetyuk, Olga V. Bautina,
Anton V. Chinarov, and Andrey A. Azhmyakov

Benefits of Circular Agriculture for the Environment: International Experience of Using Digitalization and Higher Education Development 139
Svetlana E. Karpushova, Aliia M. Bazieva, Natalia M. Fomenko,
and Elena S. Akopova

Ecologization of Cultivating Honey Plants in the Region 149
Viktoria V. Vorobyova, Daria V. Rozhkova, and Pavel T. Avkopashvili

Growth Reserves of Agricultural Production in the Ryazan Region 159
Olga A. Bodryagina, Svetlana G. Vezlomtseva, Olesya A. Zarubina,
Evgeny E. Maslennikov, and Galina Z. Cibulskya

Assessment of Export Prospects of Russian Agricultural Producers 167
Alexander A. Dubovitski, Elena A. Yakovleva, Olga Yu. Smyslova,
Gayane A. Kochyan, and Elena V. Zelenkina

The Current State of the Organic Market in Russia 181
Irina V. Chernyaeva, Larisa V. Shirshova, and Natalia V. Lashchinskaya

**Farm Size in Organic Agriculture: Analysis of European Countries
and Russia** 189
Natalia Yu. Nesterenko, Alexander V. Kolyshkin,
and Tamara V. Iakovleva

**On the Issue of Food Security of the EAEU Countries During
a Pandemic** 201
Olga B. Digilina, Andrey O. Zlobin, and Andrey A. Chekushov

**Self-Sufficiency in a Highly Productive Seed and Breeding Base
as a Factor in the Sustainability of the Food Security of the Russian
Federation in the Context of the Transformation of the World
Food System** 209
Vera A. Tikhomirova

**Biochemical Indicators of Grain Sorghum Varieties in the Rostov
Region** 217
Olesya A. Nekrasova, Elena V. Ionova, Nina S. Kravchenko,
Natalya G. Ignateva, and Vladimir V. Kovtunov

Technology for Sustainable Agriculture

**Study of the Labor Resources of Peasant (Farm) Households
by Production Type** 229
Anna V. Ukolova and Bayarma Sh. Dashieva

Rural Housing Development Potential 243
Andrey N. Baidakov, Olga S. Zvyagintseva, Olga N. Babkina,
Diana S. Kenina, and Dmitriy V. Zaporozhets

**Expanded Reproduction as the Basis for Agricultural
Sustainability: Marketing, Digital Economy, and Smart
Technologies** 255
Egor V. Dudukalov, Elena V. Patsyuk, Olga A. Pecherskaya,
and Yelena S. Petrenko

**Innovative Model of the Functioning of Consumer Cooperation
as an Incentive for Developing the Regional Food Market
and Increasing Population Welfare** 265
Natalia I. Morozova, Galina N. Dudukalova, Vladimir V. Dudukalov,
Tatiana V. Opeykina, and Larisa V. Obyedkova

**Application of Micro Preparations as an Element
of Agrobiotechnology for Soybean Cultivation in the Conditions
of the Central Federal District** 273
Kristina Yu. Zubareva, Pavel Vas. Yatchuk, Irina L. Tychinskaya,
and Evgeny Yu. Korolev

**Marker-Assisted Selection of Pea Interspecific Hybrids
with Introgressive Alleles of Convicilin** 283
Sergey V. Bobkov and Tatyana N. Selikhova

**Changes in the Agrochemical Indicators of Light Gray Forest
Soil and the Yield of Grain Crops During the Rotation of Crop
Rotation Under the Influence of Various Systems of Its Processing
in the Conditions of the South-East of the Volga-Vyatka Region** 295
Alexey V. Ivenin, Yulia A. Bogomolova, and Alexander P. Sakov

**The Influence of Agrotechnology on Grain Quality and Yield
of Winter Wheat of the Yuka Variety in the Conditions
of the Western Ciscaucasia** 309
Irina V. Shabanova, Nikolay N. Neshchadim, and Aleksandr P. Boyko

**Changes in the Fertility of Agrogenic Soil During Chemical
Reclamation** 319
Olga V. Gladysheva, Elena V. Gureeva, and Vera A. Svirina

Adaptive Capacity of Strawberries in Autumn 329
Zoya E. Ozherelieva, Pavel S. Prudnikov, Marina I. Zubkova,
and Sergey D. Knyazev

**Comprehensive Assessment of Promising Soybean Lines
of the Northern Ecotype for Cultivation in a Mixture with Corn** 337
Inna I. Nikiforova, Andrey A. Fadeev, and Inga Yu. Ivanova

**Management of Agricultural Enterprises for Sustainable
Development**

**System of Effective Financial Planning in the Sustainable
Development of Agro-industrial Organizations** 347
Liudmila I. Khoruzhy, Yuriy N. Katkov, Valeriy I. Khoruzhy,
and Ekaterina A. Katkova

Improving the Price Mechanism in Milk Production and Processing 359
Olga A. Stolyarova, Lyubov B. Vinnichek, and Yulia V. Reshetkina

**Peculiarities and Prospects of the Application of the Unified
Agricultural Tax** 371
Lidiya A. Ovsyanko, Natalia I. Pyzhikova, Georgy N. Kutsuri,
Kristina V. Chepeleva, and Tatiana A. Borodina

Biologization of Spring Wheat Cultivation with Application of Sugar Beet Waste to the Soil 381
Irina V. Gefke, Larisa M. Lysenko, and Olga V. Bychkova

Study of the Ratio of Heat and Electrical Energy Expended in Microwave-Convective Drying of Grain 391
Dmitry A. Budnikov

Clustering of Agribusiness and Its Role in Forming the New Architecture of Rural Areas (Case Study of the Republic of Bashkortostan) 403
Vilyur Ya. Akhmetov, Rinat F. Gataullin, Razit N. Galikeev, Salima Sh. Aslaeva, and Ramil M. Sadykov

Assessment of Efficiency and Production Risks in Crop Production Innovative Development 411
Tatiana N. Kostyuchenko, Darya O. Gracheva, Natalia N. Telnova, Alexander V. Tenishchev, and Marina B. Cheremnykh

Peculiarities of Organizational and Economic Interaction of Organizations in the System of Product Subcomplexes 419
Lidia A. Golovina and Olga V. Logacheva

Stock Analysis as an Element of Financial Flows Optimization in Enterprises of the Agricultural Sector of Economics 429
Anna A. Babich, Anna A. Ter-Grigor'yants, Elena S. Mezentseva, and Tatiana A. Kulagovskaya

Improving the Labor Remuneration System in Agriculture as the Basis for the Food Security of the Country 439
Aktalina B. Torogeldieva, Kseniya A. Melekhova, and Elena A. Yaitskaya

Innovative Calculating of Products in Industry Enterprises 449
Igor E. Mizikovsky, Elena P. Polikarpova, Victor P. Kuznetsov, Ekaterina P. Garina, and Elena V. Romanovskaya

A Critical Look at Circular Agriculture from a Perspective of Sustainable Development (Conclusion) 459

About the Editors

Elena G. Popkova Doctor of Science (Economics), is Founder and President of the Institute of Scientific Communications (Russia) and Leading Researcher of the Center for Applied Research of the chair "Economic policy and public-private partnership" of MGIMO University (Moscow, Russia). Her scientific interests include the theory of economic growth, sustainable development, globalization, humanization of economic growth, emerging markets, social entrepreneurship, and the digital economy, and Industry 4.0. Elena G. Popkova organizes all-Russian and international scientific and practical conferences and is Editor and Author of collective monographs, and she serves as Guest Editor of international scientific journals. She has published more than 300 works in Russian and foreign peer-reviewed scientific journals and books.

Bruno S. Sergi, Ph.D. is Professor of International Economics, University of Messina, and Associate, Davis Center for Russian and Eurasian Studies, Harvard University. Bruno S. Sergi teaches at the Harvard Extension School on the economics of emerging markets and the political economy of Russia and China. Sergi is Associate of Harvard University's Davis Center for Russian and Eurasian Studies and the Harvard Ukrainian Research Institute. He also teaches political economy and international finance at the University of Messina, Italy. He is Series Editor of Cambridge's Elements in the Economics of Emerging Markets (Cambridge University Press), as well as Editor for Entrepreneurship and Global Economic Growth and Co-Series Editor of Lab for Entrepreneurship and Development (Emerald Publishing). He is Founder and Editor-in-Chief of the International Journal of Trade and Global Markets, the International Journal of Economic Policy in Emerging Economies, and the International Journal of Monetary Economics and Finance. He is Associate Editor of The American Economist. He has published several articles in scholarly journals and many books as Author, Co-Author, Editor, or Co-Editor. Sergi's academic career and advisory roles have established him as a frequent guest and a commentator on matters of contemporary developments in political economies and emerging markets in a wide range of media. Sergi holds a Ph.D. in Economics from the University of Greenwich Business School, London.

Contribution of Agriculture to Sustainable Development and Food Security

Scientific Foundations for the Formation of the Organizational Structure of the Grain Market

Roman R. Araslanov⬥, Andrey F. Korolkov⬥, and Rafail R. Mukhametzyanov⬥

Abstract The paper presents the organizational structure of the grain market as a system of markets of different types with the function of the information environment, strictly allocated to each type of market structure. The fundamental basis is the information environment of the first level, which provides economic agents with objective market prices, production volumes, and trades. Based on these data, the government can regulate the market by administrative or economic methods. The information environment of the second level is a conductor of innovation and analytical information about the market. The information environment of the third level is protective and provides insurance of financial risks of market participants from sharp price fluctuations. Each information environment is marked with certain structural elements and dependencies on each other. To create a model of the organizational structure of the grain market, the authors study the essence of the market exchange, the main stages of its development, and classification of market structures, based on the works of classical, Marxist, institutional, and neo-institutional economic schools. After that, the authors apply the selected classification to the organizational structure of the grain market and introduce the concept of information environment as an aggregating indicator to schematically depict the relationship between different types of market structures through this attribute.

Keywords Market · Grain · Exchange · Organizational structure · Information environment

JEL codes B41 · B49

R. R. Araslanov (✉) · A. F. Korolkov · R. R. Mukhametzyanov
Russian State Agrarian University – Moscow Timiryazev Agricultural Academy, Moscow, Russia
e-mail: araslanovroman1991@mail.ru

A. F. Korolkov
e-mail: akorolkov@rgau-msha.ru

R. R. Mukhametzyanov
e-mail: mrafailr@yandex.ru

© The Author(s), under exclusive license to Springer Nature Singapore Pte Ltd. 2022
E. G. Popkova and B. S. Sergi (eds.), *Sustainable Agriculture*,
Environmental Footprints and Eco-design of Products and Processes,
https://doi.org/10.1007/978-981-16-8731-0_1

1 Introduction

There are many discussions in the scientific community on the issues of understanding the essence of market relations and the definition of the concept of market. There are plenty of scientific theories to explain the essence of the market. In this research, we look at different scientific approaches to find something common in them and try to put these scientific approaches together like parts of a puzzle to identify the essence of the organizational structure of the grain market and the way it is formed.

2 Materials and Methods

The research object is the grain market. The research subject is the organizational structure of the grain market. The research objective is to create a model of the organizational structure of the grain market. The research tasks are as follows:

- Describe the essence of the market exchange process;
- Study the main stages of development of market exchange;
- Study the existing classifications of markets on various grounds;
- Choose one classification and apply it to the model.

Theoretical results of the research are obtained by analyzing the works of different scholars, representatives of classical, Marxist, institutional, and neo-institutional scientific schools.

The research uses several research methods: monographic, retrospective, expert, and system analysis, generalization, modeling, etc. The empirical basis for the research was materials published in international and Russian scientific literature and periodicals.

3 Results

The fundamental basis for studying the market structures was laid in the classical works of Smith [19, pp. 79–82]. According to them, the limits of the possibility of exchange (the market size) set the degree of labor division. According to Marx's theory of expanded reproduction [11, pp. 91–100], the market covers all stages of the reproductive cycle, divided into four phases: production, distribution, exchange, and consumption. However, the works of A. Smith do not show the factors affecting the limits of the possibility of exchange. In our opinion, the limits of the possibility of exchange are wider the more complex the rules of exchange, which allow us to structure the exchange within a single market transaction [5] by concluding a contract [7]. Contracts can be classified into three types: classical, neoclassical, and attitudinal [9]. During the development of contract law, there was a gradual complication of exchange rules, new types of contracts appeared. The exchange itself became more impersonal due to the gradual transition from bilateral to multilateral reputation mechanism, which led to expanding the limits of exchange and the formation of new types of market structures [20, pp. 15–20]. The classification of stages in the development of the market exchange is given in the works of D. North. According to it, there are three main stages: the stage of personified (personal) exchange, the stage of impersonal exchange without the protection of contracts by a third party, and the stage of impersonal exchange with the protection of contracts by a third party [13, pp. 54–60]. There is no substitution in the transition from one type of exchange to another; we can observe their mutual coexistence in different parts of the world.

Different types of market structures were formed at each stage of the development of the market exchange. There are many classifications of markets on various grounds. Depending on the exchange subject, markets are divided into markets for goods (finished products and resources) and markets for services. Depending on the number of sellers, buyers, and barriers to entry and exit, markets are divided into monopolies, oligopolies, and perfect and monopolistic competition markets. Within the framework of the research topic, we applied the classification of markets according to the variability of exchange rules and the nature of sales described in the works of A. A. Auzan and F. Braudel and expanded and supplemented by us [2, pp. 133–136; 4, pp. 150–320]. There are the following types of retail markets: open public market, covered public market, craft shop, and department store. The wholesale markets are fair, exchange, and wholesale food markets.

The open public market is the simplest form of market exchange. It is marked with simplicity and visibility of the exchange, the ease of breaking relations, and the absence of credit and other complex forms of service trade turnover. With the development of formal trading rules established by the town community and supervised by market inspectors, the town authorities began to build specialized indoor markets.

The intermittent nature of trading in public markets created certain difficulties for both buyers and sellers. The need for daily trade led to the emergence of craft shops, their further development in the form of workshops of the artisan population, and then the emergence of shopkeepers not engaged in production but trading in finished products, according to the formula of Marx [10, pp. 150–200]: money–commodity–money. In this form of trade, the quality of exchange is guaranteed by the seller, the government, which levies taxes on the seller and carries out periodic inspections, and by a third party in the form of professional guilds and associations. With the development of the network form of trade organization, there appeared networks of craft shops involved in the circulation of certain types of goods and services (medicinal herbs, clothing, sports equipment, cab services, and food delivery). We also include such popular platforms as Instagram, TikTok, and YouTube to craft shop networks, since it is the place for sharing information and content produced daily by millions of users (craft shops). We can demonstrate the effect of A. Smith's formula on the example of Instagram. The algorithms of this platform (the basis–photos with posts) made the rules for sharing information more complex, allowing users who were among the first to take advantage of this platform to create a personal brand in the shortest time possible. Other users followed in their footsteps, greatly expanding

the limits of exchange and contributing to the development of labor specialization–
the emergence of a dozen new professions (story maker, account designer, Instagram manager, various kinds of technical specialists engaged in account support,
creating landing pages, chatbots, connecting payments, or setting up webinar rooms).
YouTube and TikTok work similarly, contributing to the development of the blogging
profession.

The evolution of retail markets can be considered another form of centralized
market exchange (e.g., the department store marked with a fixed price and a wide
range of products offered). The development of the transportation system of cities
has greatly expanded the range of potential customers of the stores. In turn, the
development of advertising and the creation of national brands allowed to attract
customers, which gave impetus to the widespread distribution of department stores.
The guarantors of the quality of exchange in this type of market are the management
of the department store and the government since the contract between the buyer, and
the seller is made in the form of a receipt. With the development of the network form of
trade organization, there occurred the development of networks of department stores
and their electronic hybrids–universal platforms (marketplaces) and aggregators of
many craft shops (Amazon, Wildberries, and Ozon).

The development of wholesale markets began with the emergence of fairs, organized and supported by the city authorities or by private individuals with the support
of the authorities. Fairs can significantly reduce the cost of finding counterparties due
to various sellers of various goods. In this case, the quality of exchange and compliance with the trade rules is guaranteed by the organizers of the fairs through the
fair court. With the development of financial instruments (bills of exchange), there
started the formation of national and global networks of fairs and specialized bill fairs
[6, pp. 57–61]. With the development of commodity exchanges and the emergence
of wholesale food markets, the functions of fairs were reoriented from the organization of large commodity turnover to the distribution of advanced (innovative)
goods and services. Modern fairs are short term, specially organized events (usually
in exhibition centers) attended by potential buyers and companies who present their
products.

The seasonality of fairs could not meet the daily need for exchange. Thus, the first
commodity exchanges were established. The emergence of joint-stock ownership
in the seventeenth century contributed to the development of the stock market and
foreign exchange markets [12, pp. 100–153]. Nevertheless, exchanges continued to
be a tool for organizing large commodity turnover for a long time until the placement of futures and options contracts into circulation, after which the exchange
function was refocused on financial risk insurance (transactions on the spot market
are rarely concluded, and only 2–3% of futures transactions are brought to execution)
[8, pp. 300–392].

The function of the development of large commodity turnover was transferred to
wholesale food markets (Rungis, Mercasa, wholesale markets in the USA, Poland,
and the Netherlands). Wholesale food markets (WFMs) are state-managed and regulated commodity distribution mechanisms with a system of transparent and fair
pricing. The mechanism of WFMs covers all distribution channels of agricultural

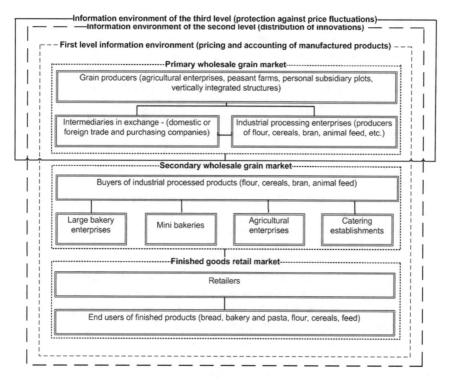

Fig. 1 Organizational structure of the grain market. *Source* Compiled by the authors

products, allowing the government to conduct an effective support policy with respect to all participants in this chain [3].

Based on the considered classification, the grain market is a set of different types of market structures, which differ in functionality (Fig. 1).

According to the studied classification of market structures, the organizational structure of the grain market consists of two interrelated types of markets (Fig. 1): over-the-counter (wholesale food in conjunction with the retail market of finished products and wholesale–retail innovation market consisting of trade fairs and craft benches of scientific institutions) and stock markets.

Each of them forms its own information environment (IE) different in functionality. The IE of the first level is formed based on WFMs and retail markets of finished products. It has a system-forming nature and provides players with objective market prices and volumes and quality of manufactured and traded products, which serve as the basis for effective regulation on the part of the government. The IE of the second level facilitates the spread of innovation. The IE of the third level is built based on the IE of the first level and aims to protect it from the effects of external and internal factors (price fluctuations). The concept of information environment was introduced as an aggregating indicator to schematically depict the relationship between different types of market structures through this attribute.

Information environment of the first level. Relying on the methodology of A. N. Osipov and E. V. Strelkov, who studied the grain market, we believe that the wholesale food market consists of a system of primary and secondary wholesale markets of grain and grain products and is closely connected with the retail market of finished products [1, pp. 60–88], which together form the information environment of the first level. Its quality depends on the speed of reading and reliability (in relation to all market participants) of the data obtained in their subsequent processing.

The reliability of the data depends on the level of provision of equal competition conditions by the government on all three types of markets. These conditions are provided by developing institutional elements limiting and stimulating the behavior of market participants, which cover all stages of the commodity supply chain (institution of licensing of organizations in the sphere of grain processing and circulation, quantitative and qualitative grain accounting, taxation, antimonopoly legislation, specification of property rights, and government support). The level of data reliability depends on the development of Supply Chain Management and E-commerce systems (e-commerce under the rules of B2B, B2C, and B2G), electronic accounting, and document management since they allow to achieve transaction transparency, reduce the market transaction costs of producers and buyers, and expand the limits of exchange [17, pp. 308–384]. The speed of reading and processing data is determined by the development of telecommunications and IT staff capacity in agriculture. While ensuring a high degree of reliability and speed of reading and processing, the data obtained is based on information about market objective prices and production and trade volumes, which leads to improved efficiency of public administration of the entire organizational structure in the application of administrative and economic methods.

Information environment of the second level. In current conditions, a high level of competitiveness of economic agents is achieved by introducing innovations in production. According to the classical definition of J. Schumpeter [18, pp. 120–203], innovation is the result of a change in the form of the production function, which consists in the creation of an innovative product during the innovation process. According to the classification of M. Porter and G. Bond, innovation is divided into upstream and downstream. Upstream innovation is associated with scientific research, which creates technological opportunities for the subsequent commercialization of scientific achievements in the form of new production functions. The exchange of upstream innovation can occur through the specially created offices of scientific institutions (i.e., a retail market–a craftsman's shop). During the exchange, upstream innovation is transformed into downstream innovation due to its commercialization. Further, it is brought to the mass consumer through a special tool (fairs and exhibitions) directly or through intermediaries. Craft shops of scientific institutions, wholesale markets for innovative products, fairs, and exhibitions, form the information environment of the second level. Nevertheless, innovation is not the only main structural component of this environment. One of the most important elements is analytical information, which allows one to correctly assess the economic situation and select the innovation in a constantly changing environment. This can be information about the current state of the market (external and internal) or the forecast of its development for a certain period. Such information can be aggregated by industry, inter-industry unions, analytical organizations, or information and advisory services [14]. An example of the successful formation of this environment is the American experience after the Bayh-Dole Act of December 1980, which abolished the state monopoly right to federal inventions. The strength of the established links between producers and consumers of downstream innovations also depends on the development of the country's exhibition industry: its infrastructure, material and technical base, availability of qualified personnel, and system of promotion and information through digital platforms. In our opinion, information and advisory services are an intermediary that brings downstream innovation from the craft shop of a scientific institution to the final consumer–the farmer in the rural area. The chain itself does not change, and only an additional element is built into it. Nevertheless, this tool has not yet received its proper development in the country [15, 16].

Information environment of the third level. The aggregate of commodity exchanges, their regional branches, and representative offices form the information environment of the third level. Its quality depends on the reliability of the data about objective market prices obtained from the information environment of the first level. The reliability of the data is also influenced by the development of the system of double warehouse certificates (depends on the level of digitalization of this process, since, in this case, the risks of document forgery are minimized, which allows for transparency of relations between the exchange and the elevators). The speed of data exchange and processing within and between environments depends on the development of telecommunications. The primary function of this environment is protective; it provides insurance of financial risks of economic agents from price fluctuations under the influence of internal and external factors. In this case, its construction is possible only based on the information environment of the primary WFMs, since grain is a classic exchange commodity. As for the peculiarities of pricing on the exchange commodity markets, in our opinion, the price of the underlying asset (regardless of the type of exchange transaction) has a real link to the dynamics of pricing on the WFMs of this asset. In this case, the data for the formation of an exchange quotation is collected through the information environment of the first level and directly through the registration of over-the-counter transactions in exchange commodities (oil, grain) through the exchange with the subsequent transfer of registers to the state body regulating the production of this commodity. In this way, the most objective information is obtained.

4 Discussion

The study examined and described the essence of the market exchange. When combined with the classical and neo-institutional theories, the market is a particular type of exchange. The exchange rules set limits on the possibility of such exchange and affect the degree of labor division. We can try to describe the market in terms of Marxist theory. Since the market covers all stages of the reproduction cycle, the exchange rules will depend on the choice of marketing channel (exchange or over-the-counter). Accordingly, there will be different limits to the possibility of exchange and the level of specialization of labor for each channel.

The authors studied the stages of development of the market exchange process: the gradual transition from a personal type of exchange to impersonal, the decreasing role of the bilateral reputation mechanism, and the increasing role of the multilateral reputation mechanism. The authors studied the classifications of markets on various grounds and chose a classification according to the variability of exchange rules and the nature of sales. During the research, the authors conducted a retrospective analysis for each type of market structure and identified their main functions at the current stage of development. In accordance with this, the authors compiled the organizational structure of the grain market as a set of markets of different types. The sign of the information environment was introduced for the convenience of displaying the relationships between market structures on the diagram. The authors provide recommendations for developing each information environment. Thus, the development of the information environment of the first level lies in the improvement of the institutions of licensing, quantitative and qualitative accounting of grain, taxation, government support, antimonopoly legislation, and the specification of property rights. It is crucial to promote the development of e-commerce systems, which can significantly reduce the market transaction costs of economic agents and increase the limits of exchange, contributing to the development of labor specialization and the whole market. The development of the information environment of the second level depends on the development of craft shops of scientific institutions and the density of their connections with consumers of downstream innovations, which also includes the development of marketing tools in the agricultural market (Instagram, TikTok, YouTube, etc.). The level of financing of domestic science (R&D costs) and promotion of information and advisory services, industry and inter-industry unions, and analytical centers is also important. Effective development of the information environment of the third level is possible only after the development of the first two levels since the exchange quotation depends on the data of objective market prices of the WFMs and is sensitive to the negative consequences of cardinal measures of state regulation (e.g., the announcement of an embargo). In turn, the embargo announcement is a consequence of the ineffectiveness of other methods of market regulation (e.g., interventions). The intervention price is based on two main parameters: the data of the objective market price and the volume of grain production. It is possible to obtain these data by developing institutions of licensing and quantitative–qualitative accounting, which can effectively apply the intervention mechanism differentiated by regions. However, the most critical factor in the effectiveness of their application is the volume of commodity or purchase interventions. The higher it is, the higher the level of market regulation. Therefore, it makes sense to start with creating the information environment of the third level when the government has sufficient financial capacity for regulation. Otherwise, the liquidity of transactions concluded on the exchange grain market will be extremely low.

5 Conclusion

The obtained results contribute to the development of methodological foundations for the formation of commodity food markets. They can serve as a basis for further research on the organizational structure of commodity markets and create the necessary tools for this analysis.

References

1. Altukhov AI, Magomedov A-ND, Osipov AN (2005) State regulation of the food market. Russia, Moscow
2. Auzan AA, Doroshenko MA, Ivanov VV (2011) Institutional economics: a new institutional economics theory. INFRA-M, Moscow, Russia
3. Avarsky ND, Prolygina NA (2015) Theory and methodology of the infrastructure of commodity circulation in the agri-food market. Agro-Ind Complex Econ Manag 10:73–79
4. Braudel F (1988) The wheels of commerce (Trans. from French; foreword Kubbel, L.E). Progress, Moscow, Russia. (Original work published 1979)
5. Commons JR (1931) Institutional economics. Am Econ Rev 21(4):648–657
6. Ferguson N (2008) The ascent of money. A financial history of the world. The Penguin Press, New York, NY
7. Greif A (1993) Contract enforceability and economic institutions in early trade: the Magribi traders' coalition. Am Econ Rev 83(3):525–548
8. Hull JC (2009) Options, futures, and other derivatives. University of Toronto, USA, NJ
9. Macneil IR (1985) Reflections on relational contract. J Inst Theor Econ 41:541–546
10. Marx K (1952) Capital. Criticism of political economy (I.I. Stepanova-Skvortsova Trans. from English), vol. 1. Gospolitizdat, Moscow, Russia
11. Marx K, Engels F (1955) Works, vol. 4. Gospolitizdat, Moscow, Russia
12. Moshensky SZ (2016) The emergence of financial capitalism. Pre-industrial securities market. Planet, Kyiv, Ukraine
13. North D (1990) Institutions, institutional change, and economic performance. Cambridge University Press, Cambridge, UK. https://doi.org/10.1017/CBO9780511808678
14. Osipov AN, Gnezdova YuV (2016) Public-private partnership as a model for the development of an innovative economy of the country. Agro-Ind Complex Econ Manag 1:26–33
15. Paptsov AG (2009) Features of information support of the agro-industrial complex abroad. Agro-Ind Complex Econ Manag 3:84–87
16. Paptsov AG, Guseva ES (2018) Investment abroad as a direction of agrarian policy in China. Agro-Ind Complex Econ Manag 12:106–112
17. Ross D (2002) Introduction to e-Supply chain management engaging technology to build market-winning business partnerships. St. Lucie Press, New York, NY
18. Schumpeter JA (1982) The theory of economic development: an inquiry into profits, capital, credit, interest, and the business cycle (Trans. from English; foreword Avtonomova V. S.). Progress, Moscow, Russia. (Original work published 1934)
19. Smith A (2007) An inquiry into the nature and causes of the wealth of nations (Trans. from English; foreword Afanasyeva V.S.). Eksmo, Moscow, Russia. (Original work published 1776)
20. Smith MB (2004) History of the global stock market. From Ancient Rome to Silicon Valley. University of Chicago Press, Chicago, IL

Priorities and Efficiency of Government Support for the Agricultural Sector of Ukraine

Leonid D. Tulush⬤, **Oksana D. Radchenko**⬤, **and Maryna I. Lanovaya**⬤

Abstract The paper summarizes approaches to the definition of priorities and effectiveness of government support of the agrarian sector of Ukraine. The authors use general scientific and specific methods of financial theory. The paper generalizes international studies of government support, which highlight scenarios typical for countries with transition economies. The paper presents macro indicators to assess the effectiveness of government support, particularly the share of the agricultural sector in GDP and budget costs (which, in Ukraine, averaged 10% and 1.2% over the past decade), as well as the level of investment, tax burden, the dynamics of production and productivity, pricing policy, and export quotas. The authors select methods for analyzing the effectiveness of government support through the indicators of social effectiveness of new government programs and definition of the share of the agricultural sector's costs in the economic sector's budget. The practical significance of this research consists in clarifying the priorities of government support in relation to global trends and developing a methodology for assessing the effectiveness of budget programs that will contribute to the formation of the strategy and concept of sustainable development of agriculture.

Keywords Support · Agrarian budget · Budget programs · Effectiveness · Government regulation · Ukraine

JEL codes G38 · Q14 · Q18 · H72

L. D. Tulush · O. D. Radchenko (✉)
National Science Center "Institute of Agrarian Economics", Kyiv, Ukraine
e-mail: oxanarad@ukr.net

L. D. Tulush
e-mail: tulush@ukr.net

M. I. Lanovaya
Academy of Urban Environment Management, Urban Planning and Printing, St. Petersburg, Russia
e-mail: lanovaya.m@mail.ru

1 Introduction

Government support for agriculture in Ukraine provides a system of measures aimed at reproducing the material and technical base, achieving the necessary level of profitability, and supporting social and environmental programs. Currently, government support undergoes significant changes and, in the context of European integration after joining the WTO, seeks to approve strategies adopted by the Common Agricultural Policy of the EU. However, this process raises many questions about the effectiveness of its mechanism and tools.

Traditionally, government support in Ukraine was formed, taking into account the significant share of the sector in the reproduction of GDP (25% at the beginning of agrarian reforms in 1990). However, government support currently observes a reduction (to 10% in 2020). In developed countries (e.g., Germany, Italy, and France), agricultural GDP is about 3%, while in newly industrialized and post-Soviet countries, it is between 6 and 10%. Accordingly, the funding for agriculture changes as well, although current expenditures are linked to the possibilities of the national budget and do not always coincide with the investment of the agricultural sector in the growth of the economy.

The gaps in scientific approaches are that they are limited by the scope of the problem posed and the choice of the country and the period. Simultaneously, it is necessary to detail the key priorities of the study on the effectiveness of government support in the agricultural sector, using the example of developing countries, to determine the path of development of national state support. Radchenko, Tulush, and Hryshchenko [19], as well as Tulush and Radchenko [22], generally identified the main advantages and disadvantages of the current mechanism of government support in Ukraine. The authors conclude that the mechanism is being under reform and has a positive momentum but poorly considers the practical needs, focusing exclusively on the budget capabilities. The problem of government support of the agricultural sector is very multifaceted and requires constant scientific monitoring and multicriteria evaluation of its effectiveness.

Summarizing the directions of research on government support, it seems possible to put forward the following hypotheses of this problem for Ukraine:

- Selection of priorities for the areas of government support;
- Selection of indicators to assess the effectiveness of government support;
- Analysis of the completeness of government programs and their compliance with the goals and strategy of sustainable development.

2 Materials and Methods

The research used the following methods:

- Financial theory, comparisons, and statistical analysis to disclose the dynamics of state support and its components;

- Method of correlation relationships to calculate the effectiveness;
- Method of hierarchies to disclose the mechanism of government support in the agricultural sector;
- Expert evaluations to adapt foreign experience.

The current state of international scientific knowledge on the subject of our research suggests different points of view on the problems of government support. Munk [16] pointed out that policymakers in all developed countries (including the EU) consistently chose to support farmers through the market, slowing the adaptation of the agricultural sector to lower prices that would have occurred otherwise. Gómez-Limón and Atance [5] indicated that the Common Agricultural Policy (CAP) was widely debated from a budgetary standpoint and optimally accommodated public goals. Lundell et al. [13] analyzed the impact of agricultural subsidy reform through government actions and pointed to the link between policy and the amount of support. Valdés [23] found that fluctuations in the income of the agricultural business in transition countries are largely determined by government policy through price regulation and are less dependent on market regulation. Analyzing government spending on agriculture and its contribution to GDP, Lawal [12] established the low efficiency of state regulation. Garmann [4] established the common paradigm that support for agriculture is provided to ensure the food security of countries and is, therefore, subject to the strict intervention of government.

Looking for a reason why industrialized countries resist cutting government support to the agricultural sector, Hee Park and Jensen [8] examined electoral competition and support for agriculture in OECD countries. They found that support in industrialized countries benefits only agricultural producers and the firms supplying agricultural products. Exploring agricultural support policies in Canada, the researchers Eagle, Rude, and Boxall [2] tried to understand why the government continues financial support of agriculture when it is profitable. Winters [26] uncovered the economic consequences of supporting agriculture and found that government intervention is only desirable in certain cases.

The policy of government support has country-specific features. Thus, in the example of Nigeria, Okolo [17] showed that the technical factors cited as the reason for the ineffectiveness of many government support programs are the cases of misuse of funds, insufficient investment, and willful political decisions in the technical issues of industry development. Adofu, Abula, and Agama [1] noted that many sub-Saharan African countries have recently pledged to increase government support for agriculture. In the example of China, Gale [3] showed the experience of developing government support through the transition from taxation to agribusiness financing. Gouin [6] provided the example of adjusting support to the agricultural sector after removing government subsidies in New Zealand through extensive measures to eliminate the deficit. Kuznetsov et al. [11] conducted a comprehensive study of the process of supporting the agricultural sector in the Russian economy. The study by Morkunas and Labukas [15] aims to identify and assess the negative effects of implementing the financial support mechanism for direct payments under the CAP on the sustainability of rural areas in Lithuania. Pardey, Kang, and Elliott [18] revealed the structure of government support for national agricultural research systems through the methods of political economy. They assessed changes in government support indexed by the intensity ratio of agricultural research. Kozlovskyi et al. [10] carried out modeling and forecasting of the level of government stimulation of agricultural production in Ukraine based on fuzzy logic theory. Ibok and Bassey [9] use the example of Wagner's Law in the agricultural sector (1961–2012) to establish a long-term relationship between government spending and industry revenues. Govereh et al. [7] showed that in 1991 the budget share of agriculture in Zambia was 26%. By 1999, it had fallen to 4.4%, which was reflected in a decline in the agricultural sector, deteriorating quality of research, falling levels of knowledge, and declining services of institutional government support.

3   Results

The modern policy of government support of the Ukrainian agricultural sector has features of protectionism, which is manifested in the measures of government regulation for the sake of competition, promoting the inflow of investment, saving material costs, and protecting jobs to maintain a stable environment for development. The legal framework of the mechanism is regulated by the Laws of Ukraine "On state support for agriculture in Ukraine" and "On the state budget of Ukraine." The expenditure of budgetary funds is determined annually by separate resolutions of the Cabinet of Ministers of Ukraine. The basics of state policy in the budget, credit, and price, regulatory, and other areas of public administration are defined in Article 1 of the Law on State Support as measures to stimulate agricultural production and development of the agricultural market and ensure food security of the population.

Table 1 shows the overall dynamics of expenditures on agriculture from the budget by functional classification of types of using budgetary funds to perform the basic functions of the government.

Over the studied period, there has been a 1.9-fold increase in the amount of funding for agriculture from the state budget and a 0.99-fold decrease in its share of the budget in 2020 compared to 2011. This increase is due to the replacement of the preferential VAT taxation of agricultural businesses (about 45 UAH bln) (existed until 2017) by direct monthly budget payments ("quasi-accumulation"), which depend on the amount of VAT paid into the budget.

The dynamics of budget expenditures on agriculture show that this industry is financed fairly high (22.39% in 2017) among other economic sectors. Nevertheless, before 2020, there was a sharp decline (7.98%) in budget expenditures on agriculture caused by the budget deficit and priority funding of socially protected expenditures. The growth potential of the absolute values of expenditures and their share in the structure of the national budget is quite significant. According to polynomial data distributions (R^2 value), the first indicator has a 77% probability, and the second has a 66% probability.

Table 1 Dynamics of expenditures of the national and consolidated budgets for the agricultural sector of Ukraine by functional classification in 2011–2020, UAH bln

Indicators	2011	2012	2013	2014	2015	2016	2017	2018	2019	2020
National budget expenditures on agriculture	6776	6541	6776	5135	4143	4075	10,532	13,054	13,020	13,496
Share of agriculture in the national budget, %	2.03	1.65	1.67	1.19	0.71	0.59	1.25	1.32	1.21	1.04
Share of agriculture in the expenditures of the national budget on the economy, %	15.13	13.24	16.40	14.92	11.15	12.97	22.39	20.52	18.00	7.98
Consolidated budget expenditures on agriculture	6930	6 660	614	5245	5461	5479	12,920	13,880	14,401	13,838
Share of agriculture in the consolidated budget, %	1.66	1.35	1.37	1.00	0.80	0.65	1.17	2.12	1.04	0.87
Share of agriculture in the expenditures of the consolidated budget on the economy, %	12.13	10.68	13.67	12.01	9.71	8.27	12.45	9.50	9.33	5.26

Source Compiled by the author based on [21]

Since the government budget is a system of monetary relations for the formation and use of the centralized monetary fund of the country, it is necessary to consider the share of spending on agriculture in the national and local budgets for several years by common indicators, primarily GDP redistribution by budgetary levers.

With the COVID-19 pandemic aftermaths, the main risks posed by the accompanying financial crisis to the agricultural industry include a deterioration in the price environment, reduced access to financial resources, and problems with logistics and sales, land market, investment climate, and a coherent government agricultural policy. According to the existing challenges, adequate food security and development measures are required.

The dynamics of macro indicators of development in 2015–2020 provide an assessment of the effectiveness of government support (Table 2), taking into account direct and indirect methods of government impact on the agricultural sector of Ukraine.

The analysis of the effectiveness of government support of the agricultural sector shows positive shifts toward improving support for lending, small-scale farming, and strengthening the material and technical resources by compensating the cost of machinery and equipment. Nevertheless, other envisaged programs such as support for the introduction of the land market, the creation of a mortgage bank, and the division of support for groups of agricultural and rural support have not yet been implemented. Thus, the sphere of government support of the agricultural sector experiences a need to increase the amount of funding and improve the institutional framework for its provision.

4 Discussion

The research confirms the effectiveness of government support and the importance of government regulation. It is necessary to introduce socially oriented programs contributing to a balanced development of the agricultural sector and rural areas, including the following:

- Support for strategically important production;
- Support for income and forms of farming;
- Creation of jobs;
- Strengthening the demography of rural settlements;
- Development of local self-government;
- Streamlining the engineering and road network, the educational, cultural, and medical components of rural development.

Currently, the sphere of government support of the agricultural sector of Ukraine has formed the following directions (Table 3): compensation of loans and the cost of purchasing domestic agricultural machinery, support for livestock and crop production, financing of farms, and agricultural insurance.

Table 2 Social and economic effect of government support of the agricultural sector by sectors and objects for 2015–2019

Indicators	2015	2016	2017	2018	2019	2019 to 2015, %, +
Number of business entities, units	79,284	74,620	76,593	76,328	75,450	95.16
Number of employed, thousand people	2870.6	2866.5	2860.7	2937.6	3010.4	104.87
Gross value–added (actual prices) scheme, UAH mln	239,806	279,701	303,949	361,173	358,072	149.32
Share of agriculture in total GVA, %	14.2	13.8	12.1	12.0	10.5	−3.70
Indices of agricultural products, %	95.2	106.3	97.8	108.2	101.4	+ 6.20
Gross value-added per employee (in prices of 2016), thousand UAH	624	765	755	868	928	148.72
Volume of products sold, UAH mln	362,309	403,645	454,380	525,096	556,325	153.55
Net profit, UAH mln	102,849	90,613	68,858	71,002	93,255	90.67
Profitability, %	29.5	24.7	16.0	13.7	16.1	−13.40
Capital agricultural investments, UAH mln	30,155	50,484	64,243	66,104	59,130	196.09
Share of agriculture in total investment, %	11.0	14.1	14.3	11.4	9.5	−1.50
Budget expenditures on agriculture, UAH bln	5461	5479	12,920	13,880	14,401	263.71
Share of agriculture in budget expenditures, %	0.80	0.65	1.17	2.12	1.04	+ 0.24

Source Compiled by the author based on [20, 21]

The variety of programs shows the lack of a comprehensive approach to the financing of agricultural producers, which indicates the lack of definition of common goals and the low effectiveness of government influence. Therefore, on April 7, 2021, the government adopted the decree for the year 2021, which expands government support of agricultural producers, supplementing it with new programs, for which the relevant ministry has calculated their socio-economic impact (Table 4).

In terms of adapting international experience, Ukraine has launched the "Acceleration of private investment in agriculture" program [25]; A $200 million loan agreement for the corresponding project was signed in 2019 with the International Bank for Reconstruction and Development. The preset agrarian components of the project include the following:

Table 3 Government programs of support for the agro-industrial complex for 2021

Sector	Objects of government support
Loans	Partial compensation of interest rates for up to 12 months for current expenses and 36 months for capital expenses; Cheaper loans (compensation of 1.5% of the discount rate: up to one year – up to 500 thousand UAH, up to three years – up to 9 thousand UAH)
Machinery and equipment	Compensation of the cost of purchased machinery: • 25% at the expense of "Financial Support of Agricultural Producers"; • 15% at the expense of "Financial support for the development of farms."
Farming	Payments to young farmers, payments on farms, and payment on farmland; Additional payments to socially insured persons of peasant farms
Livestock	Subsidies for maintaining cattle in the amount of 900 UAH per animal; Partial reimbursement of the cost of pedigree animals (50%); Compensation for the construction of livestock complexes (up to 25% of credit funds) for up to five years; Support for breeding young does, does, young ewes, and ewes (1000 UAH); Payments for the increase in the herd of cows of own reproduction (30 thousand UAH)
Crop production	80% compensation for purchasing domestic seeds and planting material; Compensation for the construction of the facilities for product storage; Subsidies (60,000 UAH) for the created farm (per 1 hectare and 1 person); Support for production on reclaimed land; Support for producers of organic products and potatoes
Insurance	Compensation for losses from damage to crops caused by emergencies

Source Compiled by the author based on [24]

- Harmonization of Ukrainian legislation with EU requirements;
- Improvement of the system of government support for agriculture;
- Encouraging the diversification of production in the agricultural sector;
- Development of rural areas;
- Strengthening food safety;
- Logistics solutions;
- Improving the quality of land use;
- Improving water management;
- Improving access to agricultural resources;
- Improving access to financial resources and risk management tools.

Table 4 Assessment of the social and economic effect of government support for farmers in 2021

Program	Volume of support, mln UAH	Workplace standards	Number of jobs
Partial reimbursement of the cost of agricultural equipment	1000.0	1	1000
Making loans cheaper	1200.0	8	9600
Financial support for the development of horticulture	450.0	20	9000
Subsidy per unit of cultivated land	60.0	20	1200
Additional payment of the unified social contribution to insured people in peasant (farm) enterprises	25.0	50	1250
Support for livestock facilities	1150.0	x	9010
Support for organic production	100.0	10	1000
Support for potato farming	60.0	9	540
Other programs	455.0	x	190
Total	4500.0	x	32,790

Source Compiled by the author based on [14]

The program's implementation will allow Ukraine to mitigate certain limitations on the growth of the participation of the private sector (particularly small and medium-sized enterprises) in the agricultural input market.

5 Conclusion

For agriculture, crises have become a kind of stimulus opening up new opportunities and acting as a catalyst for urgent changes in the industry. This is facilitated by the Ukrainian policy of supporting the agricultural sector to regulate competition, promote the inflow of investment, save material costs, and protect jobs. The assessment of the effectiveness of government financial support is characterized by the dynamics of macro indicators of the development of the agricultural sector in Ukraine in 2015–2020, highlighting the impact of direct and indirect methods of impact.

It is established that the government's actions influence the reforms of subsidization of the agricultural sector. The fluctuations in the industry's income are primarily determined by government policy at the expense of price regulation. The decline in the share of budget costs in the agricultural sector is reflected by the decline in the production of labor-intensive industries.

The comparison of the results with the purpose and objectives of the study shows that the hypotheses put forward to select the priorities of government support, the selection of indicators to assess the effectiveness and completeness of the programs, and their compliance with the goals and strategies of the industry have been implemented.

Prospects for further research in the framework of the stated problem of choosing priorities and effectiveness of government support for the agricultural sector of Ukraine consists in the need to develop an appropriate methodology for analyzing the effectiveness of all government support programs.

Acknowledgements The authors would like to express their gratitude to the Director of the National Research Center "Institute of Agrarian Economics" Yu. A. Lupenko, who assisted in the preparation of this research. The research was conducted under the budget research of the National Academy of Agrarian Sciences of Ukraine "Theoretical and methodological support of financial regulation of sustainable development of the agricultural sector and rural areas" for 2021–2025.

References

1. Adofu I, Abula M, Agama JE (2012) The effects of government budgetary allocation to agricultural output in Nigeria. Sky J Agric Res 1(1):1–5
2. Eagle AJ, Rude J, Boxall PC (2015) Agricultural support policy in Canada: What are the environmental consequences? Environ Rev 24(1):13–24. https://doi.org/10.1139/er-2015-0050
3. Gale HF (2013) Growth and evolution in China's agricultural support policies (Economic Research Report 153). USDA-ERS, Washington, DC
4. Garmann S (2014) Does globalization influence protectionism? Empirical evidence from agricultural support. Food Policy 49:281–293. https://doi.org/10.1016/j.foodpol.2014.09.004
5. Gómez-Limón JA, Atance I (2004) Identification of public objectives related to agricultural sector support. J Policy Model 26(8–9):1045–1071
6. Gouin DM (2006) Agricultural sector adjustment following removal of government subsidies in New Zealand. Agribusiness and Economics Research Unit, Lincoln University, Lincoln, New Zealand
7. Govereh J, Shawa JJ, Malawo E, Jayne TS (2006) Raising the Productivity of Public Investments in Zambia's Agricultural Sector (No. 1093–2016–87997). Lusaka, Zambia. https://doi.org/10.22004/ag.econ.54479
8. Hee Park J, Jensen N (2007) Electoral competition and agricultural support in OECD countries. Am J Polit Sci 51(2):314–329
9. Ibok OW, Bassey NE (2014) Wagner's law revisited: the case of Nigerian agricultural sector (1961–2012). Int J Food Agric Econ (IJFAEC) 2:19–32
10. Kozlovskyi S, Mazur H, Vdovenko N, Shepel T, Kozlovskyi V (2018) Modeling and forecasting the level of state stimulation of agricultural production in Ukraine based on the theory of fuzzy logic. Montenegrin J Econ 14(3):37–53. https://doi.org/10.14254/1800-5845/2018.14-3.3
11. Kuznetsov NI, Ukolova NV, Monakhov SV, Shikhanova JA (2016) Research of agricultural sector support process in Russia's economy. J. Adv Res L Econ 7:2086
12. Lawal WA (2011) An analysis of government spending on agricultural sector and its contribution to GDP in Nigeria. Int J Bus Soc Sci 2(20):244–250

13. Lundell M, Lampietti J, Pertev R, Pohlmeier L, Akder H, Ocek E, Jha S (2004) A review of the impact of the reform of agricultural sector subsidization. The World Bank, Washington, DC
14. Ministry of Agrarian Policy and Food of Ukraine (n.d.) Official website. Accessed from https://minagro.gov.ua/ua
15. Morkunas M, Labukas P (2020) The evaluation of negative factors of direct payments under common agricultural policy from a viewpoint of sustainability of rural regions of the new EU member states: Evidence from Lithuania. Agriculture 10(6):228. https://doi.org/10.3390/agriculture10060228
16. Munk KJ (1989) Price support to the EC agricultural sector: an optimal policy? Oxf Rev Econ Policy 5(2):76–89. https://doi.org/10.1093/oxrep/5.2.76
17. Okolo DDA (2004) Regional study on agricultural support, Nigeria's Case. Evolution 3:3
18. Pardey PG, Kang MS, Elliott H (1989) Structure of public support for national agricultural research systems: a political economy perspective. Agric Econ 3(4):261–278
19. Radchenko O, Tulush L, Hryshchenko O (2021) The choice of indicators for monitoring financial regulation of sustainable development of agricultural regions: the example of Ukraine. SHS Web Conf 106:01029. https://doi.org/10.1051/shsconf/202110601029
20. State Statistics Service of Ukraine (n.d.) Official website. Accessed from http://www.ukrstat.gov.ua
21. State Treasury Service of Ukraine (n.d.) Official website. Accessed from https://www.treasury.gov.ua/ua
22. Tulush L, Radchenko O (2020) Financial support for the agricultural sector of Ukraine in overcoming the COVID-19 pandemic. In Organizational-economic mechanism of the agro-industrial complex: state, problems and prospects, vol. 2. Far Eastern State Agrarian University, Blagoveshchensk, Russia, pp 67–79
23. Valdés A (ed) (2000) Agricultural support policies in transition economies (Vol. 23). The World Bank Publications, Washington, DC
24. Verkhovna Rada of Ukraine (2004) Law "On state support of agriculture of Ukraine." (June 4, 2004 No. 1877-IV, as amended June 24, 2021). Kyiv, Ukraine. Accessed from https://zakon.rada.gov.ua/laws/show/1877-15#Text
25. Verkhovna Rada of Ukraine (2019) Loan Agreement (Program "Acceleration of Private Investments in Agriculture") between Ukraine and the International Bank for Reconstruction and Development (August 27, 2019 No. 8973-UA). Accessed from https://zakon.rada.gov.ua/laws/show/996_005-19?lang=en#Text
26. Winters LA (1987) The economic consequences of agricultural support: a survey. OECD Econ Stud 9:7–54

Development of Methods of Revision Control of Financial and Economic Activities of Agricultural Consumer Cooperatives

Oxana V. Boyko⬤, Tatiana V. Ostapchuk⬤, and Liubov V. Postnikova⬤

Abstract It is necessary to transform the revision of the financial and economic activities of agricultural consumer cooperatives since the current set of revision methods does not meet the needs of economic development of the system of agricultural cooperation. The development of support measures for the system of agricultural consumer cooperation in Russia formed a vicious practice of creating "pseudo-cooperatives." Thus, one of the directions of developing the methods of revision control is the use of methods aimed at establishing the cooperative identity of an agricultural consumer cooperative. The paper examines the application of cooperative identification and authentication methods to determine cooperative identity.

Keywords Revision · Revision control · Agricultural consumer cooperative · Authentication method · Cooperative identity

JEL code Q130

1 Introduction

The essence of the agricultural consumer cooperative is expressed in the purpose of its creation, its compliance with cooperative principles, and its real assistance to the cooperative members. The noncommercial nature of these cooperatives imposes certain systemic requirements for the formation and implementation of revision control.

O. V. Boyko (✉) · T. V. Ostapchuk · L. V. Postnikova
Russian State Agrarian University – Moscow Timiryazev Agricultural Academy, Moscow, Russia
e-mail: boyko_oksana@mail.ru

T. V. Ostapchuk
e-mail: buzonik@mail.ru

L. V. Postnikova
e-mail: lpostnikova@rgau-msha.ru

© The Author(s), under exclusive license to Springer Nature Singapore Pte Ltd. 2022
E. G. Popkova and B. S. Sergi (eds.), *Sustainable Agriculture*,
Environmental Footprints and Eco-design of Products and Processes,
https://doi.org/10.1007/978-981-16-8731-0_3

The development of agricultural consumer cooperation in Russia is associated with some negative phenomena, one of which is the development of the so-called "pseudo-cooperation."

Cooperative identity is what distinguishes the cooperatives from other organizational forms in the agro-industrial complex (AIC).

A revision of cooperative identity allows us to find whether a cooperative is a real one or not. An affirmative answer to this question indicates the cooperative's integrity towards its members and the counterparties with whom it interacts.

The existing methods of revision control do not allow us to give a detailed answer to this question. Therefore, the authors propose to use identification and authentication methods to determine cooperative identity.

2 Materials and Methods

This research is conducted on the materials of the Revision unions that are members of the Russian self-regulatory organization of Revision unions of agricultural cooperatives "Agrocontrol" (RSO "Agrocontrol"). The works of Russian scientists in the field of cooperation and economic control also served as the basis for the conducted research. During the research of methods of revision control of agricultural consumer cooperatives, the authors used methods of analysis and synthesis, methods of comparison and generalization, graphic method, and method of expert evaluation.

3 Results

The financial support of the development of agricultural consumer cooperation has revealed the development threat, which requires the use of new methods of revision control.

Impressive financial support from the government has led to the fact that the creation of agricultural consumer cooperatives (ACC) has become aimed more at quantitative than qualitative indicators, which does not contribute to the goal of creating a sustainable system of agricultural consumer cooperation in Russia in the future.

The general incompetence of the ACC members, the employees of the Competence Center, and the employees of government agencies contributes to the appearance of "pseudo-cooperatives" aiming to receive grants.

The Revision union plays an important role during the creation, formation, and development of a new ACC. This union can become a coordinator and a kind of "filter" that does not allow the financial support of organizations mimicking agricultural consumer cooperatives.

This position is based on the provisions of Federal Law No. 193-FZ "On Agricultural Cooperation" [1], which entrusts the Revision union with the verification of the ACC compliance with the principles of establishment and operation of cooperatives in accordance with this law and its statutes.

In general, the following groups of methods can be used during revision control [3–6, 10–13]:

- Methods of analytical review;
- Methods of documentary revision;
- Methods of actual revision;
- Methods of economic analysis.

The methods of analytical review can be used to identify the main directions of control activities to reduce labor and time costs to identify weaknesses and problem areas of the Revision union, which should be given more attention.

Methods of documentary revision allow establishing the legality, reliability, and appropriateness of the financial and economic activities of the inspected cooperative based on primary accounting documents, accounting registers, accounting (financial) statements, and other types of existing documentation in ACC.

The methods of the actual revision are based on the actual inspection of the condition and availability of property and the determination of the correctness of its evaluation.

The use of methods of economic analysis for revision control allows evaluating the financial condition and the impact of various factors on the results of economic and financial activities of the Revision union.

Additionally, we can allocate various methods related to economic control (e.g., methods of financial control) [9].

However, the process of confirming the cooperative identity of the agricultural consumer cooperative is directly related to the definition of the cooperative as a specific form of a nonprofit organization based on membership, whose activities are based on the cooperative principles and are aimed at achieving the interests of its members [2, 7, 8].

The study of methods of revision control of agricultural consumer cooperatives revealed their insufficient adaptability for determining the cooperative identity.

The Revision union can confirm the cooperative identity of an ACC in two stages (Fig. 1).

In the first stage of the preliminary control, the confirmation of the cooperative identity occurs by applying the method of identification. This method should be implemented during the admission to membership in the Revision union of a particular cooperative.

The method of identification allows one to recognize an object by comparing the available parameters with predetermined ones. Within the framework of the revision control system, the following definition of the identification method can be given.

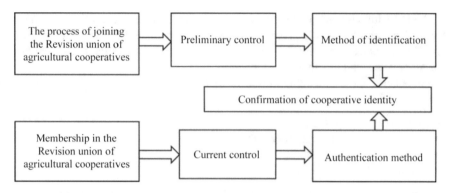

Fig. 1 Stages of confirming cooperative identity. *Source* Compiled by O. V. Boyko

The method of identifying an agricultural consumer cooperative is the process of recognizing a declared cooperative as a nonprofit organization, whose activities are based on membership, meet cooperative principles, and aim to achieve the goals for which it was created.

At the second stage of the current control, the confirmation of the cooperative identity is carried out during the revision control using the method of authentication.

It is possible to use the method of authentication during the current revision control of financial and economic activity (during a regular or extraordinary revision) to confirm the cooperative identity of the inspected ACC. The application of this method is based on the use of a mechanism for comparing the benchmark indicators of economic activity of the true agricultural consumer cooperative with the indicators of the inspected cooperative. Based on this comparison, the Revision union can either confirm or deny the cooperative identity of the cooperative assessed.

For revision, the following definition of the authentication method can be given.

The method of authentication of an ACC is the process of recognizing a Revision union, which has features of cooperative identity, namely the features of a nonprofit organization whose activities are based on membership, meet the cooperative principles, and aim to achieve the goals for which it was created. This process aims to reconcile the parameters of the economic activity of a Revision union with the established indicators of the financial and economic activity of the organization.

Thus, at the stage of preliminary control by the Revision union, it is necessary to determine the extent to which the declared cooperative is an agricultural consumer cooperative. At the same time, the Revision union does not thoroughly analyze documents. That is, only superficial data, which can be checked without interfering in the activities of the cooperative, is analyzed.

The method of authentication involves a deeper, more detailed analysis of documents, which is possible only in the course of current revision control of the ACC carried out during its revision inspection.

Summarizing the above, the authors conclude that, when applying the methods of identification and authentication, any Revision union creates a kind of checklist for checking the cooperative identity, which should be based on the principles of agricultural consumer cooperatives.

4 Discussion

Based on the current legislation on agricultural cooperation in the Russian Federation, we can identify the following principles of agricultural consumer cooperatives:

- Management, based on democratic principles, by the general meeting of members of the cooperative, with the creation of an internal supervisory body–the supervisory board;
- Voluntary membership of agricultural producers with the activity similar to the activity of a cooperative to benefit from participating in the cooperative's economic activities;
- Rendering services of the cooperative predominantly by its members (not less than 50%) and limitation of participation of nonmembers in economic activities of the cooperative, including in the amount of dividends on share contributions of members and share contributions of associated members of the cooperative;
- Distribution of profits (this possibility is limited by the Civil Code of the Russian Federation) and losses among members in accordance with their participation in business activities.

The definition of the verifiable parameters of cooperative identity may be based on the given principles of activity. It is impossible to verify compliance with the last principle during the preliminary inspection due to the lack of the necessary documents. Therefore, the main groups of tested parameters of cooperative identity, established by the method of identification, will include three groups of questions to determine the following:

- Whether the type of the cooperative's activity complies with the law and the interests of its members;
- Whether the cooperative's members participate in the cooperative's business activities;
- Whether the cooperative complies with cooperative principles.

The answer to these questions can be obtained at the stage of "acquaintance" of the Revision union and the cooperative, when the latter becomes a member of the former.

Usually, the Revision union obliges the cooperative to submit an application to join, accompanied by a list of documents (e.g., Articles of association, OGRN Certificate (Primary State Registration Number), Federal Tax ID, Copy of minutes of the general organizational meeting, etc.). These documents can serve as a basis for checking the cooperative's identity.

Table 1 presents parameters and examples of documents that can serve as a basis for establishing the relevance of the indicators of cooperative identity studied by the method of identification.

When determining the cooperative identity in the process of current control, all principles of cooperatives can be tested in their entirety. In our opinion, the verifiable parameters of cooperative identity determined by the method of authentication can be represented as four groups of questions.

Table 1 Parameters of the verification of cooperative identity by the method of identification at the preliminary control stage

No	Tested parameter	Documents that can serve as a basis for verifying the cooperative identity	A cooperative has a cooperative identity if:
1	Economic activities of the cooperative	Minutes of the general organizational meeting; Extract from the minutes of the general meeting of members; Articles of association; Register of members and associate members	• General meetings of the members are held; • The direction of economic activity of the cooperative is reflected in its name and described in the provisions of the Articles of association; • Election of the governing bodies is made in accordance with the requirements of the law; • The supervisory board of the cooperative was elected; • The membership base of the cooperative includes existing agricultural producers
2	Composition of the membership base and its participation in the economic activities of the cooperative	Register of members and associate members; Extract from the minutes of the general meeting of members	• The members of the cooperative are agricultural producers with the same (similar) type of activity; • There is a formed register of members and associate members; • The number of members and associate members in the register on the date of entry into the Revision union and the minutes of the general meeting at which it was decided to join the Revision union has the same number

(continued)

Table 1 (continued)

No	Tested parameter	Documents that can serve as a basis for verifying the cooperative identity	A cooperative has a cooperative identity if:
3	Holistic adherence to cooperative principles	Minutes of the general organizational meeting; Articles of association; Extract from the minutes of the general meeting of members	• The cooperative provides services to its members; • The scope of services is proportional to the size of the share fee and is fixed in the Articles of association or in the Minutes of the general organizational meeting; • The Minutes of the general meeting reflect the participation of members of the cooperative without exception • Voting at the general meeting is performed on a democratic basis on principle "one member – one vote."

Source Compiled by O. V. Boyko

Table 2 provides a list of parameters for verifying cooperative identity by authentication method and examples of documents that can serve as a basis for establishing compliance with the studied parameters of cooperative identity.

The examples provided in the tables above serve as a basis for drawing up checklists for inspection of agricultural cooperatives by Revision unions. They can use it in their practical activities.

5 Conclusion

Thus, eliminating nonviable and mimicry organizations is essential for laying a solid foundation for the development of the entire system of agricultural consumer cooperation.

One of the main filters in the creation and further development of the created ACC is the Revision union, which helps coordinate the cooperative's activities or prevent "pseudo-cooperatives" from receiving public funds.

The method of identification at the preliminary control stage allows us to immediately determine that the cooperative is not really a cooperative and does not have sufficient signs of cooperative identity (i.e., it is a "pseudo-cooperative").

Table 2 Parameters of the verification of cooperative identity by authentication method at the stage of current control

No	Tested parameter	Documents that can serve as a basis for verifying the cooperative identity	A cooperative has a cooperative identity if:
1	Cooperative management	Minutes of the general meeting, meetings of the management board, and supervisory board; Register of members and associate members	• General meetings are held regularly, at least once a year, there is a quorum; • Meetings of the management board and the supervisory board are held regularly in accordance with the requirements of the law and the Articles of association; • Election of governing bodies in accordance with the law
2	Membership base and its participation in the cooperative' economic activities	Register of members and associate members; Applications for membership in the cooperative; Income and expense estimates; Report on the execution of estimates	• The members of the cooperative are agricultural producers with the same (similar) type of activity; • Membership cards are filled out completely • The register of members is formed and updated regularly; • The participation of nonmembers of the cooperative is restricted; • Dividend payments on share contributions of members and associate members of the cooperative are limited
3	Economic activity	Articles of association; Contracts; Accounting documents; Accounting (financial) statements; Tax returns	• The participation of the members of the cooperative in the economic activities is fixed in the Articles of association; • The share of members' participation in business activities is at least 50% of its total volume; • The direction of the activity is fixed in the name and described in the provisions of the Articles of association and bylaws (if necessary); • The cooperative provides the services in the approximately same amount to all members, which is expressed as an economic advantage for them

(continued)

Table 2 (continued)

No	Tested parameter	Documents that can serve as a basis for verifying the cooperative identity	A cooperative has a cooperative identity if:
4	Allocation of profits and losses	Minutes of the General Meeting; Estimate of Income and Expenses; Report on the execution of the income and expenditure estimate	• Regular general meetings with the distribution of profits or coverage of losses

Source Compiled by O. V. Boyko

The cooperative identity confirmed in the course of the current control allows one to verify the honesty of the cooperative intentions toward its members and counterparties with whom it will interact, including the government.

Thus, the notion of cooperative identity is one of the main areas of focus in examining an agricultural consumer cooperative by the Revision union before this cooperative receives state funds. The use of the authentication method allows the Revision union to assess, confirm, or deny the cooperative identity of the ACC in the course of a regular or unscheduled revision.

References

1. Russian Federation (1995) Federal Law "On agricultural cooperation" (December 8, 1995 No. 193-FZ). Moscow, Russia.
2. Antsyferov AN (2019) A Course in cooperation: a scientific publication. In: Sobolev AV (ed). Kantsler, Yaroslavl-Moscow, Russia
3. Belov NG (2006) Control and revision in agriculture: textbook, 4th edn. Finance and Statistics, Moscow, Russia
4. Boyko OV (2011) Methods of revision of agricultural consumer cooperatives. Eur Soc Sci J 13(16):430–435
5. Butynets FF (1976) The subject and objects of control in agricultural enterprises. Naukova Dumka, Kiev, USSR
6. Butynets FF (1979) Control and revision in agricultural enterprises. Higher School, Kiev, USSR
7. Chayanov AV (1925) A quick course in cooperation. Central Partnership "Cooperative Publishing House", Moscow, USSR
8. Emelyanov IV (2020) The economic theory of cooperation. The economic structure of cooperative organizations. Rosinformagrotech, Moscow, Russia
9. Kolesnikov SI (2011) About the main methods of financial control. Financ Credit 36(468):60–64
10. Kolesnikova EN (2011) Revision and control in agricultural production cooperatives: theory, methodology, and practice. Ryazan branch of the Moscow University of the Ministry of Internal Affairs of Russia, Ryazan, Russia

11. Kramarovsky LM (1988) Revision and control: Textbook for Higher Education Institutions in Specialty "Accounting and Analysis of Economic Activities". Finance and Statistics, Moscow, USSR
12. Melnik MV, Panteleev AS, Zvezdin AA (2009) Revision and control: textbook. KnoRus, Moscow, Russia
13. Ovsiychuk MF (2007) Control and revision: textbook, 5th edn. KnoRus, Moscow, Russia

Sustainable Agriculture for Food Security: Conceptual Framework and Benefits of Digitalization

Alexander A. Krutilin, Aliia M. Bazieva⑩, Tatiana A. Dugina, and Aydarbek T. Giyazov

Abstract The article is intended to answer the research question of whether today's agriculture and food systems can meet the needs of a global population that is projected to reach more than 9 billion. Can we increase the required production, even as the pressures on already scarce land and water resources and the negative impacts of climate change intensify? The study shows that a transformed agribusiness should evolve into an enticing movement that contributes to the alignment and achievement of financial, natural, and social manageability goals. Through cutting-edge calculations and clever computerized innovations, farming business measurements at all levels and between all value chain partners are expected to be linked. Business situations such as brilliant and ideal water system and assurance, food control and safety, soil and variety insurance, astute homestead management, and creature creation should be recognized through such businesses, all to achieve natural, financial, and social sustainability. Additionally, increased adaptability of horticultural property, a new approach to work and participation, and a strengthened culture should allow for variation in the event of anticipated disruptions in financial, natural, and social streams.

Keywords Sustainable agriculture · Food security · Digitalization · Digital economy · Agribusiness

A. A. Krutilin
Volgograd State Technical University (Sebryakov Branch), Mikhaylovka, Russia
e-mail: kotyra84@bk.ru

A. M. Bazieva (✉)
Kyzyl-Kiya Institute of Technology, Economy and Law of Batken State University, Kyzyl-Kiya, Kyrgyzstan

T. A. Dugina
Volgograd State Agricultural University, Volgograd, Russia
e-mail: deisi79@mail.ru

A. T. Giyazov
Batken State University, Batken, Kyrgyzstan
e-mail: aziret-81@mail.ru

JEL Codes Q01 · Q56 · M15 · O14

1 Introduction

Agricultural production more than tripled over the last decade, partly owing to productivity-enhancing technologies and a significant expansion in the use of land, water, and other natural resources for agricultural purposes. The same period witnessed a remarkable process of industrialization and globalization of food and agriculture. Food supply chains have lengthened dramatically as the physical distance from farm to plate has increased. The consumption of processed, packed, and prepared foods has grown in all but the most isolated rural communities [1].

Nevertheless, persistent and widespread hunger and malnutrition remain a huge challenge in many parts of the world. The current rate of progress will not be enough to eradicate hunger by 2030, and not even by 2050. At the same time, the evolution of food systems has responded to and driven changing dietary preferences and patterns of overconsumption, which is reflected in the staggering increases in the prevalence of overweight and obesity around the world. Expanding food production and economic growth have often come at a high cost to the natural environment [2, 3].

Looking ahead, the core question is whether today's agriculture and food systems can meet the needs of a global population that is projected to reach more than 9 billion. Can we increase the required production, even as the pressures on already scarce land and water resources and the negative impacts of climate change intensify? The consensus view is that current systems are likely capable of producing enough food, but inclusively and sustainably will require significant transformations.

2 Materials and Methodology

Sustainable Agriculture for food security is considered a scientific concept in the works of [4–6]. Conceptual framework and benefits of agriculture digitalization formulated in the studies of [7–9].

Historically, agricultural techniques were evaluated using a small number of models, including benefit yields and ranch usefulness. Nowadays, agriculture is considered sustainable when it can meet the current and future demands without jeopardizing financial, natural, social, or political requirements. Sustainable Agriculture is one translation that focuses on specific types of innovation, most notably procedures that reduce reliance on nonrenewable or environmentally damaging data sources.

Agricultural sustainability goes beyond a particular cultivating framework. In horticultural frameworks, supportability is defined as the framework's ability to cradle shocks and stresses while remaining productive. It refers to the capacity to adapt and change in response to changing external and internal conditions. The calculated boundaries have evolved from an initial emphasis on ecological factors to include financial and then more extensive social and political metrics. The primary goal is to minimize negative natural and human welfare externalities, costs, or benefits that affect a party that did not choose to cause them, to enhance and maximize the value of the local biological system's assets and conserve biodiversity. Subsequent concerns include a greater appreciation for horticulture's beneficial environmental externalities. Reasonable agriculture not only produces food and other marketable goods but also provides public goods such as clean water and flood protection.

Financial perspectives on horticultural viability seek to assign a monetary value to natural resources and to include a longer time horizon in economic analysis. Additionally, they include sponsorships that promote resource depletion or out-of-step competition with other frameworks for creation. Agrarian supportive frameworks may have various beneficial side effects, including assisting in the development of average capital and reinforcing social capital.

Additionally, this concept does not imply making decisions about technological advancements or practices based on philosophical concepts. If an innovation aims to increase utility without causing undue harm to the environment, it may generate various supportability benefits. In this regard, critical sustainability standards rely on reconciliation and biological cycles in food production, such as predation, supplement cycling, parasitism, and nitrogen obsession. The elimination of non-endless data sources that is detrimental to the environment or the well-being of ranchers and shoppers. Additionally, it is based on the concept of maximizing ranchers' knowledge and abilities, thereby increasing their independence and displacing expensive external data sources with human resources. Finally, the concept is based on individuals collaborating to address common horticultural and common asset issues such as watershed, water system, and forest management.

3   Results

The concept of sustainable development takes into account economic, biological, and social factors. If appropriate strategies for estimating these various segments are available, horticulture can make reasonable progress. There are currently several approaches to quantifying Sustainable Agriculture using indicators and thus making it implementable on an individual ranch in a horticultural setting.

To determine the current contribution of digitalization to sustainable agricultural development and food security, statistics on digitalization and sustainability of agriculture in 2020 in the leading countries in terms of agricultural sustainability were collected (Fig. 1).

According to Fig. 1, the highest level of digitalization is observed in Sweden (95.146 points) and the Netherlands (92.567 points), and the highest agricultural sustainability is in Norway (73.5 points) and Finland (73.2 points). Correlation analysis of data from Fig. 1 revealed a weak connection between the considered statistical

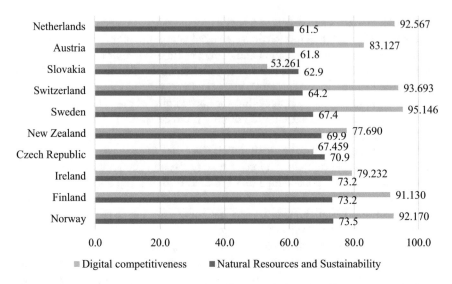

Fig. 1 Statistics of digitalization and sustainability of agriculture in 2020, scores 1–100 *Source* built by the authors based on materials from [10, 11]

indicators—the correlation was 9.63%. Consequently, digitalization currently makes little contribution to sustainable agricultural development and food security.

When surveying cultivating frameworks, the productivity of various rural land uses, expressed as yield per unit area, should be a critical indicator. The information base is adequate at all scale levels. Regardless, detailed data on location and, in an ideal world, weather-specific yield potential are required for comprehension, which must be accumulated in either information banks or yield models. Additionally, yield levels must be easily decipherable whenever data on development practices, precisely the extent of treatment and the use of plant protection products, are discovered. It is the most effective method for determining whether the yield execution to cost ratio is acceptable [12].

At all levels of perception, the example of the revolution and the recurrence of development can be viewed as similarly appropriate indicators. Models don't need to portray themselves in mind-boggling ways. Acquiring information, recognizing logic, and translating the recurrence of development are all nearly effortless. Additionally, the two pointers can provide information about biodiversity.

Hereditary variety and selection decision-making can be used as fundamental indicators of manageability. In any case, they should be located in the pivot example and development recurrence pointer chain. Although the cost of obtaining information may be prohibitively high depending on the size of perception, it is typically accomplished using data provided by plot card lists or the rural exchanging local area.

Adjusting supplements or compost is a critical indicator of a farm's viability. Proper preparation is necessary for high utility and yield quality. Simultaneously, treatment is essential for maintaining and expanding soil diversity [13]. The examination of supplement and manure changes will also provide information on soil fertility advancement and aid in evaluating natural effects. In terms of site and execution, evaluation could provide insight into the effectiveness of supplement inputs and the anticipated threat to the climate.

The energy assessment of cultivating frameworks is a focal point with a high degree of similarity and applicability. Net energy acquisition and/or energy productivity can be used as critical indicators for energy appraisal, with energy acquisition used to evaluate execution and energy proficiency used to evaluate a framework's natural effect. At all scales, the information base based on yield measurements and information on manufacturing techniques is sufficiently precise. As with the efficiency pointer, specific data on the area explicit yield potential must be gathered either through information banks or through yield models to decipher the information.

The resource-based view of a farm can be depicted by computing the benefit of the various creation measures, establishing commitment edges, and selecting the ideal explicit power. The ranch-level information base is fundamentally excellent. Regardless of whether the rules are at the territorial or public level, distinguishing proof of proficiency is risky. A determination of edge esteems only possible with territorial and homestead type-explicit correlation data.

With the reservations depicted above, results can be applied to all boundaries referenced thus far on a fundamental level. However, it should be noted that a reasonable translation may be possible if the entire framework is broken down and the collaborations between various development practices from one perspective and the site-explicit yield potential from the other are adequately reflected.

Horticulture can now be viewed as the foundation for natural, economic, and social sustainability. Significant methodology, constraints, technological advancements, client centricity, hierarchical culture, and development will all play a role in this, and one of the most valuable assets will be advanced information, which is viewed as a critical factor in rural development.

Most importantly, we must begin with horticulture's most frequently referenced primary asset: information. The agricultural industry generates enormous amounts of data that are currently understudied and underutilized. Later on, it is customary to combine produced agrarian data with open data on climate, spatial data, and other data, resulting in advanced calculations useful for dynamic, programmed measure control and ensuring management and security in horticultural activities. For example, in the production of animals or leafy foods, field observations of soil dampness, pH, precipitation, growth rate, and development can assist ranchers in determining the optimal method of preparation, water system, or other business cycles such as collecting, security, and so on. The application of such advanced calculations would contribute to the conservation of water assets in the water system, as well as to the reduction of pesticide, herbicide, and manure use, as well as to time and cost savings.

Additionally, one of the critical issues addressed will be the security and confirmation of soil biodiversity and manageability. The future period will be defined by the development of clever policies that attempt to address a portion of these issues. For instance, the role of soil in agriculture is critical, as it serves as a vital dynamic link in the interaction that is the basis of life on Earth. The establishment of a sophisticated biomonitoring framework capable of identifying and eliminating sources of contamination to maintain maximum biological potential is critical and significant not only for soil quality but also for biodiversity conservation and human health. Concerns about manageability will accelerate the development of business arrangements based on computerized innovations aimed at increasing the productivity of biomass supply chains, from precise development to prudent stockpiling and coordination to optimal end-user use. Additionally, the use of natural pesticides will become more closely linked to agricultural cycles, allowing for the preservation of healthy soil conditions and reducing soil contamination, all of which benefit human health, land use, and financial proficiency [14].

Apart from that, the critical direction, advancement, and traditional culture, as well as the limit, client centricity, and focus on all partners, are not insignificant. Later on, an adequate emphasis has been placed on ensuring manageability through agribusiness approaches and objectives. Nonetheless, all partners are accountable for cultivating advancement, participation, and a hospitable environment within the agricultural sector [15]. Agribusiness can benefit from advanced tools and platforms for communication, experience exchange, information distribution and exchange, product development, and product development. This can aid in expanding into new business sectors, meeting the needs of industries where certain rural items are scarce, and developing imaginative things and the foundation for business participation, all of which can eventually contribute to the turn of events and development of supportability. For instance, if certain rural items are scarce, manufacturers can ensure financial viability by providing individuals with things that may not be readily available to them.

Additionally, numerous specialized farming products can be developed to address the needs of various business sectors. Ducks are an example of a model. Duck meat is widely consumed in the European market, whereas duck legs are frequently used for food and duck feathers for other creative purposes in the Eastern world. There are numerous such items whose development can address the needs of various business sectors while also ensuring sustainability, which will be accomplished through the use of stages and associated value chains.

4 Conclusion

Global issues such as environmental change and dangerous atmospheric depletion, extreme climate disasters, and unexpected disruptions are becoming an increasing concern for the economy and private frameworks. This is precisely why it is necessary to investigate novel, advanced conceivable outcomes of technological advancements. Additionally, global pioneers' approaches address this issue and aim to develop event reversal and manageability development methodologies. Sustainability is a concept that emphasizes the three fundamental perspectives of financial, natural, and social sustainability. Horticulture is particularly relevant in these three perspectives, as it is one of the few activities that can directly affect them, either positively or negatively. This concept stimulated research into the impact of horticulture on manageability and the role of computerized innovations in achieving supportability in an agribusiness setting.

The focus will now be on the needs of the entire environment, including ranchers and manufacturers from various industries, policymakers and state governments, and any remaining partners. The primary asset will no longer be information, but rather advanced calculations and information that will aid in the management of rural cycles. Computerized advancements will be focused on monitoring and tending to environmental change and dangerous atmospheric devastation, as well as on responding to radical climate changes and other unsettling economic influences. A transformed agribusiness should evolve into an enticing movement that contributes to the alignment and achievement of financial, natural, and social manageability goals.

Through cutting-edge calculations and clever computerized innovations, farming business measurements at all levels and between all value chain partners are expected to be linked. Business situations such as brilliant and ideal water system and assurance, food control and safety, soil and variety insurance, astute homestead management, and creature creation should be recognized through such businesses, all to achieve natural, financial, and social sustainability. Additionally, increased adaptability of horticultural property, a new approach to work and participation, and a strengthened culture should allow for variation in the event of anticipated disruptions in financial, natural, and social streams.

References

1. Inshakov OV, Bogachkova LY, Popkova EG (2019) The transformation of the global energy markets and the problem of ensuring the sustainability of their development. In: Energy sector: a systemic analysis of economy, foreign trade and legal regulations. Springer, Cham, pp 135–148
2. Sergi BS, Popkova EG, Bogoviz AV, Ragulina JV (2019) Costs and profits of technological growth in Russia. In: Tech, smart cities, and regional development in contemporary Russia. Emerald Publishing Limited
3. Sergi BS, Popkova EG, Borzenko KV, Przhedetskaya NV (2019) Public-private partnerships as a mechanism of financing sustainable development. In: Financing sustainable development. Palgrave Macmillan, Cham, pp 313–339
4. Kumar A, Sreedharan S, Singh P, Achigan-Dako EG, Ramchiary N (2021) Improvement of a traditional Orphan Food Crop, Portulaca oleracea L. (Purslane) using genomics for sustainable food security and climate-resilient agriculture. Front Sustain Food Syst 5. https://doi.org/10.3389/fsufs.2021.711820
5. Lombardi GV, Parrini S, Atzori R, Gastaldi M, Liu G (2021) Sustainable agriculture, food security and diet diversity. The case study of Tuscany, Italy. Ecol Model 458. https://doi.org/10.1016/j.ecolmodel.2021.109702
6. Rasul G (2021) A framework for addressing the twin challenges of COVID-19 and climate change for sustainable agriculture and food security in South Asia. Front Sustain Food Syst 5. https://doi.org/10.3389/fsufs.2021.679037

7. Ebrahimi HP, Sandra Schillo R, Bronson K (2021) Systematic stakeholder inclusion in digital agriculture: a framework and application to Canada. Sustainability (Switzerland), 13(12): 6879. https://doi.org/10.3390/su13126879
8. Rijswijk K, Klerkx L, Bacco M, Scotti I, Brunori G (2021) Digital transformation of agriculture and rural areas: a socio-cyber-physical system framework to support responsibilisation. J Rural Stud 85:79–90. https://doi.org/10.1016/j.jrurstud.2021.05.003
9. Visser O, Sippel SR, Thiemann L (2021) Imprecision farming? Examining the (in)accuracy and risks of digital agriculture. J Rural Stud 86:623–632. https://doi.org/10.1016/j.jrurstud.2021.07.024
10. The Economist Intelligence Unit Limited (2021) Sustainable Development Index 2020. https://foodsecurityindex.eiu.com/Index. Accessed 11 Sept 2021
11. IMD (2021) World Digital Competitiveness Report 2020. https://www.imd.org/centers/world-competitiveness-center/rankings/. Accessed 11 Sept 2021
12. Hayati D, Ranjbar Z, Karami E (2010) Measuring agricultural sustainability. Biodiversity, biofuels, agroforestry and conservation agriculture, pp 73–100
13. Gómez-Limón JA, Sanchez-Fernandez G (2010) Empirical evaluation of agricultural sustainability using composite indicators. Ecol Econ 69(5):1062–1075
14. Bharti VK, Bhan S (2018) Impact of artificial intelligence for agricultural sustainability. J Soil Water Conserv 17(4):393–399
15. Khaled R, Hammas L (2016) Technological innovation and agricultural sustainability: What compatibility for the mechanization? Int J Innov Digit Econ (IJIDE) 7(2):1–14

Methodological Approach to the Polycriteria Assessment of Agricultural Sustainability: Digitalization, International Experience, Problems, and Challenges for Higher Education in Russia

Elena V. Patsyuk, Alexander A. Krutilin, Nadezhda K. Savelyeva◉, and Karina A. Chernitsova◉

Abstract This chapter developed a proprietary scientific and methodological approach to the polycriteria assessment of agriculture's sustainability. The advantage of the new approach is, first, the possibility for a precise quantitative evaluation of agriculture's sustainability based on the data of the official statistics. Second, a systemic consideration of all criteria of agriculture's sustainability, including stability, expanded reproduction, circularity, and environmental and energy efficiency. Third, the largest precision of the assessment results, due to the use of Saaty's hierarchy process. This method is used to determine the contribution of various criteria of agriculture's sustainability to food security. This allows assigning weight coefficients to the distinguished criteria, which reflect—quantitatively—their contribution. Due to this, the sustainability of agriculture correctly describes its contribution to the provision of food security. The approbation of the developed methodological approach is performed, which allows studying the international experience of sustainable development of agriculture and determining the disproportion in the sustainability of agriculture in the world economic system. A systemic contribution of digitalization and higher education to the sustainability of agriculture is determined, and challenges of the strategy of sustainable development of agriculture for higher education and the digital economy of Russia are described.

E. V. Patsyuk (✉) · A. A. Krutilin
Volgograd State Technical University (Sebryakov Branch), Mikhaylovka, Russia
e-mail: elenapatsyuk@yandex.ru

A. A. Krutilin
e-mail: kotyra84@bk.ru

N. K. Savelyeva
Vyatka State University, Kirov, Russia
e-mail: nk_saveleva@vyatsu.ru

K. A. Chernitsova
Plekhanov Russian University of Economics, Moscow, Russia
e-mail: Karinaaa2004@mail.ru

Keywords Sustainability of agriculture · Food security · Digitalization · International experience · Problems · Challenges of higher education · Russia

JEL Codes F52 · I25 · I28 · Q01 · Q13 · Q16 · Q17

1 Introduction

The amount of arable land available is unlikely to change, and as the population grows, it will become necessary to increase harvests from existing assets. By 2050, the world's population will reach 9 billion, up from 7.7 billion today, necessitating a 70% increase in food production to keep up with population growth [1]. Without abandoning traditional agricultural practices, the capacity of people to develop and exploit arable land will dwindle in the future. Previously, the emphasis was on making the best use of available land and producing as many harvests as possible. There was little emphasis on harvesting appropriately or protecting land, and a comparative lack of emphasis on benefit and financial savings in terms of area and assets [2]. These two are inextricably linked, and the arrangements that result will revolutionize agriculture [3–6].

Crop production should be continuously increased by 60–90% by 2050 to meet the nutritional needs of an expanding human population. Frameworks for yield development are required that produce more food with a higher nutritional content while having a negligible climatic impact. Horticultural advancement in the twentieth century was heavily reliant on compost, pesticides, and irrigation systems, all of which had a significant environmental impact. These changes occurred as a result of the Green Revolution, which secured food for a portion of the population. However, the twenty-first century's challenges are unique, and conserving soil and water will be critical to ensuring food security [7]. Achievable precision farming combined with a changing environment will not result in new effects that accelerate environmental development. As part of reasonable agribusiness, cutting-edge trimming frameworks that integrate biological innovations, precision horticulture, and precision protection should be developed to reduce manure, pesticide, and water inputs while increasing preservation adequacy to keep up with field horticulture and maintainability across a watershed. Utilizing modern breeding biotechnology methods, high-yield cultivars with improved nutritional content and resistance to abiotic stresses should be developed. These improved cultivars will almost certainly upend the horticultural sector's current state.

2 Materials and Method

In recent years, information technology has acted as a disruptive force in business, eradicating market inefficiencies through mechanization and superior choice assistance apparatuses that take residents and clients into account simultaneously [8]. Horticulture, like all other industries, has been harmed by the frequent blackouts. Regardless, late-calculating propellers, structure, and sophisticated computations indicate an impending paradigm shift, necessitating information flow from multiple sources [9].

Food's growing popularity, as well as its enormous biological influence, necessitates employment in horticulture. Long-term environmental impacts on practical agriculture should be investigated, as should data sources and resources. Dynamic cycles of streamlining and evaluation necessitate familiarity with a small number of data sources, outputs, and external consequences [10]. Numerous frameworks for information security and the board of directors have been developed to enable accurate agribusiness. Agriculture with accuracy is the application of technologies and standards to improve production execution and natural manageability. Through the application of cutting-edge technologies for precise detecting, observing, examining, planning, and control, shrewd farming improves agribusiness accuracy and dynamic capabilities [11]. Setting, context, and area awareness all contribute to the quality of the information. Continuous sensors collect a variety of data, whereas continuous actuators instantly adjust the creation bounds. In agricultural creation, there is a significant requirement for data storage and processing. Through web administrations, farming groups would communicate with and receive data from a central online programme. This web application collects, stores, and cycles data before making it available to clients or another framework. In essence, a few data processing use cases are intended to assist ranchers in managing their dynamic processes. Continuous innovation, such as the Internet of Things (IoT), enables automated data security and, consequently, intelligent agriculture. Numerous recent experiments in brilliant growing and precision farming have been conducted. Industry 4.0 is now exerting a significant influence on measures of innovative development. Numerous components of agricultural frameworks incorporate Industry 4.0 advancements, including the Internet of Things, big data, edge computing, 3D printing, expanded reality, community-oriented mechanical technology, information science, distributed computing, digital actual frameworks, computerized twins, network security, and continuous improvement.

To recognize functional competence, complete robotization, and high efficiency in these frameworks, various types of data are collected via multiple sensors from practical frameworks, stored in massive data frameworks, and generated using AI and deep learning approaches. Conventional data management procedures and frameworks are insufficient to manage this volume of data; thus, large-scale data foundations and frameworks have been created and deployed [12]. To design frameworks that can deal with the complexities of this massive amount of data, it is necessary to examine various data components. Numerous types of data have been earmarked for executives' reference structures to date. Several strategies for sustainable agriculture can help conserve the environment, improve the soil's ripeness, and increase regular assets. Horticulture has been shown to affect soil decomposition, water quality, human health, and fertilizer administration. In this capacity, practical agribusiness is critical to mitigating the negative impacts of farmland creation. Because each entertainer has a unique responsibility within this framework and the success of this interaction is highly dependent on the performance of each entertainer, sustainable horticulture requires an iterative cycle.

For the empirical part of this study, statistics of agricultural sustainability in the G7 and BRICS countries in 2020 were collected (Table 1).

We have developed our own scientific and methodological approach to the poly-criteria assessment of agricultural sustainability. The advantage of the new approach lies, first, in the possibility of an accurate quantitative assessment of agricultural sustainability based on official statistics. Second, the systematic consideration of all criteria for the sustainability of agriculture, including stability, expanded reproduction, isolation, environmental and energy efficiency. Third, the greatest accuracy of the assessment results through the use of the Saaty hierarchy process. This method is used to determine the contribution of various agricultural sustainability criteria to food security. This allows assigning weights to the selected criteria that quantitatively reflect their contribution. As a result, agricultural sustainability correctly describes its contribution to food security. Approbation of the developed methodological approach is made in Table 2.

The evaluation results are shown graphically in Fig. 1.

The polycriteria assessment of agricultural sustainability in the G7 and BRICS countries in 2020 showed that, in general, agricultural sustainability in developed countries (0.79) is slightly higher than in developing countries (0.77), and in both categories of countries it is quite high (there is practically no disproportion).

3 Results

Russia has quietly resolved the country's long-standing food deficit. The sophisticated agri-food sector in the country is one of the fastest growing segments of the public economy. Crop production is approaching historical levels (sunflower, sugar beet). Previously a consistent shipper of staple food items, the country has grown to become a significant supplier to the global market. Russia has surpassed the United States as the world leader in wheat and buckwheat cuisine and has climbed into the top ten harvest countries. Additionally, it has begun shipping domestic animal products and food products with added value. Recent years have seen significant advancements in the food quality and security sectors, which have been recognized globally.

State support for agriculture has consistently been set at levels comparable to those in the European Union and the United States, even though various assistance programmes are largely ineffective at achieving their stated goals [14]. There are emerging indicators of insufficiently executed areas (such as yields per hectare, per

Table 1 Agricultural sustainability statistics in the G7 and BRICS countries in 2020

Countries		Land for grain production	Quality and safety of food	Quantitative availability of food	Rural population	Agriculture, forestry, and fisheries, value-added	Grain yield	Price availability of food	Food Security Index
G7	Germany	6,266,500	79.8	79.1	22.69	81	7,269.9	84.9	81.5
	France	9,380,922	87.1	74.8	19.56	78	6,875.2	83.8	80.4
	Japan	1,802,951	76.7	71	8.38	67	6,049.2	82.4	76.5
	Italy	3,140,579	79.7	68.3	29.56	84	5,171.4	82.5	75.8
	Canada	13,929,520	86.7	80	18.59	20	4,042.5	83.3	82.4
	USA	53,149,164	89.1	78.3	17.74	n/d	8,280.8	87.4	83.7
	UK	3,181,631	80.9	74.4	16.6	87	7,229	83.6	79.1
BRICS	India	99,220,000	47	58.4	65.97	100	3,160.8	64.2	58.9
	Russia	44,240,971	70.9	60.1	25.57	27	2,964.3	79.8	69.7
	Brazil	22,613,700	84	58.8	13.43	15	5,208.5	77	70.1
	South Arica	3,347,247	66.2	64.5	33.65	39	5,648.2	70.8	67.3
	China	102,493,146	72.6	66.9	40.85	36	6,029	74.8	71

Source Calculated and compiled by the authors based on materials [13]

Table 2 Multicriteria assessment of agricultural sustainability in the G7 and BRICS countries in 2020

Indicator		Land for grain production	Quality and safety of food	Quantitative availability of food	Rural population	Agriculture, forestry, and fisheries, value-added	Grain yield	Price availability of food
Correlation with the food security, %		−52.78	89.07	91.97	−74.74	2.63	66.28	94.98
Weight coefficient		0.11	0.19	0.19	0.16	0.01	0.14	0.20
G7	Maximum value in the sample	53,149,164.00	89.10	80.00	29.56	87.00	8,280.80	87.40
	Sample arithmetic mean	12,978,752.43	82.86	75.13	19.02	69.50	6,416.86	83.99
	Ratio to the maximum	0.24	0.93	0.94	0.64	0.80	0.77	0.96
	Ratio between maximum and the weight coefficient	0.03	0.18	0.18	0.10	0.00	0.11	0.19

(continued)

Table 2 (continued)

Indicator		Land for grain production	Quality and safety of food	Quantitative availability of food	Rural population	Agriculture, forestry, and fisheries, value-added	Grain yield	Price availability of food
BRICS	Maximum value in the sample	102,493,146.00	84.00	66.90	65.97	100.00	6,029.00	79.80
	Sample arithmetic mean	54,383,012.80	68.14	61.74	35.89	43.40	4,602.16	73.32
	Ratio to the maximum	0.53	0.81	0.92	0.54	0.43	0.76	0.92
	Ratio between maximum and the weight coefficient	0.06	0.15	0.18	0.09	0.00	0.11	0.18

Source Calculated and compiled by the authors

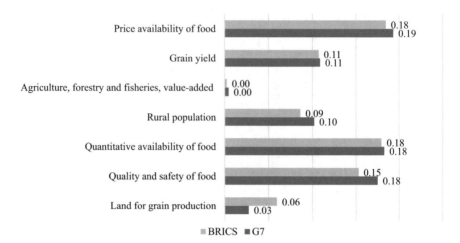

Fig. 1 Multicriteria assessment of agricultural sustainability in the G7 and BRICS countries in 2020 *Source* Calculated and constructed by the authors

head, and labor efficiency), as well as absolute factor utility (TFP) [15]. Production advances primarily due to the cost of increasing factors. Manufacturers of modular homes make extensive use of cutting-edge technology. On traditional measures of food security, Russia ranks in the top third of the world's nations.

The 1998 emergency, which halted imports and triggered a wave of domestic investment, initially in the food industry, then in critical agribusiness and upstream, was a watershed moment in the region's development. Thus, the primary factors influencing development were venture development and comparing changes like the board. 2008 was a comparable follow-up effort. In 2014, another attempt was made to secure domestic manufacturers through the presentation of authorized adversaries. By the way, development variables like speculation and executives have been virtually depleted. Russia's agricultural sector is currently being tested by new growth factors, which will be discussed in greater detail below.

Agriculture is no longer a marginal industry in the modern world; it has become an integral part (albeit a minor one) of food systems. Today's competition is based on the development of novel food products with unique characteristics aimed at specific market segments [16]. Food production has grown to be one of the world's most significant industries. Russia urgently needs to develop an innovative strategy for developing its agri-food sector in order to compete and strengthen its position in both domestic and international commercial industries.

Russian production is highly variable, particularly agricultural production (the unpredictability of yields of principal crops surpasses a similar pointer commonly in Canada, which is like Russia as far as its agri-climatic conditions and extent of agricultural creation). It is a numerical representation of mechanical slack. Another indicator of Russian agriculture's lack of innovation is the country's extreme reliance on imported duplicating materials. To summarize a significant statement made by a government official, it is possible to import all innovations at one time; it is possible to import a few innovations regularly, but it is impractical to expect to continuously import all creations. As a result, inventiveness lags and the relative decline in insincerity is exacerbated.

4 Conclusion

The primary impediment to rational agricultural expansion in Russia is, predictably, the "asset revile": the country's abundant land and water resources, as well as relative biodiversity, do not yet represent a critical demand for conservation. Russia continues to be the world's natural benefactor. Additionally, some immediate practical concerns should be addressed in the medium term.

To begin, there is now a concern about soil ripeness. In the country, there is virtually no framework for public inspection of soil quality and condition. Their depreciation must be computed using fractional master appraisals [17]. Simultaneously, privately owned firms' objectively brief organizational skylines dissuade them from investing in soil ripeness maintenance, which also has a significantly extended payback period. Similarly, reformist approaches to soil treatment and water conservation are uncommon in the country.

Increased utility per hectare has resulted in some progress toward preserving the country's biodiversity, owing to the reduction in the agricultural production area as a result of increased utility per hectare. Despite opposition, a decision was made in 2019 to surrender to the state 1 million hectares of newly removed agricultural land. As a result, ozone-depleting chemical emissions would significantly increase (contrary to our pledges under the recently concluded Paris Agreement). The country's biodiversity would suffer a significant loss [18].

Second, environmental constraints on agricultural development have been successfully implemented in some regions across the country. Animal production is coexisting with a high level of household waste in the Belgorod region, which, in the most extreme case, can enter the underground spring. The permitted amount of sunflower in crop revolutions has been exceeded in several southern locales. Data on permitted overfishing are available. Rapid hydroponics development in Russia is not accompanied by an acceptable level of ecological sustainability, which could spell the enterprise's demise.

Third, the absence of a public mechanism or even a desire to reduce food loss and waste significantly impairs the achievement of reasonable agrarian outcomes. Because there is no formal authority in Russia that monitors FLW, we must rely on the competent judgments of market participants. Mistakes in the fundamental components of the agri-food business can account for up to 40% of the yield, implying an inefficient use of a diverse array of assets. The current state of the world's experience demonstrates a plethora of strategies for minimizing food waste. Specifically, one of the most widely adopted techniques for reducing food waste has been the exchange of commodities nearing expiration for charitable causes in all countries worldwide. Russia already has several companies capable of conducting such transactions, but the country's taxation framework precludes them.

Fourth, as stated previously, advanced food frameworks should be more targeted toward specific purchaser groups. The progressive working class throughout the world, including Russia, is concerned with cost-effective food production processes. Increasing this training is becoming a necessary component of food sector competition. The concept of agriculture sustainability was launched at the all-Russian agri-food forum [19]. If the brand is properly positioned and supported by government policies, it has the potential to become a primary method of promoting economically viable agriculture in Russia.

References

1. Sundmaeker H, Verdouw CN, Wolfert J, Freire LP (2016) Internet of food and farm 2020. In: Digitising the industry, vol 49, pp 129–150. River Publishers
2. Dariusz K (2014) Modernization of agriculture vs. sustainable agriculture. Scientific papers. Ser Manag Econ Eng Agric Rural Dev 14(1):171–178
3. Popkova EG, Popova EV, Sergi BS (2018) Clusters and innovational networks toward sustainable growth. In: Exploring the future of Russia's economy and markets. Emerald Publishing Limited
4. Sergi BS, Popkova EG, Bogoviz AV, Ragulina JV (2019a) Costs and profits of technological growth in Russia. In: Tech, smart cities, and regional development in Contemporary Russia. Emerald Publishing Limited
5. Sergi BS, Popkova EG, Sozinova AA, Fetisova OV (2019b) Modelling Russian industrial, tech, and financial cooperation with the Asia-Pacific region. In: Tech, smart cities, and regional development in Contemporary Russia. Emerald Publishing Limited
6. Sergi BS, Popkova EG, Borzenko KV, Przhedetskaya NV (2019c) Public-private partnerships as a mechanism of financing sustainable development. In: Financing sustainable development, pp 313–339. Palgrave Macmillan, Cham
7. Mishra B, Gyawali BR, Paudel KP, Poudyal NC, Simon MF, Dasgupta S, Antonious G (2018) Adoption of sustainable agriculture practices among farmers in Kentucky, USA. Environ Manage 62(6):1060–1072
8. Waga D (2013) Environmental conditions' big data management and cloud computing analytics for sustainable agriculture. SSRN 2349238
9. Poppe KJ, Wolfert S, Verdouw C, Verwaart T (2013) Information and communication technology as a driver for change in agri-food chains. EuroChoices 12(1):60–65
10. Nesterenko NY, Pakhomova NV, Richter KK (2020) Sustainable development of organic agriculture: strategies of Russia and its regions in the context of the application of digital economy technologies
11. Wolfert S, Goense D, Sørensen CAG (2014) A future internet collaboration platform for safe and healthy food from farm to fork. In: 2014 annual SRII global conference, pp 266–273. IEEE
12. Wolfert J, Kempenaar CORNÉ (2012) The role of ICT for future agriculture and the role of agriculture for future ICT
13. Institute of Scientific Organizations (2021) Dataset "Corporate social responsibility, sustainable development and fight against climate change: imitation modelling and neural network analysis in regions of the world—2020" (2020). https://iscvolga.ru/dataset-climate-change
14. Kheyfets BA, Chernova VY (2019) Sustainable agriculture in Russia: Research on the dynamics of innovation activity and labour productivity. Entrep Sustain Issues 7(2):814

15. Afanasyeva L, Belousova L, Tkacheva T (2021) Formation of a system of indicators of balanced economic development in the context of globalization: space-time analysis in order to ensure economic security. In: SHS web of conferences, p 92. EDP Sciences
16. Kassie M, Zikhali P, Manjur K, Edwards S (2009) Adoption of sustainable agriculture practices: evidence from a semi-arid region of Ethiopia. Nat Resour Forum 33(3):189–198. (Oxford, UK: Blackwell Publishing Ltd.)
17. Yakushev VP, Yakushev VV (2018) Prospects for "Smart Agriculture" in Russia. Her Russ Acad Sci 88(5):330–340
18. Serova E (2020) Challenges for the development of the Russian agricultural sector in the mid-term. Russ J Econ 6:1
19. Wegren SK (2021) Prospects for sustainable agriculture in Russia. Eur Countryside 13(1):193–207

Environmental and Energy Efficiency as a Criterion for Sustainable Agriculture

Svetlana E. Karpushova◉, Igor V. Denisov◉,
A. L.-Muttar Mohammed Yousif Oudah, and Yuliya I. Dubova◉

Abstract This chapter studies such characteristics of agriculture as environmental (from the positions of waste reduction and the contribution to the fight against climate change) and energy (from the positions of energy efficiency) efficiency as the criteria of agriculture's sustainability. Based on the dataset "Corporate social responsibility, sustainable development and fight against climate change: imitation modelling and neural network analysis in regions of the world—2020," the authors determine the environmental and energy efficiency of agriculture for developed (G7) and developing (BRICS) countries. Based on the results of the evaluation, the authors determine the importance of the problem of agriculture's sustainability in developed and developing countries from the positions of environmental and energy efficiency. The method of regression analysis is used to determine the impact of the measures of state regulation (e.g., energy efficiency regulation and environment-related treaties in force) on the environmental and energy efficiency of agriculture (and, as a result, on its sustainability). Framework recommendations for state regulation of environmental and energy sustainability of agriculture are developed.

Keywords Environmental efficiency · Efficiency · Sustainability of agriculture · Dataset modeling · Corporate environmental responsibility · Developed countries · Developing countries

JEL Codes O13 · P48 · Q01 · Q15 · Q48

S. E. Karpushova (✉)
Volgograd State Technical University (Sebryakov Branch), Mikhaylovka, Russia
e-mail: sfkse@yandex.ru

I. V. Denisov (✉)
Plekhanov Russian University of Economics, Moscow, Russia

A. L.-M. M. Y. Oudah
Al-Ayen University, Dhi Qar, Nasiriyah, Iraq
e-mail: mohammed.yousif@mail.ru

Y. I. Dubova
Volgograd State Technical University, Volgograd, Russia
e-mail: dubovaui@mail.ru

1   Introduction

Pesticides have significantly increased agricultural productivity and product quality; however, once in the environment, pesticides can accumulate in soil and water, harming greenery as concentrations in industrialized ways of life reach levels that are harmful to untamed life. Additionally, pesticide accumulations degrade the quality of drinking water, contaminate food intended for human consumption, and cause adverse health effects when pesticides are applied directly to farm laborers. Simultaneously, some pesticides contain bromide intensifiers that, when volatilized, convert to stratospheric ozone-depleting gases. Difficulty in defining agricultural pesticide use rules is that pesticides vary significantly in their toxicity, persistence, and mobility, depending on the type and convergence of their dynamic fixes.

Similarly, when fewer but less dangerous pesticides are used, an increase in pesticide use may result in a decrease in ecological impact, and vice versa, highlighting the critical nature of pesticide use hazard evaluation. Additionally, the amount of pesticides absorbed by soil and water is affected by soil properties and temperature, waste, yield type, environment, application technique, and time and recurrence [1]. Additionally, when pesticides are used in conjunction with specific vermin, such as coordinated irritation of the board, the climate, pesticide clients, and food consumers may suffer negative consequences [2–5].

2   Materials and Method

Pesticides have contributed significantly to agriculture's high utility and output. Water scarcity can stymie agricultural development and have a detrimental effect on marine habitats and wildlife. Horticulture utilizes both surface and groundwater, as well as rainfall. To encourage agriculture to make better use of surface and groundwater resources, the amount of water extracted from these sources per ton of biomass/domesticated animals produced should be reduced. Maintaining and restoring the "normal" state of water assets is critical for effective water management and acceptable agricultural practices [6].

Rural development has accelerated the depletion of several countries' limited surface and groundwater resources. Similarly, CEOs' poor land management practices, such as tree removal on a rural property, can result in "abundant" water problems, such as salinization and flooding caused by rising water tables [7]. Apart from horticulture, the increased competition for water resources across the economy is a major source of concern for strategy producers in some OECD countries; however, once in the environment.

To determine the contribution of the factors of government regulation of sustainable agriculture, environmental and energy efficiency in the G7 and BRICS countries, the corresponding statistics for 2020 were collected, which are systematized in Table 1.

Table 1 Statistics on environmental and energy efficiency and government regulation of sustainable agriculture in the G7 and BRICS countries in 2020

Countries	Pollution index	Climate index	The energy trilemma index	Total natural resources rents	Industrial profile of the economy	Renewable energy regulation	Energy efficiency regulation	Environment-related treaties in force
	y_1	y_2	y_3	y_4	y_5	x_1	x_2	x_3
BRICS data								
Brazil	54.98	97.16	71.6	3.5320	28	70.9	51.8	86
China	80.77	79.19	63.7	1.4962	70	66.4	73.5	83
Russia	62.79	40.36	71.2	10.7020	44	59.9	59.2	66
India	78.87	64.87	50.3	2.1430	42	87.3	66.4	90
South Africa	57.30	95.25	58.9	5.1403	38	76.1	76.2	83
G7 data								
Germany	29.03	83.00	79.4	0.0689	47	96.6	84.5	100
Italy	55.63	92.27	76.8	0.0712	36	84.1	89.2	83
Canada	27.83	50.57	78.0	1.7297	n/d	82.3	87.7	69
UK	40.56	87.62	81.5	0.4396	27	90.6	84.2	97
USA	36.88	77.54	77.5	0.4731	29	58.4	82.0	55
France	43.56	90.25	80.8	0.0424	27	86.3	72.4	97
Japan	39.59	84.79	73.8	0.0280	50	77.4	68.5	90

Source Compiled by the authors based on materials from the Institute of Scientific Communications [8]

Based on the collected data, the following results of their regression analysis were obtained (Table 2).

According to the data from Table 2, the factors of government regulation of sustainable agriculture have a contradictory effect on environmental and energy efficiency, although in general the correlation of the considered indicators is quite high (from 41.62–72.90%). Based on the results of regression analysis, the prospects for optimizing state regulation of sustainable agriculture in the interests of environmental and energy efficiency are determined. They are illustrated in Fig. 1.

According to Fig. 1, optimization of government regulation of sustainable agriculture in the interests of environmental and energy efficiency allows to reduce the

Table 2 Regression analysis of the dependence of environmental and energy efficiency on government regulation factors sustainable agriculture in the G7 and BRICS countries in 2020

	y_1	y_2	y_3	y_4	y_5
r^2	54.59	69.70	41.62	70.50	72.90
Invariable	92.26	−13.17	43.83	22.34	10.01
Regression coefficient at variable x_1	−0.85	−1.52	−0.11	0.06	−2.08
Regression coefficient at variable x_2	−0.34	0.84	0.39	−0.18	0.56
Regression coefficient at variable x_3	0.61	1.78	0.09	−0.14	1.77

Source Calculated and compiled by the authors

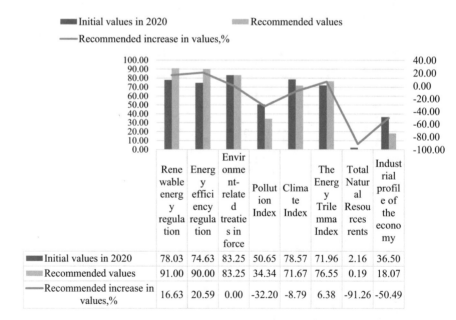

Fig. 1 Prospects for optimizing government regulation of sustainable agriculture in the interests of environmental and energy efficiency *Source* Calculated and constructed by the authors

Pollution Index by 32.20%, the Energy Trilemma Index increases by 6.38%, the Total natural resources rents decreases by 91.26%, and the Industrial profile of the economy decreases by 50.49%. However, the Climate Index is down by 8.79%. Consequently, government regulation of sustainable agriculture alone is not enough to ensure environmental and energy efficiency, although this regulation plays a very important role in this.

3 Results

Carbon dioxide (CO_2), methane (CH_4), and nitrous oxide are the three ozone-depleting compounds that are most frequently emitted as a result of rural migration (N_2O). These gases have a fluctuating and potentially harmful deviation potential in the atmosphere, which is expressed in CO_2 reciprocals. Rural CO_2 outflows occur as a result of oxidized soil matter being influenced by development or wind disintegration. CH_4 is primarily derived from the intestinal maturation and wastes of ruminant animals, paddy rice fields, and biomass consumption. Composting, animal waste, the disposal of trash capacity, the use of biomass, and petroleum derivatives all contribute to N_2O emissions [9].

Agriculture is also a GHG sink, with soil absorbing significant amounts of CO_2 through harvest and field land. Simultaneously, the soil has a high capacity for converting CH_4 to less dynamic CO_2, but soil obsession with N_2O receives less attention. Similarly, crop production and wood production on arable land both contribute to the increase in CO_2 photosynthetic obsession. However, it will be critical to distinguish farming's unique employment as a source and sink of GHGs from other sectors of the economy [10].

Agriculture has a significant impact on water quality due to high nitrogen and phosphate levels, heavy metals, dynamic pesticide fixes, caustic chemicals, and soil residue. Compost containing excessive nitrogen and phosphorus contributes to eutrophication, which can have a detrimental effect on fish populations. Unbelievably high levels of heavy metals in composted water can permeate the human evolved way of life through fish consumption. Pesticide-contaminated water can enter the system via filtration or directly through showering in near-surface water. Water fermentation can be triggered by the use of manure and petroleum compounds, as well as the consumption of biomass.

Wind- and rain-borne soil residue from agriculture and overgrazed fields can contaminate turbid water, reduce sunlight, and deplete the oxygen available to aquatic plants and fish, resulting in population declines of fish and shellfish. Additionally, residue runoff impairs the storage capacity of lakes and reservoirs, clogs streams and seepage channels, increases the frequency and severity of flooding, and degrades water conveyance frameworks.

Stockpiling feed components for creatures' feeding regimens accounts for a sizable portion of domestic animal production's energy requirements. Site conditions have a significant impact on the energy contribution of feed. Increased asset effectiveness in feed production translates into increased animal energy productivity [1]. As a result, it is critical to take the board of feed creation into account when displaying it. Because nutrition affects the energy efficiency of domesticated animals used for food production, it is critical to replace supplements with an exhibition-focused eating regimen for the creatures to increase the energy productivity of domesticated animals used for food production.

Increased animal execution results in increased energy productivity; however, this effect is only temporary; for example, as milk production increases, this effect diminishes due to the higher energy inputs required by the higher performing animals. To generate a product from the energy spent on the homestead, simple management procedures must be developed that takes a holistic view of the behavior of each animal.

Additionally, an animal's capacity to expand alternative activities affects its energy effectiveness. With fewer useful lives, the energy requirements for replacement animals increase, affecting the energy productivity of the livestock farming system. Changes in animal husbandry and care, as well as explicit reproduction processes, can be used to extend the useful life of dairy cows, thereby improving energy efficiency.

Utilizing agro-deposits for compound, feed, material, or energy generation can assist in increasing energy efficiency across the horticultural interaction. For example, straws can be used to bio-energize raw materials used in the synthesis of a variety of polymers [11]. Additionally, waste generated by animal agriculture can be used to generate biogas or materials for nonfood applications. The ability to reuse additives is a critical feature of this group of models, as it results in increased energy efficiency in arable farming.

Energy efficiency is a political objective that is guided by two principles: limited access to petroleum derivatives and consideration of environmental consequences. As a result, needs should be guided by the possibility of reducing both energy consumption and the associated negative environmental impacts [12]. Additionally, energy consumption consumes a significant portion of a ranch's total expense budget. Increased energy costs will have a variety of consequences for various farming frameworks, with those that are more energy efficient benefiting the most, which may also affect the development of agrarian creation.

Farmers believe that energy proficiency measures should be desirable enough to consider when deciding whether or not to pursue a particular invention. This means that for energy efficiency measures to be effective, they must be financially viable. Financial assistance or guidance should be considered initially as the aptitude to assist the action. The market for a specific action expands, and economies of scale lower the cost of those actions.

4 Conclusion

Global-warming-related greenhouse gas emissions endangering the ozone layer, biodiversity loss, excessive nitrogen and phosphorus use, and marine fermentation have all reached alarming levels. These factors, combined with declining access to new water, increased land corruption and deforestation, and a dearth of solutions, are jeopardizing the livelihoods of millions of genuinely developing people, particularly those living in extreme poverty.

These issues are exacerbated further by population growth. It has grown significantly larger than the seven billion-person footprint and is expected to exceed 9 billion by the middle of this century. To meet the food needs of 9 billion people, rural yields must be increased by 60% or food loss must be eliminated, while waste must be reduced by 60% [13]. Expanded food production will put every regular asset, including scarce horticultural land, woods, water, and the environment, under increasing strain. Indeed, numerous authoritative studies have concluded that horticulture will almost certainly be incapable of providing enough food to sustain the world's growing population in a healthy and robust state of living.

Simultaneously, agriculture broadly defined—which includes harvesting and domesticated animal creation, fisheries, and ranger service—provides revenue, employment, food, and other labor and products to the vast majority of today's impoverished people. In aggregate, nonfarming development is twice as effective at alleviating need as farming development and multiple times more compelling than development in resource-scarce low-income countries [14]. Thus, horticulture's future success will be contingent on its ability to provide not only a plentiful supply of nutritious food, a living wage, and attractive employment opportunities but also on its ability to address a broad range of ecological challenges. To address these numerous issues, it is necessary to transition to more sustainable forms of horticulture and to develop comprehensive techniques to aid in this process [15].

Additionally, the increase in food production aided significantly in preserving delicate, minimal, and primitive handles that would have evolved for food crops spread across the country [16]. A significant portion of this decline could be attributed to horticultural innovation, particularly in smallholder farming systems, as well as lower food costs and rising country incomes. Simultaneously, it has been linked to high energy consumption levels. Excessive agrochemical use and monocropping have resulted in natural corruption in some areas, including inefficient water use and undeniable levels of manure runoff, pesticide impacts, loss of agrobiodiversity, soil pollution, and land degradation. In this way, agricultural intensification has been both a boon and a bane, emphasizing the critical importance of including maintainability in any subsequent intensification plan [17].

By and large, expanding agribusiness has increased global food production and enabled increased average per capita food consumption in many parts of the world—despite current slowing efficiency growth rates. Simultaneously, agriculture in other parts of the world has continued to perform below its potential due to a lack of utilization of diverse sources of knowledge. This is true throughout much of Africa, particularly in areas where horticultural efficiency has advanced slowly or not at all, with a few notable exceptions. When combined with rapid population growth, Africa's low rates of horticultural improvement have resulted in many countries shifting from net food exporters to net food merchants.

Agriculture's future challenges—food production, domesticated animals, fisheries, and ranger service—are mind-boggling. Agricultural frameworks must improve in order to meet the growing demand for food, feed, fuel, and fiber. They must ensure equal economic opportunities for ranchers, particularly landless and pursued agricultural professionals, and establish rural labor standards [18]. They should be more efficient and viable in how they utilize and impact the traditional asset base. They should be more shock and change-resistant, as well as better equipped to deal with more severe climate shocks and rising temperatures. They must drastically reduce their greenhouse gas emissions. Additionally, they must provide critical environmental services such as water management, fertilization, flood and disease protection, and assistance with soil ripeness. They must reduce their reliance on petroleum products: sustainable horticulture is largely reliant on long-term, environmentally friendly electricity, and increased energy efficiency. Finally, food waste should be reduced.

We develop framework recommendations for state regulation of agriculture's environmental and energy sustainability.

- Research on the natural and financial effectiveness of best management techniques is encouraged on a local level. Ranchers are frequently skeptical of projects that imply public ideals in areas that have never been inhabited. Elective approaches that are novel to a region should be field-tested to establish their immediate ecological and economic benefits to ranchers.
- Establish a robust data collection and training programme for farmers on nearby natural issues and the effects of their actions on these issues; make results available to the local community in order to improve state-funded education and contribute to more compelling future water quality management. Expanding the existing agricultural information structure would be a beneficial instructional tool (local horticultural consultancy office). Assume ranchers are unaware of a water quality problem or believe it is the result of agribusiness. In that case, education programmes must both educate them about the issue and assist them in comprehending how their cultivating practices contribute to the problem.
- Offer a broad range of educational, specialized, and financial services.
- There are some impediments to ranchers adopting alternative management practices that cannot be addressed entirely through a single type of assistance. Schooling can educate makers about creative practices [19], specialized assistance reduces the private cost of obtaining information about a specific method on a specific ranch, augments administrative ability that may be lacking, and monetary assistance overcomes a short planning horizon, enabling the rancher to recognize more severe dangers beyond the short run, and acts as a motivator.
- Make recommendations for ways to improve inter-office collaboration, such as collaboration between natural insurance organizations, rural consulting firms,

rural directorates, ranchers associations, neighborhood specialists, and nearby nongovernmental organizations. Neighborhood groups assess asset status and requirements regularly, identify ecological needs and available assets, solicit bids for needed areas, and make programme strategy recommendations. By incorporating ranchers' site-specific data into territorial-level measures, we can strengthen the connection between ranch-level activities and provincial goals [20].

- Modify horticultural structures in ways that contribute to reducing variation in the consumption and nature of water assets and enhancing the environmental benefits of agribusiness water use.
- Develop simple water management strategies that disentangle the financial, ecological, and social costs and benefits of horticulture's water consumption, as well as any associated trades between ranchers, citizens, and customers [21].
- Strengthen existing data on agri-ecological cycles associated with the connections between horticulture, water, and climate, for example, by funding innovative public and private work; amass a superior collection of information at the public and dynamic levels about the hydrological and natural components of water frameworks, as well as the relationship between water assets and water quality and pollution.

References

1. Orlove B, Shwom R, Markowitz E, Cheong SM (2020) Climate decision-making. Annu Rev Environ Resour 45:271–303
2. Bogoviz AV, Sergi BS (2018) Will the circular economy be the future of Russia's growth model? In: Sergi BS (ed) Exploring the future of Russia's economy and markets: towards sustainable economic development. Emerald Publishing, Bingley, UK, pp 125–141
3. Popkova EG, Sergi BS (2018) Will industry 4.0 and other innovations impact Russia's development? In: Sergi BS (ed) Exploring the future of Russia's economy and markets: towards sustainable economic development. Emerald Publishing, Bingley, UK, pp 51–68
4. Popkova EG, Popova EV, Sergi BS (2018) Clusters and innovational networks toward sustainable growth. In: Sergi BS (ed) Exploring the future of Russia's economy and markets: towards sustainable economic development. Emerald Publishing, Bingley, UK, pp 107–124
5. Sergi BS, Popkova EG, Bogoviz AV, Ragulina JV (2019) Entrepreneurship and economic growth: the experience of developed and developing countries. In: Sergi BS, Scanlon CC (eds) Entrepreneurship and development in the 21st century. Emerald Publishing Limited, Bingley, UK, pp 3–32
6. Metcalf L, Eddy HP, Tchobanoglous G (1991) Wastewater engineering: treatment, disposal, and reuse, vol 4. McGraw-Hill, New York
7. Schmidhuber J, Tubiello FN (2007) Global food security under climate change. Proc Natl Acad Sci 104(50):19703–19708
8. Institute of Scientific Organizations (2021) Dataset "Corporate social responsibility, sustainable development and fight against climate change: imitation modelling and neural network analysis in regions of the world—2020" (2020). https://iscvolga.ru/dataset-climate-change
9. Vermeulen SJ, Campbell BM, Ingram JS (2012) Climate change and food systems. Annu Rev Environ Resour 37:195–222

10. Janzen HH, Angers DA, Boehm M, Bolinder M, Desjardins RL, Dyer J, Wang H (2006) A proposed approach to estimate and reduce net greenhouse gas emissions from whole farms. Can J Soil Sci 86(3):401–418
11. Antle JM, Capalbo SM (1991) Physical and economic model integration for measurement of the environmental impacts of agricultural chemical use. Northeast J Agric Resour Econ 20(1):68–82
12. Adger WN, Huq S, Brown K, Conway D, Hulme M (2003) Adaptation to climate change in the developing world. Prog Dev Stud 3(3):179–195
13. Godfray HCJ, Beddington JR, Crute IR, Haddad L, Lawrence D, Muir JF, Toulmin C (2010) Food security: the challenge of feeding 9 billion people. Science 327(5967):812–818
14. Tilman D, Balzer C, Hill J, Befort BL (2011) Global food demand and the sustainable intensification of agriculture. Proc Natl Acad Sci 108(50):20260–20264
15. Gatzweiler FW, Backman S, Sipilainen T, Zellei A (2001) Analyzing institutions, policies, & farming systems for sustainable agriculture in central and eastern European countries in transition (No. 541-2016-38665)
16. Aerts S (2012) Agriculture's 6 Fs and the need for more intensive agriculture. In Climate change and sustainable development, pp. 192–195. Wageningen Academic Publishers, Wageningen
17. Parfitt J, Barthel M, Macnaughton S (2010) Food waste within food supply chains: quantification and potential for change to 2050. Philos Trans R Soc B Biol Sci 365(1554):3065–3081
18. Giovannucci D, Scherr SJ, Nierenberg D, Hebebrand C, Shapiro J, Milder J, Wheeler K (2012) Food and agriculture: the future of sustainability. In: The sustainable development in the 21st century (SD21) Report for Rio, p 20
19. Aakkula J, Kröger L, Kuokkanen K, Vihinen H (2006) Implementation of policies for sustainable development in the context of CAP. In: New challenges for research, pp 1–19. MTT Economic Research
20. Toma L (2003) Policy recommendations for pursuing sustainable agriculture in a small rural community in Romania
21. Foley JA, Ramankutty N, Brauman KA, Cassidy ES, Gerber JS, Johnston M, Zaks DP (2011) Solutions for a cultivated planet. Nature 478(7369):337–342

Agriculture in Developing Countries: Cultural Differences, Vectors of Sustainable Development, Digitalization, and International Experience

Lyudmila M. Lisina, Aydarbek T. Giyazov,
A. L.-Muttar Mohammed Yousif Oudah, and Yuliya I. Dubova ⓘ

Abstract This chapter conducts empirical research aimed at determining the specifics of agriculture in developing countries. A unique sample of developing countries from different geographical regions of the world—i.e., countries with vivid and strong cultural differences—is formed. For different geographical regions of the world, the authors determine the vectors of sustainable development of agriculture. For this, based on the method of correlation analysis is used to determine the contribution of various potential vectors (digitalization of knowledge, digitalization of technologies, digitalization of economic practices based on IMD "World Digital Competitiveness Ranking 2020", 2021) to the achievement of agriculture's sustainability in each designated geographical region of the world. As a result, the applied recommendations to increase agriculture's sustainability for each region of the world are suggested. A systemic view of the current problems of agriculture's sustainability in developing countries and the current level of this sustainability is formed.

Keywords Agriculture · Developing countries · Cultural differences · Vectors of sustainable development · Digitalization · International experience

JEL Codes O14 · O32 · Q01 · Q13 · Q17 · Q18

L. M. Lisina
Volgograd State Technical University (Sebryakov Branch), Mikhaylovka, Russia
e-mail: lisinmih@list.ru

A. T. Giyazov (✉)
Batken State University, Batken, Kyrgyzstan
e-mail: aziret-81@mail.ru

A. L.-M. M. Y. Oudah
Al-Ayen University, Dhi Qar, Nasiriyah, Iraq
e-mail: mohammed.yousif@mail.ru

Y. I. Dubova
Volgograd State Technical University, Volgograd, Russia
e-mail: dubovaui@mail.ru

© The Author(s), under exclusive license to Springer Nature Singapore Pte Ltd. 2022   65
E. G. Popkova and B. S. Sergi (eds.), *Sustainable Agriculture*,
Environmental Footprints and Eco-design of Products and Processes,
https://doi.org/10.1007/978-981-16-8731-0_7

1 Introduction

Rural output has shifted dramatically in agricultural countries over the past two decades. Since the late 1960s and early 1970s, the World Bank and its numerous agricultural research agencies have successfully advocated for the acceptance of modern (highly synthetic information-based) rural techniques [1], such as the Green Revolution's 'natural occurrence' seeds and promises of massive yields [2]. All ranchers benefited from these cutting-edge practices, including the poor. Rates were anticipated to rise, implying that incomes would follow suit. On the other hand, developing countries cannot afford to rely so heavily on foreign sources of information. The oil and debt crises of the 1970s and 1980s exacerbated this further. The expansion of loan bundles from global monetary foundations was facilitated by the economic and financial turmoil in developing countries. Subsequently, governments seeking financing were required to implement the underlying reform strategies. Since the 1980s, approximately 100 countries have been compelled to implement primary transformation packages. On the one hand, the arrangements slowed progress, while on the other hand, they hampered the transformation of indigenous horticultural output into trade goods [2].

During the preceding two decades, small ranchers from Central to South America, Africa, and Asia shared surprisingly similar experiences. Numerous individuals have been forced to abandon diverse traditional polycultures in favor of monocultures to compete in internationally competitive economic sectors [3]. Arrangements for augmentation administrations and loans, for example, were frequently contingent on ranchers tolerating innovations in trade crops [2, 4–9].

Ranchers have faced a similar constraint in shifting to export crops as nearby prices for staples and traditional yields have plummeted as a result of small subsidized imports from developed nations flooding neighborhood markets. The cycle has been one of deliberate deprivation for the majority of small ranchers. Numerous individuals have been removed from growth environments both indoors and outdoors. Rather than alleviating food scarcity, which has been repeatedly cited as a reason for public interest in rural innovation and hybrid seeds, global food surpluses are increasing, but, unexpectedly, appetite and food insecurity remain critical concerns for the most vulnerable.

2 Materials and Method

Agribusiness remains the primary source of commerce, employment, and income for between 50 and 90% of the population in developing countries. Small ranchers, who account for between 70 and 95% of the farming population, account for the majority of this rate [10]. As a result, small ranchers make up a significant portion of the population. They have historically survived solely through resource generation. Various individuals have experimented with various methods of crop distribution over the past two decades, with varying degrees of initial success but numerous tragic failures. They have reaped relatively few benefits from agricultural diversification or modernization. As a result of the market's globalization, the number of small players has decreased, as detailed below. However, they should not be overlooked financially [11]. Their underestimation has perpetuated an endless cycle of deprivation for segments of society, an unusually imbalanced turn of events, and, as a result, the inability of many non-industrial governments to achieve a level of overall turn of events that is acceptable.

Residents of rural areas are impoverished to the point of having little purchasing power. They do not create a sizable market for home industries in this manner. Since domestic business sectors are typically too small to justify significant financial activity, creation is frequently directed toward obscure economic sectors and metropolitan elites. As a result, economic interest is at an all-time low, making a broad-based, viable turn of events impossible. This results in a disproportionate reliance on unknown business sectors and a lack of underlying motivation (which can result in) increased expectations for the daily comforts of poor people. Simply put, poverty becomes an endless cycle that serves as a barrier to advancement.

As part of the empirical part of this study, we will define the specifics of agriculture in different geographic regions of the world, that is, countries with bright and strong cultural differences. The sample includes both developed and developing countries to obtain the most complete and reliable picture of the world economy as a whole. For different geographic regions of the world, we will define the vectors of sustainable agricultural development. To do this, using the method of correlation analysis, we will determine the contribution of various potential vectors (Knowledge, Digitalization of Technologies, Digitalization of Economic Practices based on the IMD "World Digital Competitiveness Ranking 2021" [12]) in achieving sustainable agriculture in each designated geographic region of the world. The data for the study are given in the Table 1.

Based on the obtained statistics, a correlation analysis was carried out, the results of which are shown in Fig. 1.

The results of the correlation analysis revealed significant differences in the impact of digitalization on the sustainability of agriculture in different geographic regions of the world. For example, in America, the contribution of Knowledge (94.09%), Digital Technologies (99.98%), and Digitalization of Economic Practices (99.53%) to the sustainability of agriculture is equally high. In Asia, the contribution of Digital Technologies (27.34%) to agricultural sustainability is small, while the impact of Knowledge (-44.26) and Digitalization of Economic Practices (-38.64%) is negative.

In Europe, the contribution of Knowledge (99.85%) is much higher than the contribution of Digital Technologies (88.46%) and Digitalization of Economic Practices (60.70%) to agricultural sustainability. In Africa and Oceania, the contribution of Knowledge (81.65%), Digital Technologies (90.20%), and Digitalization of Economic Practices (90.65%) to agricultural sustainability is moderate.

Table 1 Food security and digitalization statistics in 2020

Geographic regions of the world	Country	Food security	Knowledge	Digitalization of technologies	Digitalization of economic practices
Americas	Chile	70.2	49.501	60.318	59.236
	Mexico	66.2	48.874	45.179	44.976
	USA	77.5	97.922	89.927	98.652
Asia	Japan	77.9	70.092	71.773	67.932
	Singapore	75.7	92.031	99.504	87.123
	China	69.3	85.105	71.706	80.004
Europe	Finland	885.3	80.438	86.270	91.184
	Ireland	83.8	68.812	68.134	85.252
	Russia	73.7	67.891	51.653	44.807
Africa and Oceania	South Africa	57.8	43.055	46.216	40.289
	New Zealand	77.0	66.603	75.946	75.023
	Australia	71.3	77.848	81.766	81.302

Source Compiled by the authors based on materials from [12, 13]

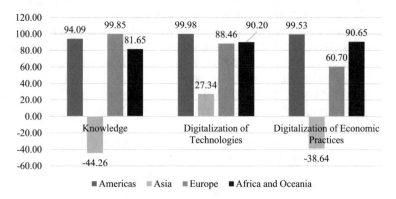

Fig. 1 Results of correlation analysis in the context of geographic regions of the world *Source* Calculated and built by the authors

3 Results

If small farmers in agricultural nations are to survive—and they must work for the development of developing countries—we need more equitable food production models that can continue to provide small farmers with a viable source of income and access to their daily food needs. Ranchers on the periphery lose out as food production systems degrade the environment and soils, while global bureaucrats relocate to greener pastures [14]. We require more environmentally friendly agricultural methods that contribute rather than relying on assets for long-term food production.

Today, agribusiness corporations tout "existence sciences" and their genetically engineered seeds and creation operations as the next revolution in resolving the world's hunger crisis. Increased output, fewer competitors, and more nutritious food varieties are all promised (for example, Nutrient A rice). Surprisingly, as with the previous upheaval, the appetite issue has been improperly prefaced, with the false premise that there will not be enough food to feed the world, rather than the more pertinent question of food transportation and availability [15].

Genetically modified (GM) seeds and production methods will replicate the benefits of the Green Revolution for small-scale ranchers. Additionally, GM crops contain a high concentration of synthetic information. Since the innovation is licensed to multinational corporations, the resulting social distortions will be comparable, if not worse. Small farmers will face similar supply chain challenges. They will be helpless against delegates, contract employees, and exporters, and will be forced to deal with the worst-case climate and harvest scenarios. They will earn comparable returns, if any, to players engaged in pre- and post-farmgate operations, such as international food, transportation, and retail collaborations [16].

On the other hand, natural hazards are almost certain to be just as destructive. Until now, all information has focused on excessive or prolonged pesticide use. Genetic resistance to pesticides is also found in weeds and insects. Since GM crops will also be grown in monocultures, natural threats to biodiversity will exist. Given the unknown metabolic consequences of hereditarily altered yields and their spread to wild cousins, as well as the unknown health consequences for consumers, the harm could be significantly worse. In comparison to current high-input frameworks, alternative agroecological strategies for food production for family and neighborhood consumption can provide stable food supplies, adequate access to food, job security, and natural manageability for long-term food production for the country's poor.

The vectors of sustainable development of agriculture are the following:

1. Input Optimization: On-ranch assets are expanded through the use of sustainable creative practices. Within implied inputs, substituted assets include family labor, serious, personal frameworks, repurposed supplements, vegetable nitrogen, crop pivots, the use of sustainable solar energy, and improved pests, soils, and woods. According to studies, these adjustments can be made while maintaining yields and frequently result in increased net homestead profit. These revenues can be used to benefit the community by expanding nearby retail options and stabilizing the fee structure [17].

2. Diversification: Sustainable agriculture places a premium on diverse pruning and animal frameworks in order to promote soil health and reduce reliance on purchased inputs. Expanding can result in more consistent ranch income by mitigating financial risk associated with the environment, difficulties, and

volatile farming industry sectors. It enables ranchers to stay on the land and protects the local economy from the shock associated with a significant decline in a single product/industry.

3. Natural Capital Conservation: Discounting capital resources is a common accounting technique. Ranchers' average capital has not historically deteriorated as a result of expanding procedures that do not track assets [18]. By and large, the misfortune is genuine, affecting yields, ranch value, and maintainability by optimizing land and water resource efficiency while improving human well-being and the climate; decent horticulture generates financial value.

4. Value-added product capture: Promoting harvests and manufactured goods is by far the most vulnerable link in ranchers' "farm to table" food system mission. Ranchers should advocate for strategies that add substantial value to the homestead to develop and sustain truly sustainable horticulture. While individual ranchers are capable of designing, measuring, and selling their products directly, many other value-added systems require a greater number of assets than a single rancher can afford. As a result, these value-added systems will necessitate the formation of a cooperative of local ranchers and collaboration with the surrounding community.

5. Community: All networks require sustainable agriculture components. To advance sustainable agriculture, we must acknowledge the rural–urban divide, the conflicts, and the enormous liberties. The benefits of a viable producing framework include a shared obligation to benefit, food security, proper food handling, open space for water recharge, a healthy natural environment that supports greenery, wildlife, and recreation, and a supportive and robust social and economic foundation for the local community. Our urban networks are currently distinct from cultivating networks in their thought processes and standard organization, most notably in their understanding of the entire food production and conveyance framework. Recognize the role of agriculture in balancing our local environment; otherwise, we will continue to obliterate our provincial texture and preferred ways of life. As a result, in order to ensure the continued viability of decent regional and metropolitan networks, we must rekindle an appreciation for our neighbors' government assistance.

Let us consider the contribution of various potential vectors to agriculture's sustainability in each of the world's designated geographical regions. Preserving critical assets for agricultural success also requires managing soil in such a way that it maintains its legitimacy as a complex and well-ordered constituent composed of mineral particles, natural matter, air, water, and living organic creatures. Ranchers concerned with long-term viability frequently place a premium on soil health, believing that healthy soil promotes healthy crops and livestock. Maintaining a consistent schedule of soil work demonstrates a commitment to preserving or, in the worst-case scenario, expanding natural soil matter. Natural matter in soil is critical as a source and sink of nutrients, a substrate for microbial action, and a buffer against changes in acridity, water content, and toxin levels. Additionally, soil natural matter development can contribute to the reduction of CO_2 expansion and, as a result, environmental modification. Natural soil matter also plays a critical role in providing a superior soil structure, which results in increased water entry, reduced overflow, improved waste, and increased strength, all of which contribute to the reduced breeze and water disintegration.

Due to a heavy reliance on synthetic manures, agroecosystems have been deprived of critical plant nutrients such as nitrogen and phosphorus internal cycling. While phosphate minerals for compost are being mined at the moment, global stockpiles will last another 50–100 years. As a result, phosphate prices are expected to increase unless new deposits are discovered and waste phosphate recovery is improved [19]. Reusing nitrogen and phosphorus (at the homestead and community level), optimizing compost application, and relying on natural supplement sources (creature and green manure) are all critical components of sustainable horticulture [20]. Supplements are reused through enhanced horticulture, which spatially integrates animals and harvest generation. As a result, broad blended yield animal frameworks, particularly in developing countries, have the potential to make significant future contributions to rural sustainability and global food security.

Social value and equity are frequently brought up in discussions about sustainable horticulture. In the majority of industrialized nations, ranch wages are so low that their horticulture industries rely heavily on temporary labor from less fortunate countries, subjecting ranchers to changing labor arrangements and putting pressure on government social administrations [21]. The ambiguous legal status of a sizable proportion of these workers also contributes to their frequently low income and standard of living, lack of professional stability, lack of advancement opportunities, and deviations from the middle of the world's linked health insurance in other endeavors. Pooling assets among numerous ranchers to improve lodging, dividing work between ranches with varying yields to reduce the irregularity of job openings, teaching laborers to obtain and work their homesteads, and experimenting with novel approaches to provide representatives with adequate medical coverage and educational liberties are largely omitted.

Ranchers lack the financial resources to negotiate higher prices for their bits of feedback and harvests as food producers, advertisers, and homestead input providers consolidate. It obliterates their revenue, leaving many ranchers with few assets to work within natural and working conditions [9]. Ranchers can bolster their financial security by forming, preparing, or promoting cooperatives. Ranchers can further increase their share of the monetary value of their harvests by processing them on-ranch before selling, developing higher value forte yields, developing direct marketing opportunities that bypass agents, and exploring specialist markets. In the long run, agreements that promote solidification can benefit ranchers as well [22].

Due to ranchers' financial strains, many rural networks have become increasingly vulnerable as homesteads and associated rural ventures close their doors. Fiscal and monetary policies that encourage increased agricultural productivity on family farms can lay the groundwork for stronger national economies [23]. Buyers can also contribute within the confines of the market system; their purchases communicate powerfully to manufacturers, merchants, and others involved in the framework what they value, such as natural quality and social worth.

Finally, some of the same financial pressures that have harmed on-ranch manageability have created social value issues for purchasers in low-wage networks, who are frequently left without access to nutritious food as conventional retailers seek to shore up their dwindling net revenues. These issues are addressed through food development and promotion programmes that include community and household nurseries, rancher advertisements, and the incorporation of new neighborhood ranch produce into school lunch programmes and local food cooperatives.

4 Conclusion

Recommendations for enhancing agriculture's sustainability in each region of the world:

1. Citizens of agricultural nations should specify precisely the arrangements for which their economies and enterprises are prepared and from which they can profit. This should be viewed as a critical component of truly differentiated and successful Special and Differential Treatment [24]. Additional new challenges should be examined not because the framework is already overburdened, but because they go beyond the realm of commerce and impinge on states' internal sway. Agricultural nations now require more space for planning, not less, to chart their course of development.

2. In agribusiness, a plurilateral structure implies that non-industrialized nations whose rural areas are unprepared for competition and whose small size means the agreement will have a significant impact on ranchers and food security will seek to withdraw until their ranchers and economies are prepared [25]. This is especially noteworthy in light of the unfair competition and dumping that characterize the trade of many non-industrialized countries with advanced economies. This is not to say that governments should completely shut down, but rather to acknowledge that massive detonation advancement has wreaked havoc on the world's most vulnerable areas and that non-industrial nations with a diverse range of development needs should be able to painstakingly time and build their integration into the global economy.

3. It is critical for small ranchers in agricultural countries to be adequately protected against novel development methods, as the market does not work in their favor, as it does for the major players [26]. Thus, trade policy should ensure the long-term viability of small ranchers and rural poor in developing countries. Additionally, they impose safeguards against dumping and unjustified competition by foreign-sponsored manufacturers. Increasing citizens' security results in more even and equitable development for nations as a whole.

References

1. Smit B, Smithers J (1993) Sustainable agriculture: interpretations, analyses and prospects. Can J Reg Sci 16(3):499–524
2. Shiva V (2000) Why industrial agriculture cannot feed the world. The Ecologist
3. Sannikova IN, Prikhodko EA, Muhitdinov AA (2021) Assessment of the universities impact on global competitiveness based on rankings. In: E3S web of conferences, vol 296, p 08009. EDP Sciences
4. Bogoviz AV, Sergi BS (2018) Will the circular economy be the future of Russia's growth model? In: Sergi BS (ed) Exploring the future of Russia's economy and markets: towards sustainable economic development. Emerald Publishing, Bingley, UK, pp 125–141
5. Sergi BS, Popkova EG, Bogoviz AV, Ragulina JV (2019a) Entrepreneurship and economic growth: the experience of developed and developing countries. In: Entrepreneurship and development in the 21st century. Emerald publishing limited
6. Sergi BS, Popkova EG, Borzenko KV, Przhedetskaya NV (2019b) Public-private partnerships as a mechanism of financing sustainable development. In: Ziolo M, Sergi BS (eds) Financing sustainable development: key challenges and prospects. Palgrave Macmillan, pp 313–339
7. Sergi BS, Popkova EG, Vovchenko NG, Ponomareva MV (2019c) Central Asia and China: financial development through cooperation with Russia. In: Sergi BS, Barnett WA (2019) (eds) Asia-Pacific contemporary finance and development. Bingley, UK, Emerald Publishing
8. Sergi BS, Popkova EG, Bogoviz AV, Ragulina Yu V (2019d) The agro-industrial complex: tendencies, scenarios, and policies. In: Sergi BS (ed) Modelling economic growth in contemporary Russia. Bingley, UK, Emerald Publishing
9. Sheail J (1995) Elements of sustainable agriculture: the UK experience, 1840–1940, pp 178–192. The Agricultural History Review
10. Kirsten J, Sartorius K (2002) Linking agribusiness and small-scale farmers in developing countries: is there a new role for contract farming? Dev South Afr 19(4):503–529
11. Arnon I (1981) Modernization of agriculture in developing countries: resources, potentials, and problems
12. IMD (2021) World digital competitiveness ranking 2021. https://www.imd.org/centers/world-competitiveness-center/rankings/world-digital-competitiveness/
13. The Economist Intelligence Unit Limited (2021) Sustainable development index 2020. https://foodsecurityindex.eiu.com/Index. Last Accessed 11 Sept 2021
14. Kingsbury N (2009) Hybrid: the history and science of plant breeding. University of Chicago Press
15. Reardon T, Barrett CB, Berdegué JA, Swinnen JF (2009) Agrifood industry transformation and small farmers in developing countries. World Dev 37(11):1717–1727
16. Spitzer S (2003) Industrial agriculture and corporate power. Glob Pestic Campaign 13(2):1
17. Francis C, Lieblein G, Gliessman S, Breland TA, Creamer N, Harwood R, Poincelot R (2003) Agroecology: the ecology of food systems. J Sustain Agric 22(3):99–118
18. Rosset PM, Altieri MA (1997) Agroecology versus input substitution: a fundamental contradiction of sustainable agriculture. Soc Nat Resour 10(3):283–295
19. MacRae RJ, Hill SB, Henning J, Mehuys GR (1989) Agricultural science and sustainable agriculture: a review of the existing scientific barriers to sustainable food production and potential solutions. Biol Agric Hortic 6(3):173–219
20. Gliessman SR, Engles E, Krieger R (1998) Agroecology: ecological processes in sustainable agriculture. CRC Press
21. Pannell DJ, Schilizzi S (1999) Sustainable agriculture: a matter of ecology, equity, economic efficiency or expedience? J Sustain Agric 13(4):57–66
22. Bryld E (2003) Potentials, problems, and policy implications for urban agriculture in developing countries. Agric Hum Values 20(1):79–86
23. McSorley R, Porazinska DL (2001) Elements of sustainable agriculture, pp 1–10. Nematropica
24. Hinrichs CC, Lyson TA (Eds) (2007) Remaking the North American food system: strategies for sustainability. U of Nebraska Press

25. Tolley GS, Thomas V, Chung MW (1982) Agricultural price policies and the developing countries. Johns Hopkins University Press
26. Diaz-Bonilla E, Robinson S (2010) Macroeconomics, macrosectoral policies, and agriculture in developing countries. Handb Agric Econ 4:3035–3213

Potential and Opportunities of Organic Agriculture in Russia

Kseniya A. Melekhova⊕, Xenia G. Yankovskaya⊕,
and Alevtina G. Demidova⊕

Abstract The main trends in the development of organic farming in Russia, Kaza-khstan, and Ukraine are revealed, the features of the formation of a system of control and certification of organic agriculture in Russia, the advantages, and potential opportunities for the development of organic production in agriculture are highlighted. The main stages of the transition from the traditional farming system to the organic farming system are described. Conclusions about contrasting organic and traditional farming in the face of changes in the world's population are formulated; in the coming decades, Russia should focus on the balanced development of both farming systems.

Keywords Organic production · Agriculture · Licensing activity · Quality control · Environmental certification · Power quality · Conversion processes

JEL Codes Q53 · Q55 · Q150

1 Introduction

Organic Agriculture is one of the promising and sustainably developing areas in all countries of the world. Square under the production of this product is constantly increasing, about 2.2 million producers of organic products are currently certified, and most of them (about 75.0%) operate in developing countries.

The concepts of "Environmental Certification" and "Organic Certification" should be distinguished. Environmental certification includes certification of the qualitative

K. A. Melekhova (✉)
Institute of Economics of the Russian Academy of Sciences, Moscow, Russia
e-mail: kschut@mail.ru

X. G. Yankovskaya
Gorno-Altai State University, Gorno-Altaysk, Russia
e-mail: ksysha.78@mail.ru

A. G. Demidova
Peoples' Friendship University of Russia, Moscow, Russia
e-mail: al.demidowa2011@yandex.ru

© The Author(s), under exclusive license to Springer Nature Singapore Pte Ltd. 2022   75
E. G. Popkova and B. S. Sergi (eds.), *Sustainable Agriculture*,
Environmental Footprints and Eco-design of Products and Processes,
https://doi.org/10.1007/978-981-16-8731-0_8

composition of raw materials and finished products; the production process should be also based on environmental principles. Organic certification includes all elements of the product life cycle; the entire chain of the goods is "from the field to the table". The period of passing certifications is also different—environmental once every three years, and organic products must be certified every year.

Nowadays, to have the right to label a product as organic, an enterprise should pass the certification of organic products and certification of organic production. The first implies testing of products and raw materials for compliance with the established regulatory requirements. The second is to check the production process of organic products or raw materials. These procedures predetermined the features of the development of Russian organic production: in contradistinction to most countries of the world, where the segment of environmentally friendly products is a niche of small forms of business, in Russia, organic agriculture is developing mainly in enterprises, which are part of large agro-holding (Economy-APK LLC, Agrivag LLC, and others), including international scale. There are many factors for the emergence of such behavior of commodity producers, including the unresolved financial aspects of the "certification" entry into the market of organic agricultural products, amounting to 300–800 thousand rubles per year [1].

2 Materials and Methods

The research is carried out in order to systematize the problems in the area of organizing organic farming in Russia. The sources of information on the area of certified agricultural land in the countries of the world, including the EU countries, Russia, Kazakhstan, and Ukraine, were the number of enterprises certified in international certification centers, for example, the Eurostat, European, and global organic farming statistics database (FiBL Statistics). The statistical data are processed using standard MS Excel functions.

3 Results

More than 1% of the world's agricultural land is used for organic production. 180 countries of the world are engaged in organic agriculture and almost half of them have their own regulatory and legal framework in the area of production and circulation of organic products, including and in Russia (Federal Law No. 280-FZ "On Organic Products" [2]). The growth rate of organic products is constantly increasing, so, this trend will continue in the future (the growth rate of organic production is 12–15%).

In 2018, compared to 2012, the total area for organic production in the EU increased by 33.7%. The largest increase occurred due to countries, such as Bulgaria (129.1%), Croatia (223.4%), Ireland (124.8%), France (97.3%), Italy (67.7%), and Hungary (60.3%). In general, for the countries of the European Union for the period

2012–2019 the area of agricultural land involved in the production of organic prod-ucts increased by 1.42 times from 10.0 million hectares to 14.2 million hectares and their share in the total area of agricultural land in the EU countries—from 5.66 to 7.92%. In some countries, the share of agricultural land allocated for the production of organic products reached 20.43% in 2019 and 25.33% in Sweden and Austria (Table 1).

The main conditions that influence the full development of the organic market can be singled out: assistance to agricultural producers from the state, private investors, including while attracting international capital; development of logistics infrastruc-ture (distribution channels, forms of retail and wholesale trade, fairs, shopping

Table 1 Area of agricultural land (excluding vegetable gardens) involved in the production of organic products in the EU countries

GEO		2012	2015	2019
European Union	ha	10,015,993	11,105,856	14,252,939
	% of total agricultural area	5.66	6.20	7.92
Spain	ha	1,756,548	1,968,570	2,354,916
	% of total agricultural area	7.49	8.24	9.66
France	ha	1,030,881	1,322,911	2,240,797
	% of total agricultural area	3.55	4.54	7.72
Italy	ha	1,167,362	1,492,571	1,993,225
	% of total agricultural area	9.30	11.79	15.16
Germany	ha	959,832	1,060,291	1,290,839
	% of total agricultural area	5.76	6.34	7.75
Austria	ha	533,230	552,141	671,703
	% of total agricultural area	18.62	20.30	25.33
Sweden	ha	477,684	518,983	613,964
	% of total agricultural area	15.76	17.14	20.43
Czechia	ha	468,670	478,033	535,185
	% of total agricultural area	13.29	13.68	15.19
Greece	ha	462 618	407 069	528 752
	% of total agricultural area	9.01	7.69	10.26
Poland	ha	655 499	580 731	507 637
	% of total agricultural area	4.51	4.03	3.49
United Kingdom	ha	590 011	495 929	459 275
	% of total agricultural area	3.41	2.89	2.62
…	…	…	…	…
Bulgaria	ha	39,138	118,552	117,779
	% of total agricultural area	0.76	2.37	2.34

Source Compiled by the authors based on Eurostat [3]

centers, goods storage system, product packaging, transportation, etc.); creation of a regulatory framework in the area of organic farming (for example, the law "On organic agriculture", which has been adopted in more than 135 countries); effective control by state authorities over the implementation of the regulatory framework; training in the marketing of organic products, their production, primary, and/or more deep processing.

In Russia, scientists began to develop research in the direction of farming on an organic basis only in the 90s. However, the problems that existed during this period (the collapse of the USSR and the rupture of economic ties with many union republics, the reforming of the agro-industrial complex economy, including a fundamental change in the system of state regulation of agricultural production, the transformation of agricultural marketing systems, and the relationship of partners in the agro-industrial complex) slowed down this process for a long time [4, 5]. And in the early 2000s, enterprises, which weren't only engaged in the production of organic products, but also sought to occupy a highly marginal and unoccupied niche in this market, began to appear.

At present, in Russia, close attention is paid to the development of organic production; however, official statistics don't keep records of organic products and lands with the conduct of organic farming and animal husbandry.

In Russian Federation, 674.4 thousand hectares of agricultural land are involved in organic agriculture, which is 20.0 times higher than the level of 2007 (the average annual growth rate exceeds 28.3%). Among the countries, which were previously part of the USSR and have a significant potential for growth in the production of organic products, Russia in 2019 was the leader in terms of land area with organic production, ahead of Ukraine (468.0 thousand hectares) and Kazakhstan (294.3 thousand hectares) (Table 2).

However, according to the research results of Kruchinina, Russia in 2015 was significantly inferior to Ukraine in terms of the number of producers (210 enterprises) and processors (110) of agricultural products [7]. According to the data for 2019, the number of Russian organic producers certified in the EU or the USA was 8.2 times lower than certified economic entities of Ukraine (57 and 410 organizations) (Fig. 1).

According to most common estimates, the share of Russia in the world market for organic products reaches 0.15–0.20% [8] and there is a significant reserve for increasing production, because of the potential for increasing this level to 10–15% [9, 10]. On January 1, there were 130 certified business entities in Russia (48 of them received Russian certificates, 70—international, 12 enterprises passed double certification), and about 30–50 organizations were at various stages of conversion. Organic products are sold on the territory of the Russian Federation and abroad.

The conversion (transitional) stage is a prerequisite for passing international and Russian certification. In general, the following stages of the transition to environmentally friendly agricultural production can be distinguished (it's considered using the example of organic farming) [9]:

Table 2 Agricultural lands of some post-Soviet countries involved in the production of organic products, ha

Years	Countries						
	Belarus	Kazakhstan	Kyrgyzstan	Russian Federation	Tajikistan	Ukraine	Uzbekistan
2007	0	2,393	15,147	33,801	0	249,872	1,854
2008	0	157,176	9,867	46,962	69	269,984	2,530
2009	0	134,861	11,415	78,448	69	270,193	324
2010	0	133,561	15,040	44,016	390	270,226	65
2011	0	196,215	15,097	126,847	460	270,320	209
2012	0	291,203	2,696	146,250	12,771	272,850	213
2013	0	291,203	2,856	144,253	98	393,400	213
2014	0	291,203	6,929	245,845	201	400,764	0
2015	0	303,381	7,565	385,139	3,800	410,550	0
2016	0	303,381	7,973	315,154	7,013	381,173	0
2017	1,338	256,741	19,327	479,828	4,920	289,000	0
2018	1,360	192,133	36,748	606,974	8,806	309,100	943
2019	1,374	294,289	19,053	674,370	10,340	467,980	931

Source Compiled by the authors based on FiBL Statistics—European and global organic farming statistics [6]

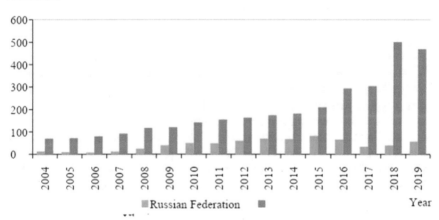

Fig. 1 The number of Russian and Ukrainian producers of organic products certified in international certification centers, units *Source* Compiled by the authors based on FiBL Statistics—European and global organic farming statistics [6]

1. Preparation for the transition (conversion): self-assessment by agricultural producers of the available internal reserves of greening the processes of soil cultivation and plant care, the duration of growing crops without the use of mineral fertilizers, genetically modified organisms, pesticides, and growth regulators. Assessment of the prospects and channels of product sales, selection of certification standards (Russia, EU, USA), assessment of the compliance of the implementation of the main elements of the system of conducting industries with the requirements of standards;

2. Conversion (transition period) from 1 to 3 years: conclusion of an agreement with a previously selected certification authority, passing some on-site checks to audit technological processes, seeds that are used, fertilizers, plant protection products, storage, and processing of products (primary and/or deeper industrial), conformity assessment by several other criteria. Obtaining a certificate in compliance with the requirements of standards;

3. Functioning in the "organic" status: posting information about the location, volumes, and prices of agricultural products on the electronic site of the certification body, which allows interaction with buyers from many countries. Prices for organic products on the world market, as a rule, are 1.5–2.0 times higher than prices for non-certified products [10–12].

The motivation for certification of agricultural products is to obtain a higher profitability of products, but, at the stage of conversion and in the status of "organic", commodity producers face a number of problems: the use of biological products requires a special technology for their use, certification authorities may limit the list of biological products permitted in organic farming, require justification of the need for their use in production processes; a decrease in the yield of agricultural crops to 30.0–40.0%, depending on the previously used technology for their cultivation (on the other hand, studies, which were conducted on cereals and soybeans in the United States, suggest, that the yield of agricultural crops obtained in organic farming is comparable to the yield in the traditional farming system, if certain conditions are met, including technological conditions [13, 14]); certification authorities evaluate not only the process of growing crops, but also the process of storage, processing of finished products, the compliance of the used circulating assets with the requirements of standards; not only the territory and infrastructure of an economic entity is subject to verification, but also territories and objects, which are located near it.

4 Conclusion

The potential for the development of organic production in Russia is determined not only by significant tracts of agricultural land but also by the established management practice of agricultural commodity producers, whose financial condition didn't allow large-scale chemicalization of crop production for a long period. As a result, the level of environmental pollution is insignificant, and the soils have passed the phase of natural self-recreation. The development of organic farming will make it possible to introduce into economic circulation unused agricultural land now, provide additional employment and income for rural residents, and improve the quality and environmental friendliness of agricultural products. However, it's premature to oppose organic and traditional farming in conditions of changing the global population. As it has already been noted by Perfiliev [15] and Schulze [16], the main attention should be focused on the balanced development of both farming systems in the coming decades.

References

1. Alekhin VT (2019) Problems connected with the transition to organic farming. Plant Prot Quar 3:10–11
2. Federal Law of August 3, 2018, No. 280-FZ "On Organic Products". http://www.fao.org/fao lex/results/details/ru/c/LEX-FAOC178322/. Last Accessed 27 Mar 2021
3. Eurostat (2021). Organic crop area by agricultural production methods and crops (from 2012 onwards). URL: https://appsso.eurostat.ec.europa.eu/nui/show.do?dataset=org_cropar& lang=en (Accessed: 10.06.2021).
4. Avarsky ND, Astrakhantseva EY (2017) Methodological aspects of the development of organic agriculture in Russia. AIC Econ Manag 8:38–56. https://doi.org/10.33305/178-38
5. Vorobyova VV, Vorobyov SP, Shlegel SV (2021) Economic viability of the specialized agricultural micro-enterprises in the Altai Krai. In: IOP conference series: earth and environmental acience, p 670. https://doi.org/10.1088/1755-1315/670/1/012051
6. FiBL Statistics—European and global organic farming statistics. The World of Organic Agriculture 2021. https://www.organic-world.net/yearbook/yearbook-2021.html. Last Accessed 10 June 2021
7. Kruchinina FS (2017) State regulation of organic products market in Russia. Bull Voronezh State Univ Eng Technol 79:296–305. https://doi.org/10.20914/2310-1202-2017-2-296-305
8. Mironenko OV (2017) The organic market in Russia: results and prospects. Meat Technol 8:38–43
9. Korshunov SA, Lyubovedskaya AA, Asaturova AM, Ismailov VY, Konovalenko LY (2019) Organic agriculture: innovative technologies, experience, prospects. Rosinformagrotech
10. Union of Organic Agriculture (2021) The project organic agriculture—new opportunities. The system and practices of responsible land use, sustainable development of rural areas. https://soz.bio/wp-content/uploads/2021/02/metodicheskie-rekomendacii-itogovyy-sbo rnik-may-2021.pdf. Last Accessed 10 June 2021
11. Rozhkova DV (2019) Organic production as a priority direction of the development of the "green" economy. Bull Nizhny Novgorod State Eng Econ Univ 2:59–68
12. Vorobyova VV, Vorobyov SP, Shmakov AA (2019) The Factors Determining Profitability of Grain Production in a Region. In: Environment. technology. Resources: proceedings of the 12th international scientific and practical conference. https://doi.org/10.17770/etr2019vol1.4036
13. Bundesministerium für Ernährung, Landwirtschaft und Verbraucherschutz (2021). Die wirtschaftliche Lage der landwirtschaftlichen Betriebe. Buchführungsergebnisse der Testbetriebe des Wirtschaftsjahr 2017/2018. https://www.bmel-statistik.de/fileadmin/daten/BFB-011 1001-2018.pdf. Last Accessed 27 Mar 2021

14. Criticism and frequent misconceptions about organic agriculture: the counter-arguments (2008). http://infohub.ifoam.bio/sites/default/files/page/files/misconceptions_compiled.pdf. Last Accessed 22 Oct 2015
15. Porfiriev BN (2015) Development of "green agro-economics" in Russia – long-term response to sanctions and strategic line of national agro-industry modernization. Russ Econ J 1:110–116
16. Schulze E, Pakhomova NV, Nesterenko NY, Krylova JV, Richter KK (2015) Traditional and organic agriculture: analysis of comparative efficiency from the position of the sustainable development concept. St Petersburg Univ J Econ Stud 4:4–39

Agflation as a Threat to Food Security: Analysis of Inflationary Factors

Oleg P. Chekmarev⬤, Pavel M. Lukichev⬤, and Alexander N. Manilov⬤

Abstract The paper aims to identify the key factors of agflation processes in the Russian economy and the formation of applications for long-term stabilization of prices of agricultural and food products. The authors use a market equilibrium and food balance model to identify inflationary factors. The empirical basis of the research includes the official sources of statistical information of Russia, other countries, and international organizations. The research shows that the main factors in developing agflation in Russia are imperfect planning of the volume of sown areas, production, and exports. The shortage of migrant labor contributes to the reduction of the potential supply of agricultural products. The paper proposes long-term directions for smoothing agflation in Russia.

Keywords Agflation · Food security · Labor migration · COVID-19 · Factors of price growth

JEL Codes Q11

1 Introduction

The COVID-19 pandemic has had and continues to have a strong impact on agricultural development. This impact is most evident in the growth of food prices (agflation) and the deterioration of national food security. The term agflation was introduced during the global financial crisis of 2007–2009 by H. Rasko and R. Bernstein,

O. P. Chekmarev (✉) · A. N. Manilov
Saint-Petersburg State Agrarian University, Saint Petersburg, Russia
e-mail: oleg1412@mail.ru

A. N. Manilov
e-mail: manilov_alex@mail.ru

P. M. Lukichev
Baltic State Technical University "VOENMEH", Saint Petersburg, Russia
e-mail: loukitchev20@mail.ru

© The Author(s), under exclusive license to Springer Nature Singapore Pte Ltd. 2022   83
E. G. Popkova and B. S. Sergi (eds.), *Sustainable Agriculture*,
Environmental Footprints and Eco-design of Products and Processes,
https://doi.org/10.1007/978-981-16-8731-0_9

the economists of the investment bank Merrill Lynch, who combined two terms—agriculture and inflation—to describe the impact of the agricultural sector on price increases [10].

The last year and the beginning of 2021 were marked by global trends of agflation due to the COVID-19 crisis. According to the UN Food and Agriculture Organization (FAO), the FAO Food Price Index (FFPI) in April 2021 was 30.8% higher than last year's corresponding period. This indicator was rising for the eleventh consecutive month [4]. Such price increases affect the affordability of food for the population. The COVID-19 pandemic also leads to a decrease in food production, which raises the issue of food supply and food availability.

The COVID-19 crisis caused a deterioration in the global and national food markets due to the breakdown of agricultural supply chains, the reduction of migrant labor in agricultural production, the localization of food production, and increased government bans on the movement of resources and products between countries. According to N. Njegovan and M. Simin, this fact increases the cost of resources of the first sphere of agro-industrial complex, global agflation processes [12], and related problems [2].

The issue of agflation is deepened since agricultural production is currently on the way to radical change [4].

Numerous studies link agflation and problems of food insecurity to changes in the migration situation. Labor migration is part of a strategy to secure livelihoods for the countries of origin and diversify agricultural production risks for the host countries. According to International Organization for Migration (IOM), COVID-19 had a major impact on migration, leaving 272 million international migrants vulnerable [8]. As a result, poor countries, heavily dependent on food imports and remittances, suffered from malnutrition and hunger caused by the COVID-19 crisis [5, 17]. Internal migration, which is about two and a half times the scale of international migration, also suffers due to lockdown [17]. Internal migration is especially important for large countries such as the USA and Russia. Migrant workers play a crucial role in global food production and supply chains, doing more than 25% of all work [9].

The COVID-19 crisis also greatly affected agricultural production in developed countries. Quantitative estimates show that travel restrictions and diseases associated with COVID-19 resulted in a shortage of 80,000 farm workers in the UK, about 70,000–80,000 workers in Spain, and 250,000 workers in Italy [9].

In recent years, Russia has become more and more active in the world market of agricultural products. Not only are agricultural producers dependent on the supply of exported machinery, equipment, seeds, and other resources, but there is also an increase in exports of agricultural products [6]. In such a situation, the global rise in food prices also causes problems with domestic agflation [15]. According to statistics [13], the growth of food prices in Russia in 2020 and early 2021 was lower than the global average. However, this indicator differed much from earlier growth rates. In addition to the export component, Russian researchers and analysts note the following agflation factors:

- Risks of opportunistic behavior of food producers on consolidated markets (e.g., sunflower oil market) [15];
- Increased price of imported products due to the ruble's devaluation [14];
- Shortage of foreign migrant labor for agricultural production in Russia.

Despite the sufficiently large number of studies, the available studies poorly present factor analysis of the causes of agflation in Russia, taking into account the significance and direction of the influence of certain factors on the growth of food prices. For example, the factor of reduced international labor migration is recognized as a factor of global and Russian agflation, but the mechanisms and degree of its influence on inflationary processes are not assessed practically. Based on the formulated problem, the paper aims to identify the priority factors of Russian agflation during the COVID-19 pandemic and determine the appropriate directions of state policy in the stabilization of food prices in the long term.

The inflation in the Russian food market is closely related to the global situation, making it necessary to study global changes in food prices, especially for the main positions of Russian exports. It is clear that Russian agflation in a pandemic, especially in the second wave of the COVID-19 crisis (fall-winter 2020), is not associated with an increase in public demand for food, which allows us to narrow the search field to the study of supply factors in food markets. Based on this reasoning and the research purpose, we can identify the following research objectives:

- To assess the structure of the cost of agricultural production and agflation potential within its most significant elements;
- To analyze planning factors, potential, and influence of uncontrollable parameters;
- To rank the main factors of Russian agflation;
- To develop recommendations in the field of stabilizing food prices and ensuring sustainable economic access to food.

2 Materials and Methods

The basic approach used in this research to assess agflation factors is a market equilibrium model. This model determines the price dynamics by the interaction of supply and demand for agricultural products and food. Supply and demand are considered based on food balance models.

The analysis is based on statistical information published by the official statistical agencies of certain countries and international organizations. The authors analyzed the ratio of the growth rate of resource prices to food prices or retail prices of agricultural producers to identify the contribution of various supply factors to agflation. The research also considers the possible impact of resource markets on agflation through rising costs and shortages of labor, spare parts, and other production factors. However, it is rather problematic to estimate their precise contribution to agflation due to the lack of detailed statistical information.

Simultaneously, the lack of open sources of information and the limited scope of this research do not allow for a detailed study of each agflation factor. Therefore, the results of this research should be treated as primary and require additional verification with the appearance of new statistical information and its inclusion in the analysis.

3   Results

As noted above, all-Russian inflation for food products is lower than in the rest of the world. However, it is higher than in many economically developed countries. (e.g., the USA). Along with that, following global trends, in 2020 and at the beginning of 2021, Russia also experienced a considerable increase in food prices. Relative to 2019, price increases have more than doubled and are at least 40% higher than in 2018 [6, 16].

Russia is also marked with other trends. The rise in food prices significantly outpaces the rise in nonfood prices and occurs against the background of decreasing real incomes of the population, which reduces the affordability of food for low-income groups of the population.

Let us consider the cost structure of agricultural production to understand the source of supply-side agflation better. The data [6] show that, compared to the statistical average for the agricultural economy, material costs (65% of total costs) and labor costs (18% with deductions to social funds) dominate in agriculture, taking into account the high volatility in the sub-branches. Therefore, further investigation of the factors of supply inflation is conducted in these two groups of costs. According to the Federal State Statistics Service (Rosstat), the share of labor in the cost structure in agriculture is relatively higher than in the economy as a whole (higher by 12.5% on the payroll with deductions). This figure increases significantly in natural terms, given the relatively lower wages in agriculture than in other industries.

Studying the factor of the growth of material costs (Table 1), we can conclude that domestic prices of manufacturers of industrial products started to actively grow only in Q1 2021. The material costs of agricultural producers on purchasing domestic

Table 1 Price indices

Price index indicator	2018	2019	2020	January–March 2021
Producer price index of industrial goods	111.7	95.7	103.6	111.1
Producer price index of agricultural products	112.9	95.5	113.1 (incl. 110.2 for Q4 2020)	106.2
Tractors (domestic production)	93.4	107.6	110.6	–
Forage crops	106.2	112.1	111.5	–

Source Compiled by the author based on [13]

resources grew at a modest rate throughout 2020. A surge of agflation in prices of agricultural producers emerged in the fourth quarter of 2020, which accounted for almost 80% of the annual increase in the price index (110.2% out of 113.1%). Thus, inflation of material resources cannot explain the observed dynamics of prices for agricultural products. Simultaneously, price indices for some items of material costs (tractors, fodder crops, etc.) significantly increased (increased prices for fodder, the rise in import price due to the ruble's devaluation by about 25%, etc.).

The second factor of agflation in the country is the labor force. Unlike many other sectors of the economy, production in agriculture during the lockdown period did not practically stop. Therefore, we cannot speak of serious problems with the employment of the indigenous population in agriculture. However, Russian agriculture extensively uses migrant labor with differentiation by sub-branches and cultivation periods. Considering the available statistics on the prevalence of foreign migrants in the farms of the Leningrad Region, we can estimate the total pre-COVID need for foreign labor in agricultural production as at least 400–500 thousand people. This figure, based on Rosstat data on the employment structure in the economy (IOM, 2020b), amounted to 10%–15% of all people employed in the agricultural sector.

Data from the Ministry of Internal Affairs of the Russian Federation show a sharp decrease in the flow of labor migrants from abroad during the pandemic (Table 2). In 2020, the flow of migrants aiming to enter Russia to work decreased by more than two times and almost three times under officially issued patents. Statistically, these figures do not include migrants from EAEU countries. However, the citizens of these countries were also subject to severe restrictions on international movements during the pandemic.

Thus, an acute shortage of migrant workers arose in the Russian labor market. Its impact on agflation can develop in two main ways through the growth of wages and reduction of production. Wages in agriculture increased only by 9.3%, which could give no more than a 1.6% increase in prices to the total agflation. However, data from previous years show a similar increase in wages [9].

The shortage of human resources could have caused a decline in agricultural production, which, in turn, would have led to a shortage and an increase in prices due to a supply shock. In our view, this scenario is more significant in terms of its impact on inflation. Unfortunately, an accurate assessment of its impact is challenging due

Table 2 Dynamics of official labor migration

Indicator	2018	2019	2020	Q1 2020	Q1 2021
Patents issued, thousand pcs	1671.7	1767.3	1132.6	426.7	228.6
Number of valid patents at the end of the reporting period, thousand pcs	n.d	n.d	n.d	1670.3	961.9
Migration registration for the purpose of entry "Work"	5047.8	5478.2	2358.8	1203.6	1316.9

Source Compiled by the author based on [11]

to the lack of statistics on foreign migrants by industry. However, the limitation of agricultural production due to the shortage of migrants is pointed out in some studies [7]. The shortage of migrant labor in agriculture caused by closed borders has also been exacerbated by the overflow of migrants remaining in the country into alternative (higher paying) sectors of the economy. With the impossibility of entering Russia due to the COVID-19 pandemic, the distribution of migrants by industry starts to change as their share in the higher paid industries increases. Thus, within agriculture, the sectors with a high share of migrant labor (vegetable growing, potato farming, etc.) could be the most affected by the labor shortage.

Of particular interest is the issue of the influence of the planning of agricultural production on the manifestation of agflation processes. The planning of sowing areas in Russia and a significant part of the sowing campaign occurred before introducing COVID-19 restrictions. Thus, we can assess the impact of this factor on agflation in terms of compliance between the plans of agricultural production and their real implementation as the gross harvest of agricultural products.

There is a fairly strong correlation between the decline in the sown areas of several crops (sugar beets, potatoes, sunflowers, and forage crops) [3] and the rise in their prices (Table 3). In late 2020 and early 2021, for crops with sharply reduced sown areas and their processed products, the crop industry saw a price increase several times higher than the overall price dynamics for agricultural products. In the case of sugar beet and sunflower, there is a synergy of negative factors. To avoid losses from the cultivation of sugar beet (after overstocking in 2019), the area sown under this crop was reduced by 19.1%, while weather conditions led to a decrease in gross yields by more than 40% for the year [3]. The result was a sharp increase in retail prices for sugar, which forced the government to take emergency and non-market measures to curb prices. A similar situation developed with sunflower and potato. Nevertheless, sunflower and grain showed a record rate of exports due to the increase in world prices for sunflower oil and grain crops [9], which led to an even greater shortage of products on the Russian market and affected the dynamics of prices for livestock products. Table 3 shows that after the grain harvest and increased exports, prices for chicken eggs, for example, rose quite rapidly (by 32% between October and December 2020), although other inflationary components also influenced this increase. Thus, the sharp increase in exports has led to the development of price dynamics for the relevant products on the Russian market. Let us note that the planning of sowing areas and the dynamics of exports are the parameters that can be potentially controlled by the government (mechanisms of government support and customs regulation) and industry unions and associations. Nevertheless, the fluctuations in these factors play a primary role in developing a supply shock and further agflation.

It is important to note that against the background of factors contributing to the manifestation of agflation processes, factors restraining the price dynamics acted as well. While agricultural organizations have been experiencing a decline in sown areas for a decade, the development of peasant (farm) enterprises (P(F)Es) shows the opposite dynamics. From 2010 to 2020, the sown area on farms increased by more than 9 million hectares, against a drop of 3.6 million hectares in agricultural organizations. Similar dynamics are characteristic of gross yields [13]. In the pandemic

Table 3 Price indices of agricultural producers. Influence of the world conjuncture

Products/Price Indexes	To the previous month									
	2020							2021		
	Jan	Feb	Mar	Sep	Oct	Nov	Dec	Jan	Feb	Mar
Products and services of agriculture and hunting	99.5	100.24	100.06	100.49	103.01	103.1	103.81	101.79	102.61	101.68
Wheat	100.79	101.51	101.27	101.11	103.56	104.24	102.9	101.36	101.11	99.9
Corn	99.43	100.97	101.09	100.17	101.38	105.29	105.32	102.62	101.67	101.43
Sunflower seeds	100.08	101.36	101.89	101.53	111.59	114.05	112.96	104.61	103.74	104.95
Potatoes	100.42	100.67	101.91	98.05	103.69	105.9	104.89	104.97	103.43	104.45
Sugar beet and sugar beet seeds	100.99	98.24	99.89	105.73	128.44	113.41	101.37	99.82	98.76	99.86
Fresh chicken eggs	91.27	92.04	104.73	101.62	108.88	107.48	112.82	100.56	102.52	104.91

Source Compiled by the author based on [13]

year, in contrast to agricultural organizations, farmers increased the volume of sown areas for forage crops and vegetables in the open field and significantly reduced the sown areas under potatoes. Thus, we can say that farmers are a factor stabilizing agricultural production. Therefore, farmers are a stabilizing factor in agrarian processes, especially in areas of extensive labor application (due to less dependence on migrant labor). Considering other positive effects [1], collective farms and other small forms of farming in rural areas are already a significant element in ensuring food security and its sustainability.

Based on the research results, we can propose the following measures to stabilize agflation processes in Russia and improve the sustainability of economic affordability of food in Russia:

1. Development of a system of planning of cultivated areas and exports at the level of the government and industry unions (and cooperatives), taking into account the balance of agricultural products;
2. It is advisable to form state reserves and commodity interventions for long-term storage goods (cereals, sugar, frozen meat, etc.) to eliminate planning errors and smooth out uncontrollable factors (e.g., weather conditions);
3. Formation of a transparent, stable, and flexible mechanism of customs regulation preserving the market motivation of economic agents of the agrarian sphere but supporting the country's food security;
4. Development of small forms of farming in rural areas, settlement of rural areas with increased support for cooperation to stabilize production volumes and reduce dependence on migrant labor;
5. Import substitution of the raw material base of the agro-industrial complex (seeds, hatching eggs, machinery, etc.) and the formation of strategic reserves of spare parts for imported machinery and other imported resources with a significant shelf life within the framework of industry unions (upper-level cooperatives) with government support.

4 Conclusion

Priority factors of Russian agflation during the COVID-19 pandemic include imperfect planning of production and export volumes of agricultural products. Additional factors of inflation in the food market are a shortage of migrant labor, ruble devaluation, and, to a lesser extent, other components. The impact of migrant labor shortages during the pandemic had more of an impact on agflation through reductions in production relative to potential output than through higher labor costs.

The implementation of the proposed measures can contribute to the stabilization of agflation and ensure the sustainability of the country's food security in general and the economic availability of food in particular.

References

1. Chekmarev OP (2020) Small forms of farming in rural areas are a prerequisite for the sustainable development of Russia. In: Scientific support for the development of the agro-industrial complex under conditions of import substitution (pp. 61–66). Russia, St. Petersburg
2. Devereux S, Béné C, Hoddinott J (2020) Conceptualizing COVID-19's impacts on household food security [Opinion piece]. Food Secu 12:769–772. https://doi.org/10.1007/s12571-020-01085-0
3. Expert-Analytical Center of Agribusiness (2021) Planted areas and yields of major crops. Results for 2020. Retrieved from https://ab-centre.ru/news/posevnye-ploschadi-i-sbory-osnovnyh-selskohozyaystvennyh-kultur-itogi-za-2020-god
4. FAO (2017) The future of food and agriculture: Trends and challenges. Italy, Rome. Retrieved from http://www.fao.org/3/i6583e/i6583e.pdf
5. FAO (2021) FAO Food Price Index. Retrieved from http://www.fao.org/worldfoodsituation/foodpricesindex/en/
6. Federal State Statistics Service of the Russian Federation (Rosstat) (2021) Official website. Retrieved from https://www.gks.ru
7. Guryanov S (2021) Foreign countries will help us: agrarians ask to let migrants into the country. Retrieved from https://iz.ru/1123593/sergei-gurianov/zagranitca-nam-pomozhet-agrarii-prosiat-pustit-v-stranu-migrantov
8. International Organization for Migration (IOM) (2020) The impact of COVID-19 on migrants (Migration Fact Sheet No. 6). Retrieved from https://www.iom.int/sites/default/files/our_work/ICP/MPR/migration_factsheet_6_covid-19_and_migrants.pdf
9. IOM (2020) Migrants & global food supply (COVID-19 Analytical Snapshot No. 18). Retrieved from https://www.iom.int/migration-research/covid-19-analytical-snapshot
10. Lukichev PM (2008) Agflation: The emergence and influence on the development of the agrarian sphere in Russia. Bulletin of the St. Petersburg State Agrarian University 8:97–100
11. Ministry of Internal Affairs of the Russian Federation (2021) Statistical data on the migration situation. Retrieved from https://xn--b1aew.xn--p1ai/dejatelnost/statistics/migracionnaya
12. Njegovan N, Simin M (2020) Inflation and prices of agricultural products. Econ Themes 58(2):203–217. https://doi.org/10.2478/ethemes-2020-0012
13. Rosstat Unified Interdepartmental Statistical Information System (2021) Official website. Retrieved from https://www.fedstat.ru
14. Rudenko MN, Subbotina YD (2021) Food security of Russia. Proceedings of the St. Petersburg State University of Economics, 1(127), 84–90
15. Ternovsky DS, Shagaida NI (2021) Agriculture during the pandemic. Econ Dev Russia 28(1):24–28
16. USDA-ERS (2021) Food Price Outlook Changes in Consumer Food Price Indexes, 2018 through 2021. Retrieved from https://www.ers.usda.gov/data-products/food-price-outlook/summary-findings/
17. World Bank (2020) COVID-19 crisis through a migration lens (Migration and Development Brief No. 32). Washington, D.C.: World Bank. Retrieved from https://openknowledge.worldbank.org/handle/10986/33634

Food Subsystem: Innovative Technologies and Prospects for Rural Development

Svetlana A. Chernyavskaya⊙, Taisiya N. Sidorenko⊙,
Nadezhda A. Ovcharenko⊙, Elena E. Udovik⊙,
and Tatiana E. Glushchenko

Abstract The formation of a food subsystem using innovative technologies in crop and livestock production should become a priority strategic objective of federal and regional governments. Solving this problem will increase the level of food security and the region's competitiveness, create new jobs, and improve the budget revenues. Moreover, it will maximize profits, increase profitability, and improve the quality and availability of food. The Krasnodar Territory, having the necessary competencies, can become a catalyst for creating and developing the region's food subsystem based on innovative technologies. Considering the possibility and feasibility of the project on long-term concessional loans at a rate of 3% per annum and participation in the program of the Krasnodar Territory on the subsidization of the measures related to the reclamation of agricultural land, we can conclude that this project is effective and has a payback period of 3.2 years. In this case, for ten years of this project, the economic entity will obtain the following results: revenue -1.3 billion rubles; interest on the bank loan -29.8 million rubles; net profit -355 million rubles. Implementing these innovative technologies will contribute to the formation of a competitive environment and the construction of an export-oriented food subsystem of the region.

Keywords Accounting · Automation · Control · Analysis · Region · Agriculture · Food subsystem

S. A. Chernyavskaya (✉) · T. E. Glushchenko
Kuban State Agrarian University Named After I.T. Trubilin, Krasnodar, Russia

T. E. Glushchenko
e-mail: glu3630@yandex.ru

T. N. Sidorenko · N. A. Ovcharenko
Russian University of Cooperation, Mytishchi, Russia
e-mail: taisianik@yandex.ru

N. A. Ovcharenko
e-mail: nade-o@yandex.ru

E. E. Udovik
Kuban State Technological University, Krasnodar, Russia
e-mail: ydovik-ydovik@rambler.ru

JEL Codes M410

1 Introduction

The formation of a food subsystem using innovative technologies in crop and live-stock production should become a priority strategic objective of federal and regional governments [1]. Solving this problem will increase the level of food security and the region's competitiveness, create new jobs, and improve the budget revenues. More-over, it will maximize profits, increase profitability, and improve the quality and availability of food. The Krasnodar Territory, having the necessary competencies, can become a catalyst for creating and developing the region's food subsystem based on innovative technologies [9].

2 Materials and Methods

Specialization of agricultural producers in the production of crops and livestock (in particular dairy cattle), as a rule, predetermines the cultivation and harvesting of rough and succulent feed (e.g., silage, hay, haylage, corn for grain, and oats) [6]. In the structure of crop rotation, the area under fodder crops takes 40%–45%. It reduces the share of cash crops, thereby reducing the revenue and net cash flow in conditions of deficit of own working capital and high level of crediting of agricultural producers. Additionally, it is necessary to consider the existing obligations to pay rent to owners of land shares.

On the other hand, crop production in the Northern zone of the Krasnodar Territory, which is considered a zone of risky agriculture due to low average annual precipitation (400–450 mm), is not always profitable because of moisture deficit [3].

In our opinion, the main reserve for revenue and profit growth on a limited crop rotation area is introducing an innovative system of subsurface drip irrigation.

3 Results

Let us consider the possibility of implementing a project of subsurface drip irrigation system on the example of an economic entity (agricultural producer) located in the Northern zone of the Krasnodar Territory, which applies zero-tillage in crop production using a subsurface irrigation system on an area of 576 ha (1 year – 180 ha) (Fig. 1) [5].

The increase of yields of forage crops is discussed in Table 1.

Table 2 shows the release of the area occupied by forage crops due to introducing a drip irrigation system in the Northern zone of the Krasnodar Territory.

Area of arable land with an active drip irrigation system (seven-field crop rotation), total	Unit of measure	Year										
		2022	2023	2024	2025	2026	2027	2028	2029	2030	2031	2032
	ha	180	576	576	576	576	576	576	576	576	576	576
including												
soy		-	90	90	-	90	126	180	90	90	-	90
corn for grain	ha	60	180	90	90	-	90	126	180	90	90	-
alfalfa	ha	90	216	396	396	360	180	180	216	396	396	360
winter wheat	ha	30	90	-	90	126	180	90	90	-	90	126
hay corn for silage	ha	30	90	-	90	126	180	90	90	-	90	126
		Area, ha										
Field 1	30	30	30	30	30	30	30	30	30	30	30	30
Field 2	30	30	30	30	30	30	30	30	30	30	30	30
Field 3	30	30	30	30	30	30	30	30	30	30	30	30
Field 4	30	30	30	30	30	30	30	30	30	30	30	30
Field 5	30	30	30	30	30	30	30	30	30	30	30	30
Field 6	30	30	30	30	30	30	30	30	30	30	30	30
Field 7	30		30	30	30	30	30	30	30	30	30	30
Field 8	30		30	30	30	30	30	30	30	30	30	30
Field 9	30		30	30	30	30	30	30	30	30	30	30
Field 10	30		30	30	30	30	30	30	30	30	30	30
Field 11	30		30	30	30	30	30	30	30	30	30	30
Field 12	30		30	30	30	30	30	30	30	30	30	30
Field 13	30		30	30	30	30	30	30	30	30	30	30
Field 14	30		30	30	30	30	30	30	30	30	30	30
Field 15	30		30	30	30	30	30	30	30	30	30	30
Field 16	30		30	30	30	30	30	30	30	30	30	30
Field 17	30		30	30	30	30	30	30	30	30	30	30
Field 18	30		30	30	30	30	30	30	30	30	30	30
Field 19	36		36	36	36	36	36	36	36	36	36	36
	576	180	576	576	576	576	576	576	576	576	576	576

Fig. 1 Recommended structure of crop rotation when implementing subsurface drip irrigation system. *Source* Compiled by the authors

Table 1 Estimated yield of fodder crops using drip irrigation system in the Northern zone of the Krasnodar Territory

Culture	Unit of measure	Yield	
		Without irrigation	With irrigation
Soy	t/ha	1	4.5
Corn for grain	t/ha	4	14
alfalfa (or):			
Green mass	t/ha	*12.5*	*70*
Green mass dried on the haylage	t/ha	*6*	*43.8*
hay	t/ha	*4.5*	*29.2*
Winter wheat	t/ha	5.5	8
Hay corn for silage (green mass)	t/ha	–	20

Source Compiled by the authors

Table 3 shows the calculation of the cost of electricity in the cultivation of forage crops using a subsurface irrigation system in the Northern zone of the Krasnodar Territory. Table 4 shows the calculation of the estimated cost of production of fodder crops using a subsurface irrigation system in the Northern zone of the Krasnodar Territory.

Figure 2 presents the possibility of using the program of the Krasnodar Territory on the subsidization of the measures related to the reclamation of agricultural land (implementation period up to 2024) [8].

Table 2 Releasing the area of forage crops using drip irrigation system in the Northern zone of the Krasnodar Territory

Indicator	Unit of measure	Total area over ten years of project implementation
Release of the area of arable land for cash crops due to:	ha	**5292**
Introducing the corn for silage into the crop rotation	ha	**876**
Eliminating silage from the diet from May 15 to October 15 (i.e., reducing the corn silage crop)	ha	**2322**
Reducing alfalfa crops as a result of higher yields	ha	**1062**
Increasing the yield of corn on grain	ha	**332**
Excluding the sowing of annual grasses (oats and peas) for haylage		**700**

Source Compiled by the authors

Table 3 Calculation of energy costs for forage crops cultivation using subsurface irrigation system in the Northern zone of the Krasnodar Territory

Culture	Yield, t/ha	Pump power, kW	Pump capacity, cubic meters per hour	Power consumption, kWh	Water quantity per ha per season season, cubic meters	Power consumption per ha, kW	Price of 1 kW without VAT, RUB	Cost of electricity per ha, RUB
Corn for grain	14	160	500	0.32	4185	1339	8.42	11,276
Soy	4.5	160	500	0.32	4787	1532	8.42	12,898
Alfalfa	0.8	160	500	0.32	12,000	3840	8.42	32,333
Winter wheat	8	160	500	0.32	2400	768	8.42	6467
Hay corn for silage (green mass)	20	160	500	0.32	2930	937	8.42	7893

Source Compiled by the authors

Table 4 Calculation of the cost of production of fodder crops using subsurface irrigation system in the Northern zone of the Krasnodar Territory

Culture	Forecast with irrigation, 2020								
	Yield, t/ha	Production costs without irrigation per ha, RUB	Costs due to irrigation, RUB/ha					Production costs per ha, RUB	Unit cost, RUB/kg
			Depreciation per ha, RUB	Repair and maintenance costs per ha, RUB	Cost of electricity per ha, RUB	Other costs per ha (1%), RUB	Total		
Corn for grain	14	51,619	36,228	10,738	11,276	582	58,825	110,444	7.89
Soy	4.5	35,354	36,228	10,738	12,898	599	60,463	95,817	21.29
Alfalfa	0.8	48,769	36,228	10,738	32,333	793	80,092	128,861	1.84
Winter wheat	8	38,965	27,171	8054	6467	417	42,108	81,073	10.13
Hay corn for silage (green mass)	20	25,810	9057	2685	7893	196	19,831	45,641	2.28

Source Compiled by the authors

Program of the Krasnodar Territory on the subsidization of the measures related to the reclamation of agricultural land
Conditions:

Availability of the examined design and estimate documentation (examined in the government or non-government bodies)
Approval for water withdrawal from the Ministry of Natural Resources and Environment of the Russian Federation
Act of commissioning (possibly by stages) for irrigated areas
The cost of the design and estimate documentation is not considered when calculating the subsidy
The cost of work and rates in the acts of work performed and KS-2 must comply with the design and estimate documentation

Subsidy size:

products sold on the domestic market	54822.64	RUB/ha
products sold for export	99892.73	RUB/ha

Indicator	Total	2021	2022
The area of arable land for subsurface drip irrigation system, ha	576	186	390
ACTUAL PROJECT COSTS			
The cost of subsurface drip irrigation system including the entire project with 20% VAT, with delivery to the farm, installation, and commissioning works	159650396	51553774	108096622
The cost of construction of main and branch pipelines for drip irrigation (including pumping station)	49312863	15923945	33388918
The cost of the power line	10000000	4000000	6000000
The cost of deepening the water intake	3000000	1500000	1500000
The cost of design and estimate documentation	4000000	1500000	2500000
Interest on a bank loan (10 years)	29827150		
Total:	**255790408**	**74477719**	**151485540**
Actual costs per ha	**444081**		
COSTS ACCEPTED FOR THE CALCULATION OF SUBSIDY (without VAT)			
The cost of subsurface drip irrigation system including the entire project with 20% VAT, with delivery to the farm, installation, and commissioning works	133041996	42961478	90080518
The cost of construction of main and branch pipelines for drip irrigation (including pumping station)	41094052	13269954	27824098
The cost of the power line	8333333	3333333	5000000
Total:	**182469382**	**59564766**	**122904616**
Actual costs per ha considered for subsidy (excluding VAT)	**316787**	**320241**	**315140**
Possible subsidy size:			
products sold on the domestic market	**31577841**	**10197011**	**21380830**
products sold for export	**57538212**	**18580048**	**38958165**

Fig. 2 The analysis of the possibility of applying the Program of the Krasnodar Territory on the subsidization of measures related to the reclamation of agricultural land (implementation period up to 2024). *Source* Compiled by the authors

The analysis of cash flow [7] on the project of implementing a subsurface irrigation system in the Northern zone of the Krasnodar Territory on concessional lending of the Ministry of Agriculture of the Russian Federation (own funds -20%; borrowed funds -80%) is considered in Fig. 3.

4 Conclusion

Thus, when considering the possibility and feasibility of the project on long-term concessional loans at 3% per annum and participation in the program of the Krasnodar Territory on the subsidization of the measures related to the reclamation of agricultural land, we can conclude that the project is effective and a payback period equals 3.2 years [10].

Over a decade of implementing this project, the economic entity will get the following results: revenues -1.3 billion rubles; interest on the bank loan -29.8 million rubles; net profit -355 million rubles [4].

Implementing these innovative technologies will contribute to the formation of a competitive environment and the construction of an export-oriented food subsystem of the region [2].

Cash flows on the investment project, RUB

Indicator	2021	2022	2023	2024	2025	2026	2027	2028	2029	2030	2031	2032
	0	First year	Second year	Third year	Fourth year	Fifth year	Sixth year	Seventh year	Eighth year	Ninth year	Tenth year	Eleventh year
Investment activities												
Capital expenditures	-74477719	-151485540	-	-								
Balance of investment activities	-74477719	-151485540	-	-								
Operating activities												
Revenue	0	38790088	120814637	132897638	149621709	141229293	134270088	119034018	120814637	132897638	149621709	138655293
Production costs, sales costs, taxes	-	-18299778	-56607914	-60922025	-73881913	-69597213	-67164209	-55240765	-56607914	-60922025	-73881913	-68067573
Depreciation charge	-	6738439	20867425	20867425	20867425	20867425	20867425	20867425	20867425	20867425	20867425	20867425
Balance of operating activities	-	27228750	85074149	92843038	96607222	92499505	87973304	84660678	85074149	92843038	96607222	91455145
Discounted balance from operating activities	-	24750934	70271247	69725122	65982732	57442192	49616944	43430928	39729627	39365448	37290388	32009301
Total balance for the year	-74477719	-124256789	85074149	92843038	96607222	92499505	87973304	84660678	85074149	92843038	96607222	91455145
Cumulative total balance of the project	-74477719	-198734508	-113660360	-20817321	75789900	168289405	256262709	340923387	425997536	518840575	615447796	706902941
Discounting multiplier (d=10%)	1	0.909	0.826	0.751	0.683	0.621	0.564	0.513	0.467	0.424	0.386	0.35
Discounted cash flow from operating activities on an accrual basis	-74477719	-180649668	-93883457	-15633808	51764502	104507721	144532168	174893698	198940849	219988404	237562849	247416029
Financial activities												
Receipt of credit	59582175	121188432										
Repayment of the loan with interest	-	-7745683	-23321432	-22779121	-22236809	-21694497	-21152185	-20609873	-20067561	-19525250	-18982938	-12482408
Balance of financial activity	59582175	113442749	-23321432	-22779121	-22236809	-21694497	-21152185	-20609873	-20067561	-19525250	-18982938	-12482408
Total balance for the year	-14895544	-10814041	61752716	70063918	74370413	70805008	66821119	64050805	65006587	73317789	77624284	78972736
Discounted balance for the year by three types of activities	-14895544	-9829963	51007744	52618002	50794992	43969910	37687111	32858063	30358076	31086742	29962974	27640458
Cumulative total balance of the project	-14895544	-25709584	36043132	106107050	180477463	251282471	318103590	382154395	447160982	520478771	598103054	677075791
Discounting multiplier (d=10%)	1	0.909	0.826	0.751	0.683	0.621	0.564	0.513	0.467	0.424	0.386	0.35
Cumulative discounted cash flow	-14895544	-23370012	29771627	79686395	123266107	156046414	179410425	196045204	208824179	220682999	230867779	236976527
Simple payback period, years	3.22											
Discounted payback period, years	3.24											
Internal rate of return IRR												
Profitability index PI	2.34											

Fig. 3 Analysis of cash flow for the project to implement a subsurface irrigation system in the Northern zone of the Krasnodar Territory on concessional lending of the Ministry of Agriculture (own funds −20%; borrowed funds −80%). *Source* Compiled by the authors

References

1. Bzhasso AA (2014) Key objectives of anti-crisis regional economies management. Bull Adyghe State Univ Ser "Econ" 2(141):30–35
2. Chernyavskaya SA, Berkaeva AK, Iyanova SA, Kashukoev MV, Misakov VS (2020) Regional and sectoral system for integrated assessment and green supply chain management of natural resources. Int J Supply Chain Manag 9(2):714–718
3. Cozzani V, Zanelli S (2007) An approach to the assessment of domino accidents hazard in quantitative area risk analysis. J Hazard Mater 130:24–50

4. Frantsisko OYu, Ternavshchenko KO, Molchan AS, Ostaev GYa, Ovcharenko NA, Balashova IV (2020) Formation of an integrated system for monitoring the food security of the region. Amazonia Investiga 9(25):59–70
5. Gubinelli G, Cozzani V (2008) Assessment of missile hazards: evaluation of the fragment number and drag factors. J Hazard Mater 134:12–40
6. Hodarinova NV (2017) Organization and evaluation of the effectiveness of internal and external control of modern agricultural organizations. J Econ Entrepreneurship 4–1(81):536–547
7. Jensen M (1986) Agency costs of free cash flow, corporate finance, and takeovers. Amer Econ Rev 26:323
8. Jensen M (2001) Value maximization, stakeholder theory, and the corporate objective function. J Appl Corp Financ 14(3):8–21
9. Kartashov KA, Isachkova LN, Sotskaya TV, Kunakovskaya IA (2018) Digital economy as a basis of the modern world or new problems for the Russian society. Bull Adyghe State Univ Ser 5: Econ 4(230):167–173
10. Stein J, Usher S, LaGatutta D, Youngen J (2001) A comparables approach to measuring cashflow-at-risk for non-financial firms. J Appl Corp Financ 13(4):100–109

Features of Performance and Development of the Vegetable Production Market in the Context of Transition to New Economic Conditions

Gulzat K. Kantoroeva, Elena V. Kletskova⬤, and Galina G. Vukovich⬤

Abstract Vegetable production is an important subcomplex of agriculture. Its successful operation determines the supply of vegetable produce to the country's population, and thereby food security of the country. Vegetables contain indispensable vitamins, minerals, and dietary fiber, which are essential to humans and have a positive impact on their physical and mental health. The authors of the paper conducted the analysis of the current state of vegetable production and vegetable consumption in Russia, made a forecast of demand for vegetable produce until 2035, and gave recommendations for the successful development of the subcomplex under the conditions of transition to the digitalization of agriculture. The authors conclude that the transition of vegetable production to a new level of development requires the improvement of available technologies as well as enhanced training of employees' skills in the development of technical and technological innovations. Considering the fact that the main characteristics of employees determine the quality and quantity of goods produced in the agricultural sector under current conditions; special attention should be paid to changing the paradigm of agricultural education.

Keywords Vegetables · Vegetable production · Consumption · Forecast · Human capital · Education · Qualification

JEL Codes Q5 · Q52 · Q59

G. K. Kantoroeva (✉)
Kyrgyz National University Named After Zhusup Balasagyn, Bishkek, Kyrgyzstan
e-mail: kletskova_elena@mail.ru

E. V. Kletskova
Altai State University, Barnaul, Russia
e-mail: kletskova_elena@mail.ru

G. G. Vukovich
Kuban State University, Krasnodar, Russia
e-mail: kaf224@yandex.ru

1 Introduction

The primary function of agriculture is to supply healthy foods to the population.

However, a long period of economic reforms resulted in the fact that currently agriculture is financed with whatever funds remain and experiences considerable difficulties in ensuring the quantity and quality of foods produced. However, the transition to digital technologies in agriculture serves as an accelerator of sustainable development of agriculture and the supply of sufficient amount of vegetable produce of good quality to the population. Among other things, the performance of agricultural producers in the vegetable production market directly depends on the use of sophisticated technologies. The research that was conducted by a group of authors will present the analysis of the current state of production and sales of vegetable produce, as well as the definition of competitiveness of the Russian Federation in the vegetable production market.

2 Materials and Methods

According to the authors, the primary goal of the research is to examine and analyze the main trends in the development of the agro-industrial complex of the Russian Federation.

Agricultural producers and consumers of agricultural products are the objects under observation.

The main economic entities presented in the vegetable production market and involved in the production, processing, storage, and sales of finished goods are the targets of the research.

The subject of the research is a set of business relations arising in the vegetable production market.

The research is based on the works of foreign and domestic researchers in the field of vegetable production, papers dealing with the problem of development of agriculture, laws, regulations, statistical data, as well as the results of empirical observation of the authors.

The methods of statistical analysis, analogy, and comparison, as well as cause-and-effect, graphic and monographic methods have been used in this research.

The fundamental foundations of the scientific issues studied in this article are laid in the publication of Nonaka I, Takeuchi H [3].

3   Results

Under current conditions, the issues of food security improvement are quite pressing, since the supply of healthy foods to the country's population determines the health and life duration of people. Today, foreign countries often supply vegetables where the pesticide content is over the limit, resulting in an increase in morbidity of the population. This is due to the fact that most diseases are caused by malnutrition, including the consumption of contaminated foods. In order to ensure the preservation of the health of the population, one should supply people with healthy foods and increase consumption of fresh and processed vegetables [5]. As may be inferred from Fig. 1, the area of farmland intended for vegetable production has an unstable dynamic pattern, and amounted to 95.7 thousand hectares in 2019.

We would like to point out that only the major public agricultural producers providing the population with all necessary vegetables demonstrated efficient performance in the period preceding the economic reforms in agriculture [7]. Under today's conditions, subsistence farms and peasant farm enterprises carry out their activities in parallel with organizations, occupying a significant share in the overall production structure (Fig. 2).

As may be inferred from Fig. 1, today, peasant farm enterprises are of key importance in the supply of foods to the population. Thus, their percentage increased more than 4 times in 2019 and was recorded at a level of up to 12.5%. Peasant farm enterprises have a number of competitive advantages over major agricultural producers, consisting of the fact that they are able to respond more rapidly to ever-changing demands on the market. As for subsistence farming, its share in the market structure has significantly decreased in recent years due to the fact that private plot activities are associated with several difficulties coming from the unpredictability of natural

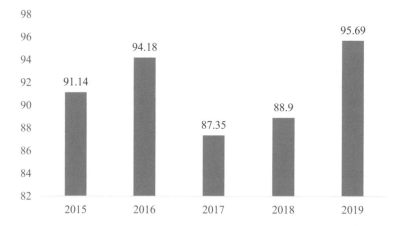

Fig. 1 The area under vegetable crops in the Russian Federation, thousand hectares. *Source* Agriculture in Russia: a statistical book (compiled by the authors)

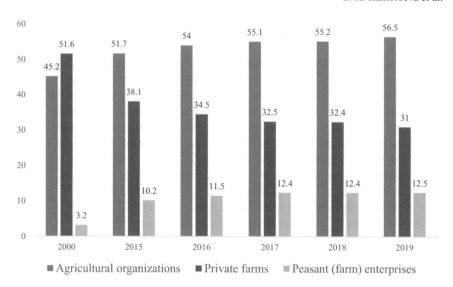

Fig. 2 The structure of supply of agricultural products by major economic categories in the Russian Federation, %. *Source* Agriculture in Russia: a statistical book (compiled by the authors)

and climatic conditions in farming [6]. At the same time, private farms are able to fully provide themselves with necessary foods.

The actual prices for agricultural products have increased more than 15 times over the recent 20 years [8–11]; however, this precludes from judging the stable development of the industry, since high inflation rates significantly reduce the incomes of agricultural producers. Thus, the financial status of producers can be judged by the index of production of agricultural products compared to the previous year (Fig. 3).

As may be inferred from Fig. 3, the index of vegetable production has a negative trend in all categories of economy management.

Further, we would like to note that a negative trend in terms of reduction of the area under outdoor vegetable crops by all major vegetables and gourds by more than 40% can be observed for the period under consideration [4, 11, 12].

In our opinion, this may be due to the introduction of elements of precision agriculture and city farming in agriculture. Their active use largely makes it possible to significantly reduce the area required for farmland [1].

The most advanced technologies that are currently used in city farming are as follows:

(1) Aeroponics, which shall be understood to mean a plant cultivation technology in an air or mist environment without any growing media;
(2) Hydroponics—plant cultivation with the use of artificial growing media and cultural solutions;
(3) Hyponics—plant cultivation by means of improved hydroponics;
(4) Vertical farms—indoor tiered plant cultivation in vitro;

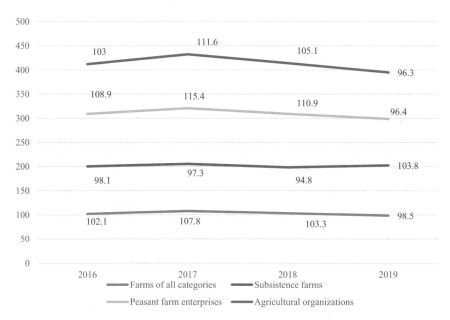

Fig. 3 Index of vegetable production subcomplex by categories of farms in the Russian Federation, %. *Source* Agriculture in Russia: a statistical book (compiled by the authors)

(5)   Robotic-aided greenhouses—a method of digital control of cultivation through differentiated application of fertilizers and irrigation of plants.

Further, we would like to point out that the use of sophisticated technologies makes it possible to grow vegetables, mushrooms, and berries all the year round. The use of photoculture made it possible to improve the yield of tomatoes to 60 kg per square meter and above, and the yield of cucumbers—to 100 kg per square meter and above.

However, the rapid increase in protected vegetable production in modern greenhouses of the fourth and fifth generations has not yet made it possible to meet the population's demand for fresh vegetables to the extent required all year round due to the high cost of digital and robotic technologies. The types of vegetable crops grown on the field and under glass are limited as well. Due to the fact that it is impossible to grow vegetables all year round in most regions for the time being, both due to natural and climatic conditions, and due to storage logistics and further processing consisting in the lack of specially equipped vegetable store cellars and warehouses. As a result, the demand for the domestic vegetable production market can be met through imports of fresh and processed vegetables (Fig. 4).

The volume of imports of vegetables was twice as high as the volume of exports in 2019. China is the major importer of vegetable produce; the volume of imports from this country in 2019 reached over 52 percent of the total cost of imported goods. The following countries with the highest volume of imports of vegetables in 2019 can be identified from non-CIS countries: China comes first with 52%, followed by

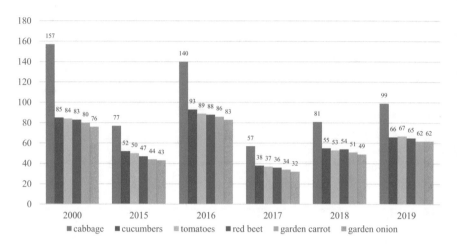

Fig. 4 The area under vegetable crops, thousand hectares. *Source* Agriculture in Russia: a statistical book (compiled by the authors)

the Netherlands with 2%. Among CIS countries, the highest amount of vegetables is imported from Kazakhstan −26% and Uzbekistan −9% [2] (Fig. 5).

The high volatility of the profitability of crop production suggests that agriculture is largely influenced by natural climatic conditions related to the geographical location of many districts. In addition, one of the reasons for a decrease in crop production is the diminishing of the fertility of farmland due to a lack of nutrients, mineral fertilizers, as well as the absence of crop protection measures. The number of units of agricultural equipment in the industry has significantly decreased over

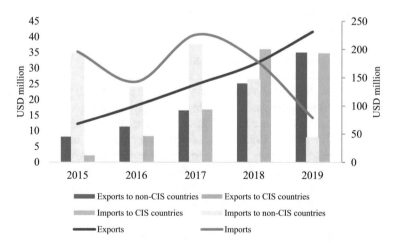

Fig. 5 The current state of exports and imports of vegetables in the Russian Federation. *Source* Agriculture in Russia: a statistical book (compiled by the authors)

the years of reforms, which results in an increase in load per unit of equipment by 2–2.5 times. This results in an increase in harvest time, resulting in a significant loss of agricultural products.

As for the rate of consumption of vegetable produce, their standard value has not been reached in any region of the Russian Federation in the recent few years. The rate of consumption of vegetables and gourds by the country's population is 15 to 50% less than the recommended standard. This is primarily associated with high market prices and the low income of the vast majority of the population. By the way, the average statistical expenditures of households on foods amount to 30 to 70% of the total income. This is significantly higher than in the developed countries of the world, where these expenditures are no more than 12 percent of the family's income. At the same time, the analysis of expenditures of the urban and non-urban population of the country shows that food expenditures of urban residents far exceed those of non-urban residents (Fig. 6).

The authors presented projected values of the necessary vegetable consumption volume at the rate of recommended standard values in an amount of 140 kg per year.

When the vegetable consumption forecast was made, the foods consumption rate approved by legislation pursuant to Order No. 614 of August 19, 2016, by the Ministry of Health of the Russian Federation, was taken into account (Fig. 7). The authors' forecast was based on predictive statistical data on the population size for the next 15 years, corrected for the coronavirus pandemic.

To sum up what has been said, we would like to emphasize the fact that under current conditions, high figures of vegetable production on a nationwide scale still do not afford an opportunity to provide the population with organic vegetables, which contradicts the achievement of the goal of food security of the country.

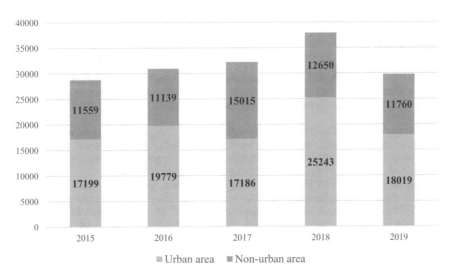

Fig. 6 The pattern of cash expenses of the population of the Russian Federation, %. *Source* Agriculture in Russia: a statistical book (compiled by the authors)

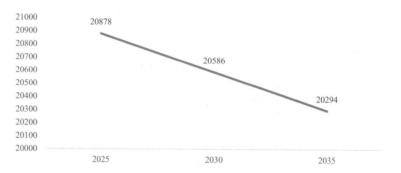

Fig. 7 Vegetables and gourds consumption forecast in the Russian Federation, million kg. *Source* Agriculture in Russia: a statistical book (compiled by the authors)

4 Conclusion

In conclusion, we would like to point out that in order to ensure the supply of high-quality vegetable produce to the population, according to medical standards, technical and technological innovations must be used in the precision agriculture system that makes it possible to significantly improve yields of vegetable crops and resolve the problem of food security of the population through self-production of goods that would correspond to medical consumption standards. Furthermore, the use of modern high-tech equipment allows resolving the problem of seasonal cultivation of vegetable produce, reclaim farmland that suffered from adverse environmental impacts and bring it back to nature's cycle.

References

1. Mensch G (2000) Technologie- und Innovationsmanagement in diversifizierten Unternehmen. In: Hinterhuber HH (ed) Die Zukunft der diversifizierten Unternehmen, Munchen, pp 185–200
2. Mezger C (2004) Humancapital - der Schlussel fur wirtschaftliches Wachstum? Ibidem-Verlag, Stuttgart, p 166
3. Nonaka I, Takeuchi H (1995) The knowledge-creating company: how Japanese companies create the dynamics of innovation. Oxford University Press, New York, Oxford, p 295
4. Sydow J, Management von Netzwerkorganisationen - Beiträge aus der Managementforschung (3. Auflage). Wiesbaden: Gabler. Westdeutscher Verlag, pp 293–354
5. Glotko A, Okagbue H, Utyuzh A, Shichiyakh R, Ponomarev E, Kuznetsova E, Structural changes in the agricultural microbusiness sector. Entrep Sustain Issues 8(1):398–412
6. Glotko A, Polyakova A, Kuznetsova M, Kovalenko K, Shichiyakh R, Melnik M (2020) Main trends of government regulation of sectoral digitalization. Entrep Sustain Issues 7(3):2181–2195
7. Kuznetsova I, Okagbue H, Plisova A, Noeva E, Mikhailova M, Meshkova G (2020) The latest transition of manufacturing agricultural production as a result of a unique generation of human capital in new economic conditions. Entrep Sustain Issues 8(1):929–944
8. Kuznetsova I, Polyakova A, Petrova L, Artemova E, Andreeva T (2020) The impact of human capital on engineering innovations. Int J Emerg Trends Eng Res 8(2):333–338

9. Rudoy E, Shelkovnikov S, Matveev D, Sycheva I, Glotko A (2015) "Green box" and innovative development of agriculture in the Altai territory of Russia. J Adv Res Law Econ VI, Winter 3(13):632–639
10. Stadnik A, Shelkovnikov S, Rudoy Y, Matveev D, Maniehovich G (2015) Increasing efficiency of breeding dairy cattle in agricultural organizations of the Russian Federation. Asian Soc Sci 11(8):201–206
11. Vittadini G, Lovaglio P (2007) Evaluation of the Dagum-Slottje method to estimate household human capital. Struct Chang Econ Dyn 18(2):270–278
12. Vries G, Timmer M, Vries K (2015) Structural Transformation in Africa: Static gains, dynamic losses. J Dev Stud 51(6):211–227

International and Practical Experience
of Sustainable Agriculture

Ensuring the Competitive Advantage of the Agricultural Sector of Kyrgyzstan (Case Study of Cultivation, Processing, and Marketing of Plum)

Saibidin T. Umarov, Kubanych K. Toktorov, and Keneshbek M. Maatov

Abstract The paper aims to develop theoretical and methodological foundations for the formation and use of conditions and factors affecting the competitiveness of agro-industrial production as a platform for economic growth. The study is conducted on the example of growing and industrial processing of fruits (plums), which allows for experimental calculations on the design, organization of production, processing, and sales of products with the definition of economic benefits in the framework of international cooperation of Kyrgyzstan with Germany. The research substantiates that to increase the competitiveness of agricultural products, it is necessary to create a competitive advantage. In the conditions of Kyrgyzstan, these competitive advantages can be formed due to the production of environmentally friendly products, optimization of production and trade costs, and the introduction of new technological solutions in the cultivation of crop and livestock products. The process of implementing new technological solutions is accompanied by modern methods and techniques inherent in trade activities, including marketing and promotional activities, preparing products for sale in a convenient form for the consumer, ensuring timely delivery, etc. The authors try to design a business plan based on growing, processing, and selling plum fruits within the framework of international cooperation Kyrgyzstan–Germany. The results of calculations proved the feasibility and profitability of this cooperation.

Keywords Agrarian sector · Competitive advantages · Competitiveness · Kyrgyz Republic · Cultivation · Industrial processing · Selling plums · Experimental calculations · Business planning · International cooperation

JEL Codes Q1 · Q10 · Q11 · Q13 · Q17 · D41

S. T. Umarov (✉) · K. M. Maatov
Osh Technological University, Osh, Kyrgyz Republic

K. M. Maatov
e-mail: kengesh2006@mail.ru

K. K. Toktorov
Osh State University, Osh, Kyrgyz Republic

1 Introduction

Economic development is provided by the production and marketing of various closely interconnected sectors of the economy. This relationship is objective. One of the examples is agro-industrial production, which belongs to different sectors of the economy (agriculture and industry) but, in fact, technologically, consumerally, and organizationally inseparable. For example, grown agricultural products are further subject to industrial processing, after which the product passes through the channels of promotion and trade until it reaches the consumer.

Under these conditions, the overall performance of economic activities of interrelated economic sectors is often better ensured due to the effective organization of intersectoral linkages compared with the overall performance of each sector of the economy separately.

Another reason for new approaches to organizational structures for ensuring close integration ties in agro-industrial production is based on common goals. According to these goals, agriculture, processing, or trade of agro-industrial products pursue the goal of the final product being in great demand and bringing as much economic benefit as possible. For this purpose, the final product must have sufficient competitiveness, which is one of the basic principles of market relations.

2 Materials and Methods

The research aims to develop theoretical and methodological foundations for the formation and use of conditions and factors affecting the competitiveness of agro-industrial production. Experimental calculations were carried out based on the example of the growing and industrial processing of fruits (plums). The authors implemented the methods of design, organization of production, and sales of products to determine economic benefits within the framework of international cooperation of Kyrgyzstan and Germany.

The research hypothesis is that its results can be used to develop strategies for agro-industrial production, transportation, sales, and consumption of agricultural products. Simultaneously, the authors consider such an important aspect of the problem as the organization of the release of agricultural and livestock products on the market. The theoretical basis of the research consists of the works [1–10].

3 Results

3.1 Fruit Cultivation

The cultivation of agricultural raw materials plays the primary role in integrating agriculture and processing. This is due to the fact that the content of the grown fruit is almost unchanged in industrial processing: drying, manufacturing jams, and grinding dried fruit as an additive to food (making cakes, pastries, bakery products, preparing dishes in catering outlets (pilaf, various soups, etc.)). In other words, processing changes the appearance of the grown crop, while its content is kept the same as in agricultural production.

Therefore, competitive advantages are provided during the growing and industrial processing of fruits, including plums. Competitive advantage can also be determined based on the demand for a particular product in a particular economic space. There is a great demand for products from plums in the Alabuka and Aksy districts of southern Kyrgyzstan in the current reality. The demand for these products has grown by 2–3 times just in 3–4 years.

Thus, according to the protocol between the German entrepreneurs and the two districts of the Kyrgyz Republic, the need for raw materials for processing equals 500 tons per year. Additionally, Indian entrepreneurs have expressed interest in buying dried fruit in the amount of 400 tons per year. Traders from the United Arab Emirates and entrepreneurs from Russia and Georgia want the same amount. As a result, about 1.5 thousand tons of dried plum fruits seem to find their buyers in countries far and near abroad.

The demand of German businesspeople is based on long-term cooperation since they have tested the demand for dried fruits from Kyrgyzstan in practice over 5–6 years. Whereas, Indian businesspeople showed such interest after the exhibition and fair, where Kyrgyz people showed their products, including fruits and plums. The fair coincided with the Indian folk festival dedicated to giving each other dried fruit (in India, there is a tradition of presenting each other dried fruit in a certain period for three days).

The interest of Arab, Russian, and other businesspeople arose due to international economic relations with these countries within the EAEU and other cooperation channels. Other countries may also show such interest.

In our opinion, the first competitive advantage in fruit and plum growing lies in the natural and climatic conditions of the Alabuka and Aksy districts, since plums grow in other regions of Kyrgyzstan, but such interests are shown to a lesser extent there. The influence is provided by the soil composition and cultivation technology used by gardeners with many decades of experience.

Relatively recently (5–7 years ago), there appeared new varieties of plum of intensive type: dwarf and semi-dwarf new *Hungarian Italian, Stanley, Renklod*, and others. These varieties were brought from abroad and adapted to local conditions due to experiments on experimental plots with elements of research work.

New varieties require more care in terms of agronomic practices and systematic cultivation along the terrain and watering of seedlings and mature trees. The yield of fruit per tree, taking into account the placement on 10,000 m^2 of land, is higher than that of extensive trees. It is possible to plant 4–5 extensive trees on 100 m^2 and twice as many dwarf and semi-dwarf trees.

Additionally, the height of dwarf trees allows for a 1.6–2 times decrease in labor consumption during harvesting. However, intensive trees require unique technological approaches to cultivation. In particular, drip irrigation, organic fertilizers instead of mineral ones, special loosening of each tree bush, chemical protection against pests, and others are considered appropriate. This allows for getting a harvest with less sugar content than in extensive trees, which comply with medical recommendations.

Nevertheless, the spread of new high-yielding plum varieties has been slow. The main reasons for this are the following:

- A variety of forms of land compatibility and the creation of small-scale peasant farms contributed to the fact that peasants were free to dispose of their land shares. Moreover, peasants were not psychologically prepared for innovation and intensified farming;
- There is a lack of experience in introducing intensive technologies, partly due to a lack of financial resources;
- Destruction of previously established irrigation networks and vast areas does not allow to coordinate and systematically grow fruit with the optimal structure of land area and logistics;
- There is a lack of an organized start of fruit sales through logistics centers and organized access to domestic and international markets, which leads to the creation of unnecessary intermediary links;
- There is no partnership between peasant farms and government in the context of low interest of local authorities in such cooperation;
- There is a low level of investment attractiveness of peasant farms in conditions of riskiness due to spontaneous sales of grown crops in markets, etc.

The problems mentioned above do not allow farmers to grow high-grade fruit trees to replace the old ones. Moreover, these problems complicate the development of new areas for sowing new varieties of trees. However, the demand for new fruit varieties can be a sufficient incentive to set strategic goals and move toward achieving them.

3.2   Fruit and Plum Processing

Processing includes several technological operations to dry fruit and prepare them for further cooking and manufacturing various food products. In processing, it is crucial to preserve the consumer properties in terms of the composition of fruits containing various substances, including glucose, fructose, sucrose, vitamins A, B1, B2, C, H, and PP, as well as essential minerals such as potassium, magnesium, zinc, copper, manganese, iron, chromium, bromine, nickel, phosphorus, and sodium. The main useful property of plums is a mild cleansing of the stomach and normalization of the gastrointestinal tract. Plums are rich in potassium, which strengthens the heart muscle and the walls of blood vessels, which is important in diseases of the cardiovascular system and hypertension.

Thus, the main requirement for processing is on the part of consumers. Since plums are used mainly in dried form, it is essential to provide high-quality drying and preserve the original color and a form convenient for the consumers.

Farmers need drying machines with a larger capacity (1.5–2 times larger than at present) to ensure an increase of up to 4 tons of fresh plums. On the other hand, drying machines, costing $60–70 thousand, are expensive for peasants and some individual entrepreneurs. Drying machines made by local craftsmen cost $25 thousand, but such equipment is much worse in drying quality.

Moreover, farmers want to add a device for automatic washing of fresh plums and delivering it in the pan to the dryer. This device adds another $10,000 to the cost of the dryer.

Thus, a modern dryer costs about $80,000. Such a device can only be acquired within a collective organizational form such as cooperatives or agro-industrial associations using corporate forms of management. There are no such organizational forms of management in the districts.

One can offer a high price of drying services to cover the cost of purchasing the dryer and make a profit. However, few people are willing to take advantage of such services. Most farmers prefer to dry fruits in primitive ways, i.e., on the asphalt or the roof of the house. Some people use the dryers by local artisans, who have a much lower service for drying. Another obstacle to innovation is ignorance of the competitive advantages of modernized drying methods, as well as the lack of motivation in the local population to innovate and use technological innovations in cultivating and drying fruits. In addition, the return on innovation takes a long time – about 4–5 years (3–4 years to grow and harvest, and one year to sell products).

A considerable obstacle to innovation is the practical lack of investment and its high cost. Thus, there is a vicious circle in which farmers want the economic benefits of growing and processing fruits and plums but encounter difficulties.

3.3   Sales of Plums and Arising Problems

Under market conditions, the final stage of production activity is the sale of products to customers, during which there is a change in the form of value from commodity to monetary form. This stage summarizes all previous activities in production, movement of goods, storage, etc. In this case, the activity's success is determined based on the profitability of transactions. In turn, the value of profit and profitability is determined based on the prevailing supply and demand ratio for a particular type of product.

Consequently, the market situation, particularly its conjuncture, plays a decisive role in the effective organization of production and sales of products, including plum products. The sale of products, especially for export, is becoming difficult due to the fierce competition among sellers, rapidly changing trends in product consumption, which depend on many factors that are sometimes not accounted for. Moreover, there is difficulty in grasping the relationship between the sales process and the changes occurring in the production and consumption of products.

Thus, there is currently globalization of the economy, which is accompanied by the integration of production processes and the strengthening of export–import operations. These processes bring together the habits and traditions in the consumption of various foods of different peoples located at considerable distances from each other. On the other hand, along with global processes, the nationalization of the economy is increasing, which hinders integration processes and the globalization of the economy. This can be seen in the example of sanctions imposed by countries against each other for political, regional, national, and other reasons. In turn, this negates the principles of world trade in goods, since in some cases, such rules do not work in practice.

For example, the WTO rules, which were developed by the efforts of many countries, do not work due to the national ambitions of several countries. This is also reflected in the sale of food products, including fruit.

Another difficulty is that agricultural products (fresh and processed) are subject to spoilage depending on seasonality, time, storage, and consumption conditions, which require adequate measures. This is also a natural difficulty in the sale of products subject to noncompliance with trade rules.

The selling price of the product plays a vital role in the sale of fruit in the domestic and international markets. As a transformed form of the value of goods, price includes the reimbursement of production costs, sale of products, and the desired level of profit. Therefore, the price is an expression of the interests of market participants (producers, buyers, intermediaries, etc.) relating to this product.

Since the price depends on producers, buyers, and the supply–demand ratio in the markets, it is difficult to predict the possible coincidence of circumstances for the prospective period. Therefore, those who predict future consumer behavior and market conditions are likely to succeed.

When analyzing price dynamics and determining their level, the following price indicators are usually used: contract prices, stock exchange quotations, reference prices, price lists, price indices, etc. The problem in determining the predicted prices of dried fruits and plums lies in the lack of systematic practice of sales to the traditional buyer inside and outside the country. Therefore, demand formation is one of the most important tasks in fruit growing and processing.

It should be noted that market forecasting has a two-step nature. The first stage involves preparing a forecast of the manifestations of the main factors affecting market conditions (permanent, cyclical, non-cyclical, and temporary). The second stage involves developing a comprehensive conjuncture forecast, the main sections of which are forecasts of commodity production, consumption, international trade, and prices.

The price and the volume of consumption largely depend on the product's competitiveness.

Competitiveness is defined as a characteristic of the product, which is reflected in its difference from the products of competitors in the degree of compliance with a particular public need and the cost of its satisfaction. Since the consumer properties of goods are inseparable from their cost characteristics, the competitiveness of goods depends on both the consumer properties of goods and their cost.

4 Discussion

In the case of plum fruit growing in the southern region of Kyrgyzstan (the Alabuki and Aksy districts), the manifestations of competitiveness are defined in two steps. The first step determines the comparative efficiency of growing plum from a unit of land area compared to other crops by comparing the costs and benefits from each type of plant, taking into account the labor intensity and duration of the periods compared. For example, dwarf plums yield in only three years, while wheat yields annually. Thus, the results of the plum crop in the third year must be divided by three years. The second step of assessing competitiveness comes down to comparing quantitative and qualitative characteristics of plum growing with the competitors growing the same plums.

Another aspect of competitiveness is the characteristic of the consumer properties. The presence of consumer properties determines the efficiency of the consumption of goods and its beneficial effect. The higher the consumer properties of a product and its possible useful effects, and the lower the cost, the higher its ability to be sold.

The ratio of the price indicator to the product's useful effect and similar indicators of other goods gives an idea of the level of its competitiveness. A competitive product does not simply withstand the competition (of other goods) but surpasses it.

There is a direct link between product competitiveness and export of goods – the higher competitiveness of a product, the more customers abroad will want to buy this product.

When selling dried fruit, especially for the prospective period, it is crucial to assess the competitors' strategy. Usually, when evaluating a strategy, the following issues are identified and solved:

- The main factors of the competitiveness of competitors' products;
- Practices of competitors firms in advertising and sales promotion;
- Practices of competitors firms regarding product naming;
- Practices of competitors regarding the movement of products (types of transport; volume of inventory; location of warehouses; types of warehouses and their costs).

As a result, we get information allowing us to: (1) conclude on the actions of competitors, the range of their products, and pricing policies, (2) calculate the sales restraints of competitors, (3) identify the competitive advantages of products, and (4) determine the costs of competitors for advertising and promotion of goods.

Each competitor is considered individually, and then a summary is compiled, which allows for determining the key success factors for each strategy.

5 Conclusion

Given the extreme importance of quantitative and qualitative measurement of the competitiveness of agro-industrial products, as well as the lack of development of these aspects in practice, the authors made clarifications on the methods of assessing the competitiveness, which allow for its objective characterization. The paper puts forward the idea that to increase the competitiveness of the agricultural sector, it is necessary to create competitive advantages. In the conditions of Kyrgyzstan, competitive advantages can be formed due to producing environmentally friendly products, establishing cheaper costs, and introducing new technological solutions in the cultivation of crop and livestock products.

Simultaneously, trade is accompanied by methods and techniques inherent in trading activities: marketing and advertising activities, preparation of products for sale in a convenient form, ensuring timely delivery, etc. The authors attempted to design a business plan based on growing, processing, and selling plum fruit within the framework of international cooperation Kyrgyzstan–Germany. The calculations show the feasibility and profitability of such cooperation for both sides of the economic partnership.

References

1. Abdymalikov K (2010) The Economy of Kyrgyzstan (in Transition): Textbook, 2nd edn. Bijiktik, Bishkek, Kyrgyz Republic
2. Beregnoy AE (2004) Size of agricultural enterprises and production efficiency. Econ Agric Process Enterp 12:18–21
3. Bolobov A (2003) Competitiveness of agricultural production. Int Agric J 3:21–25
4. Gerasimov BI, Zharikov VV, Zharikov VD (2010) Fundamentals of logistics. Moscow, Russia: INFRA-M
5. Kaliev G, Nikitina G (2008) Improving the competitiveness of agricultural products and food in Kazakhstan. Selected Works 3:9–62
6. Kultaev TCh (2011) Economic forecasting of agricultural production based on modeling. Nauka, Bishkek, Kyrgyz Republic
7. Kuzmenko NI, Salikov YuA (2016) On the issue of economic security factors in regional development strategies. In Igolkin SL (Ed), Materials of the XIX Report scientific-practical conference of the teaching staff (pp 39–43). Voronezh, Russia: Voronezh Institute of Economics and Law
8. National Statistical Committee of the Kyrgyz Republic (2013) External and mutual trade of the Kyrgyz Republic: 2008–2012. Bishkek, Kyrgyz Republic: National Statistical Committee of the Kyrgyz Republic. Retrieved from http://www.stat.kg/media/publicationarchive/51c14e5c-9948-48d6-b3c4-0fcdec32c0ee.pdf

9. Sharpe WF, Alexander GJ, Bailey JV (2001) Investments: textbook (A. N. Burenina, & A. A. Vasina Trans. from English). Moscow, Russia: INFRA-M

10. Umarov ST (2020) Competitive activity of production and sale of agricultural products of Kyrgyzstan in the context of international integration. Econ Innov Manag 2:54–60. https://doi.org/10.26730/2587-5574-2020-2-54-60

Problems of Managing Large-Scale Breeding in Russian Cattle Breeding

Vladimir I. Chinarovⓘ, **Sergey A. Shemetyuk**ⓘ, **Olga V. Bautina**ⓘ, **Anton V. Chinarov**ⓘ, **and Andrey A. Azhmyakov**

Abstract Government investments aimed at developing domestic pedigree cattle breeding allowed the formation of a wide network of breeding plants, breeding reproducers, and gene pool farms throughout the country to meet the needs of agricultural producers in pedigree products. Since 2016, dairy cattle breeding has witnessed expanded reproduction. The industry is fully provided with heifers bred in Russia for the replenishment of the herd. Breeding farms provide simple reproduction and have a sufficient number of young cattle of almost all dairy cattle breeds for sale within the country and export. The Russian market of pedigree products is marked with a positive trend to increase the sale of pedigree cattle. With high annual rates of import substitution of breeding cattle, the sales of breeding cattle increased by 62.3% in six years, which allowed for the reduction of import to 15.6%. The situation on the market for bull semen is more complicated; artificial insemination covers only two-thirds of the potential cattle herd, and import consumption is 22.1%.

Keywords Cattle breeding · Import substitution · Pedigree resources · Semen · Seed bulls · Breeds

JEL Codes O13 · Q16

V. I. Chinarov (✉) · O. V. Bautina · A. V. Chinarov
L.K. Ernst Federal Research Center for Animal Husbandry, Podolsk, Russia
e-mail: vchinarov@yandex.ru

O. V. Bautina
e-mail: Oky4@yandex.ru

A. V. Chinarov
e-mail: chinant@yandex.ru

S. A. Shemetyuk
JSC "GCW", Moscow, Russia
e-mail: shamilk@mail.ru

A. A. Azhmyakov
JSC "Udmurtplem", Izhevsk, Russia
e-mail: udrnurtplern@yandex.ru

© The Author(s), under exclusive license to Springer Nature Singapore Pte Ltd. 2022   123
E. G. Popkova and B. S. Sergi (eds.), *Sustainable Agriculture*,
Environmental Footprints and Eco-design of Products and Processes,
https://doi.org/10.1007/978-981-16-8731-0_13

1 Introduction

Over six years of implementing the "State program of agricultural development and regulation of markets of agricultural products, raw materials, and food for 2013–2020," there have been significant structural shifts in Russian cattle breeding. During these years, the production of protein for food purposes in cattle production increased by 8.9% due to an increase in the production of marketable milk (2955.1 thousand tons) and the growth of cattle (39.8 thousand tons). These results were obtained due to the dynamic development of specialized beef cattle breeding. With the continued reduction of the total number of cattle, the increase in livestock production was provided by the intensification of the industry and the realization of the genetic potential of the animals. Dairy and meat productivity increased by 15.4% and 7.5%, respectively.

The annual reduction in the number of cattle does not allow Russian cattle breeding to get out of stagnation. It is the main reason for the lack of expanded reproduction in the industry [4]. The number of dairy cattle is decreasing faster than the number of beef cattle growing due to an increase in crossbred cattle. In 2014, with the introduction of a subsidy per beef cow, the total number of cattle in beef cattle breeding increased by 116.8 thousand cows for the year, including 10.1 thousand cows due to the transfer from the dairy herd. During the analyzed period, the number of cattle in dairy farming decreased by 12.4% (i.e., 2114.1 thousand animals). In turn, the number of cattle in beef cattle breeding increased by 44.3% and amounted to 3,149,300 heads. The structure of breeding stock also changed: the share of beef cows increased by 4.4% and amounted to 15.6%. With an extremely high import dependence of the Russian market of milk (21.8%) and beef (16.0%), it is necessary to introduce an economic mechanism to exclude competition of specialized meat and dairy cattle breeding in the resource market [3]. If no urgent action is taken, these trends will continue. At 5000 kg of milk productivity, the optimal share of specialized beef cattle in the herd structure should be 40%. In fact, this indicator equals only 17.4% (Table 1).

2 Materials and Methods

The intensification of production and the lack of proper attention to the productive longevity of animals has led to the fact that many commercial farms cannot achieve even simple reproduction. Moreover, the replenishment of the main herd is carried out mainly at the expense of purchased heifers. Additionally, herds producing breeds with lower productivity do not withstand competition since their productivity level does not provide profitable farming. This reduces the number of cattle in dairy farming and creates the illusion of a rapid (100–150 kg per year) increase in productivity through the realization of the genetic potential of livestock. The solution to the problem of the dependence on technological import in breeding products and sustainably expanded

Table 1 General characteristics of cattle breeding in Russia

Indicators	2013	2014	2015	2016	2017	2018	2019
Number of cattle, thousand heads	19,272.6	18,919.9	18,620.9	18,346.1	18,294.2	18,151.4	18,126.0
Including dairy cattle breeding	17,090.8	16,507.7	16,018.5	15,735.6	15,568.1	15,272.6	14,976.7
Beef cattle breeding	2181.8	2412.2	2602.4	2610.5	2726.1	2878.8	3149.3
Share of beef cattle, %	11.3	12.7	14.0	14.2	14.9	15.9	17.4
Number of cows, thousand heads	8430.9	8263.2	8115.2	7966.0	7950.6	7942.3	7964.2
Including dairy cattle breeding	7485.9	7201.4	6952.7	6801.3	6784.9	6773,5	6732.6
Beef cattle breeding	945.0	1061.8	1162.5	1164.7	1165.7	1168.8	1231.6
Share of beef cows, %	11.2	12.8	14.3	14.6	14.7	14.7	15.6
Change in the number of cows for the year in total, thousand heads	−197.6	−167.6	−148,0	−149.2	−15.4	−8.3	22.0
Including dairy cows		−284.4	−248.7	−151.4	−16.4	−11.4	−40.9
Including beef cows		116.8	100.7	2.2	1.0	3.1	62.8
Milk production, million tons	29.9	30.0	29.9	29.8	30.2	30.6	31.4
Beef production, thousand tons	1608.0	1621.4	1617.1	1588.8	1569.3	1608.1	1625.2
Including dairy cattle breeding	1474.0	1475.5	1456.9	1406.7	1369.3	1395.9	1382.4
Beef cattle breeding	134.0	145.9	160.2	182.1	200.0	212.2	242.8
Share of beef cattle breeding, %	8.3	9.0	9.9	11.5	12.7	13.2	14.9
Average annual milk yield per dairy cow, kg	4041	4085	4223	4331	4443	4516	4664
Average daily gain, g	689	700	710	706	703	727	741
Including dairy cattle breeding	713	725	736	726	718	750	766
Beef cattle breeding	503	524	533	586	618	603	623

Source: Calculated by the authors based on the data of the Federal State Statistics Service of the Russian Federation [8]

reproduction in dairy cattle breeding is impossible while the calf crop per 100 cows is less than 85 heads.

Long-term development of Russian livestock breeding is impossible without modern pedigree material and availability of own original lines and purebred cattle. Strengthening the breeding base in the regions and stimulating the development of national holdings for reproduction to meet the growing needs of agricultural producers in high-quality breeding products and high-quality genetic material should become a priority direction, contributing to bringing dairy cattle breeding out of stagnation. In the context of globalization of world livestock breeding, unlike the general trend toward mono-breeding, Russia, according to FAO, remains one of the few countries with a diverse gene pool of animals and actively involves local breeds in agricultural production [9].

The "Head Center for Reproduction of Agricultural Animals" (AO HCR) was created to provide Russian cattle breeding with high-quality semen and breeding products for expanded reproduction based on the use of the best gene pool. The need for these products is determined by the number of cows, the structure of cattle breeding, and the current breed composition of cattle in Russia.

3 Results

The final transition of Russian cattle breeding to artificial insemination with 50% coverage of cattle kept in private farms requires at least 24.1 million doses of semen annually (Table 2).

With the existing technology for producing this amount of semen, it is necessary to have at least 3.5 thousand bulls, including 3.2 thousand main animals.

Due to the large volume of imports, Russian reproduction centers are forced to store large volumes of semen. The limited demand for bull semen from Russian producers is not caused by its quality. The main reason is the lower price from importers. About 40% of the imported semen had a price below its value on the Russian market. The high breeding, evaluation, and maintenance costs of breeders are covered only if the selling price is at least 180 rubles per semen dose. Simultaneously,

Table 2 Demand for semen in cattle breeding in Russia for artificial insemination, thousand doses

Categories of farms	Dairy cattle breeding		Beef cattle breeding		Total	
	2018	2019	2018	2019	2018	2019
Breeding farms	4391	4544	796	800	5187	5343
Commercial farms	9731	9491	4084	4353	13,814	13,844
Private farms	4957	4911			4957	4911
All categories of farms	19,078	18,946	4880	5153	23,958	24,099

Source Calculated by the authors

Table 3 Formation of the Russian market for bull semen, thousand doses

Years	2018	2019
Demand for bull semen (including exports)	23,882	26,751
Including breeding farms	5187	5342
Actual capacity of the Russian market	14,971	15,800
Unmet demand	8911	10,275
Semen produced by AO HCR	6829	7198
Share of AO HCR in demand, %	28.6	27.6
Sold semen produced by AO HCR	5686	5553
Share of AO HCR in the Russian market, %	38.0	35.1
Import	6073	3486
Including at a price lower than the selling price of agricultural producers	4627	1366
Export	303	96
Including at a price higher than on the Russian market	40	36

Source Calculated by the authors based on official data of the Federal Customs Service of the Russian Federation [7]

the government canceled the import tariff, thereby providing significant preferences to importers of pedigree products. Since October 1, 2016, a zero rate of value-added tax has been introduced to purchase imported breeding material, reducing its value by another 10%.

After amendments to the tax code and VAT refunds on imported breeding products in 2019, there were downward trends in imports of bull semen. However, import consumption is still 22.1%; out of the 3.49 million doses, 1.37 million doses were imported at dumping prices (Table 3).

The most important task facing the AO HCR is to ensure expanded reproduction in stockbreeding utilizing cattle of Russian selection. Over the seven years, the number of cows in breeding herds has increased by more than a hundred thousand heads. The growth of breeding stock was provided by the intensive introduction of first heifers into the main herd, the number of which was 1.9 times higher than the number of cows retired. Out of 2518.8 thousand retired cows in Russian breeding farms, 71.9% were replaced by heifers of own reproduction, 477.3 thousand purchased heifers of domestic selection, and only 330.8 thousand imported animals.

Against the background of a decline in the number of cattle in dairy cattle and an increase in the number of cows in breeding farms in 2019, the share of breeding animals in the industry reached 15.6%, and in beef cattle 15.0% due to the presence of a large number of crossbred cattle.

4 Discussion

Another important task for the AO HCR is preserving genetic diversity and multi-breeding in Russian cattle breeding [1].

The breed structure and economically useful and productive characteristics of cattle change annually. An essential factor in the intensification of modern cattle breeding, which depends on purposeful breeding and pedigree work, is the outstripping growth rate of livestock productivity with a reduction in the number of livestock. Specialized breeds are the most adapted to industrial technology, compared to cattle of combined production direction [10]. More intensive breeding and improvement are observed among those breeds that provide more output of the products most demanded in the domestic market per unit cost.

With its variety of natural and economic conditions, Russia's regional aspect of breed zoning is of no small importance. The structure and breed composition of cattle are different in all federal districts of Russia. The breed diversity preserved in Russian cattle breeding is an absolute and relative competitive advantage.

The competitiveness of a breed depends on how well it is adapted to zonal conditions and to what extent its genetic potential is realized. The breeding efficiency in the region is determined by the animals' productivity and the breed's ability to expand reproduction [2].

Of 24 breeds of cattle raised in dairy cattle breeding of Russia, 18 breeds have the status of breeding animals. Animals of the Caucasian Brown, Istobenskaya, Danish Red, and Tagilskaya breeds, as well as the Dagestan and Yakutian Mountain cattle, did not have breeding status and are traditionally bred in commercial and private subsidiary farms. With a considerable variety of breeds in terms of reproduction, productivity, product quality, and adaptation to zonal conditions for proportional development of regions, multi-breeding is an objective necessity for Russia.

Changes in the breed structure in dairy cattle breeding were determined mainly by the dairy productivity of the breeds. In the context of fierce internal and external competition in the livestock market, breeds with higher production efficiency, determined by the ratio of production volume to the cost of raising and maintaining the animal, are promising for breeding. Since the price of milk and beef is currently set in terms of quality, the breed used to produce a particular product is now especially important.

Comparative assessment of animals in Russian dairy and beef cattle breeding on the main indicators of reproduction, productivity, and income per cow per year showed significant differences not only among breeds but also in the efficiency of use and breeding of particular breeds, depending on the region (Table 4).

More than two-thirds of the total number of cows in Russia are four breeds:

- In dairy cattle breeding—Black-and-white and Holstein breeds;
- In beef cattle breeding—Aberdeen Angus and Kalmyk breeds.

Of 7964.2 thousand cows of 38 breeds contained in all categories of farms, ten breeds had the herd of more than 50 thousand cows in dairy cattle breeding. Only

Table 4 Ranking of breeds in the region by economically useful traits and income per cow per year (breeding stock 2019)

Breed	Nationwide		Central federal district		Northwestern federal district		Southern federal district		Volga federal district	
	*	**	*	**	*	**	*	**	*	**
Dairy cattle breeding										
Holstein black-and-white	18	2	13	6	6	1	3	2	11	2
Black-and-white	14	5	8	2	3	2	7	6	10	5
Ayrshire	15	4	11	8	5	3	4	3	5	1
Simmental	6	12	4	5			1	4	4	8
Brown Swiss	5	11	1	11			5	1	7	6
Red-and-white	13	10	5	10			6	7	8	3
Red Steppe	9	16					2	5	3	10
Kholmogorskaya	12	7	12	12	4	4			9	4
Yaroslavskaya	11	9	7	9	1	5				
Montbéliarde	16	6	9	4						
Sychevskaya	2	15	2	13						
Kostromskaya	1	3	3	3						
Jersey	17	8	10	7						
Swiss Red	10	1	6	1						
Gorbatovskaya Red	3	13							1	7
Estonian Red	7	17			2	6				
Bestuzhevskaya	4	14							2	9
Suksunskaya	8	18							6	11
Beef cattle breeding										
Hereford	5	4	3	3	4	4	4	5	2	2
Aberdeen Angus	4	6	4	4	3	3	5	6	3	4
Kazakh Whiteheaded	7	1	1	1			3	1	4	3
Kalmyk	9	2					6	2	5	1
Galloway	8	3	6	5						
Limousin	2	5	2	2	2	2			1	5
Charolais	1	8			1	1	1	4		
Simmental	6	7	7	6						
Aubrac	10	10	5	7						
Russian Komolaya	3	9					2	3		

(continued)

Table 4 (continued)

Breed	Ural federal district		Siberian federal district		Far Eastern federal district		North Caucasian federal district	
	*	**	*	**	*	**	*	**
Dairy cattle breeding								
Holstein black-and-white	3	3	6	6	3	1	6	1
Black-and-white	2	1	2	2	4	4	7	6
Ayrshire			3	1	6	6	5	4
Simmental	1	2	1	4	1	5		
Brown Swiss							2	5
Red-and-white			5	3	5	2		
Red Steppe			4	5			3	7
Kholmogorskaya					2	3		
Yaroslavskaya							4	3
Montbéliarde							1	2
Sychevskaya								
Kostromskaya								
Jersey								
Swiss Red								
Gorbatovskaya Red								
Estonian Red								
Bestuzhevskaya								
Suksunskaya								
Beef cattle breeding								
Hereford	3	3	1	2	4	4	4	4
Aberdeen Angus	1	1	4	4	5	5	1	1
Kazakh Whiteheaded			2	1	2	3	2	2
Kalmyk					3	2	3	3
Galloway			3	3	1	1		
Limousin								
Charolais								
Simmental	2	2						
Aubrac	4	4						
Russian Komolaya								

Note *–Economically useful features; **—Income per year
Source Calculated by the authors based on the results of cattle appraisal in the Russian Federation [5, 6]

four breeds had a herd of more than 50 thousand cows in beef cattle breeding. The most common breed is the black-and-white breed, which produces 47.2% of milk in Russia. The most productive breed is the Holstein breed, though it has a very low productive longevity.

Of 38 breeds of cattle bred in Russia, only 16 in dairy and 10 in beef cattle breeding are currently of production importance. The rest must be preserved and improved as carriers of unique genetic material. Five breeds of cattle (Black-and-white (2812.4 thousand animals), Holstein (1680 thousand animals), Simmental (390.1 thousand animals), Aberdeen Angus (305.4 thousand animals), and Hereford (179.1 thousand animals)) are represented in all federal districts of Russia.

The development of beef cattle breeding in Russia has a zonal character. However, the industry has recently started to actively develop mainly in regions of intensive agriculture. This fact has led to a quarter of Russia's beef cows being concentrated in the Central Federal District.

Beef cattle breeding is developing most intensively in the Northwestern, Ural, and Far Eastern Federal Districts by strengthening the breeding base, which has already brought the share of breeding cows in the beef herd to 40%–50%, with an average share of 19.2% in the whole country. On the one hand, it allowed the regions to raise the level of the industry. On the other hand, it led to a sharp imbalance in the location of beef cattle breeding and its breeding base.

The Central and Southern Federal Districts stand out in terms of the diversity of specialized beef cattle (8 breeds each). Seven breeds were bred in the Ural Federal District. The Volga and Far Eastern Federal Districts bred six breeds each. Five breeds were bred in the Siberian Federal District. Four breeds each were bred in the Northwest and North Caucasus Federal Districts.

According to the breed structure, placement of livestock, and achieved herd reproduction indices, the need for seed bulls used to obtain a sufficient amount of semen for artificial insemination is calculated for each region (Table 5).

5   Conclusion

Cattle breeding in Russia increases its potential and production volumes. It is developing in accordance with current global trends—the share of beef cattle breeding is increasing with outstripping growth of intensification of dairy cattle breeding based on a higher level of the genetic potential of domestic breeding cattle.

The intensification of dairy cattle breeding has led to a decrease in the meat industry's potential. With the growth of milk productivity of cows, the reproductive ability of animals and the duration of their economic use are reduced. The most abundant dairy breeds have low meat productivity. Therefore, with the achieved level of dairy productivity of cows in Russia, it is necessary to accelerate the development of beef cattle breeding to meet the demand in the Russian beef market.

In the Central Federal District, the most promising for breeding are Brown Swiss, Swiss Red, and Kazakh Whiteheaded breeds; in the Northwestern Federal

Table 5 The number of cows and needs of seed bulls (2019)

Breed	Central federal district		Northwestern federal district		Southern federal district		Volga federal district		Ural federal district	
	Cows	Bulls	Cows	Bulls	Cows	Bulls	Cows	Bulls	Cows	Bulls
Dairy cattle breeding										
Black-and-white	261,443	133	126,622	69	161,209	49	1,316,091	557	275,315	117
Holstein black-and-white	401,080	205	66,385	36	478,619	144	155,358	66	72,877	31
Red Steppe					88,238	27	11,132	5		
Simmental	40,347	21	477	1	8783	3	69,515	30	10,921	5
Red-and-white	53,164	28			8701	3	39,370	18	272	1
Kholmogorskaya	16,944	9	39,318	22			161,331	69		
Ayrshire	14,343	8	41,078	23	90,947	28	12,652	6		
Brown Swiss	19,276	10			3858	2	17,211	8		
Yaroslavskaya	43,918	23	5237	3						
Jersey	14,901	8			11,081	4				
Montbéliarde	9060	5					106	1		
Bestuzhev							28,273	13		
Mountain cattle of Dagestan										
Kostromskaya	10,835	6								
Sychevskaya	7863	5								
Caucasian Brown										
Suksunskaya							4418	2		
Gorbatovskaya Red	124	1					2615	2		

(continued)

Table 5 (continued)

Breed	Central federal district		Northwestern federal district		Southern federal district		Volga federal district		Ural federal district	
	Cows	Bulls	Cows	Bulls	Cows	Bulls	Cows	Bulls	Cows	Bulls
Yakutian cattle										
Istobenskaya	929	1								
Danish Red							1626	1		
Estonian Red			418	1						
Tagilskaya							353	1		
Swiss Red	371	1								
All breeds (24)	894,597	464	279,536	155	851,435	260	1,820,051	779	359,386	154
Beef cattle breeding										
Kalmyk					243,831	140	17,676	10		
Aberdeen Angus	234,598	134	28,484	17	5271	4	18,227	11	1622	1
Hereford	10,052	6	189	1	8671	6	73,993	45	14,998	9
Kazakh Whiteheaded	2322	2			19,382	12	38,078	22		
Limousin	1040	1	137	1	1336	1	7334	5	208	1
Charolais			90	1	4874	3			332	1
Galloway	2267	2								
Simmental	2699	2							1520	1
Aubrac	1779	2							2160	2
Russian Komolaya					2028	2				

(continued)

Table 5 (continued)

Breed	Central federal district		Northwestern federal district		Southern federal district		Volga federal district		Ural federal district	
	Cows	Bulls	Cows	Bulls	Cows	Bulls	Cows	Bulls	Cows	Bulls
Salers							893	1	1059	1
Mandalong specials										
Blanc Bleu Belge	643	1								
Santa Gertrudis					608	1				
All breeds (14)	255,400	150	28,900	20	286,000	169	156,200	94	21,900	16
Total (38)	1,149,997	614	308,436	175	1,137,435	429	1,976,251	873	381,286	170

Breed	Siberian federal district		Far Eastern federal district		North Caucasian federal district		Nationwide	
	Cows	Bulls	Cows	Bulls	Cows	Bulls	Cows	Bulls
Dairy cattle breeding								
Black-and-white	492,399	200	91,159	28	88,128	24	2,812,365	1177
Holstein black-and-white	104,638	43	154,529	48	246,559	68	1,680,045	641
Red Steppe	169,086	69	9263	3	303,661	84	581,381	188
Simmental	193,571	80	45,685	15	20,794	6	390,094	161
Red-and-white	217,831	90	68,001	21			387,338	161
Kholmogorskaya			65,896	21			283,488	121
Ayrshire	6980	3			8912	3	174,912	71
Brown Swiss			842	1	65,023	20	106,210	41
Yaroslavskaya					29,376	9	78,531	35

(continued)

Table 5 (continued)

Breed	Siberian federal district		Far Eastern federal district		North Caucasian federal district		Nationwide	
	Cows	Bulls	Cows	Bulls	Cows	Bulls	Cows	Bulls
Jersey	113	1			49,180	13	75,274	26
Montbéliarde					24,425	7	33,591	13
Bestuzhev							28,273	13
Mountain cattle of Dagestan					14,853	5	14,853	5
Kostromskaya							10,835	6
Sychevskaya							7863	5
Caucasian Brown					44,889	13	44,889	13
Suksunskaya							4418	2
Gorbatovskaya Red							2739	3
Yakutian cattle			2105	1			2105	1
Istobenskaya							1626	1
Danish Red							929	1
Estonian Red							418	1
Tagilskaya							353	1
Swiss Red							371	1
All breeds (24)	1,184,616	486	437,480	138	895,799	252	6,722,900	2688
Beef cattle breeding								
Kalmyk	639	1	17,837	11	53,601	32	333,584	194

(continued)

Table 5 (continued)

Breed	Siberian federal district		Far Eastern federal district		North Caucasian federal district		Nationwide	
	Cows	Bulls	Cows	Bulls	Cows	Bulls	Cows	Bulls
Aberdeen Angus	7930	5	693	1	8576	6	305,401	179
Hereford	46,977	28	6032	4	18,181	11	179,092	110
Kazakh Whiteheaded	17,941	11	11,361	7	19,342	12	108,425	66
Limousin							10,054	9
Charolais							5297	5
Galloway	1912	2	530	1			4709	5
Simmental			149	1			4368	4
Aubrac							3939	4
Russian Komolaya							2028	2
Salers							1059	1
Mandalong specials							893	1
Blanc Bleu Belge							643	1
Santa Gertrudis							608	1
All breeds (14)	75,400	47	36,600	25	99,700	61	960,100	582
Total (38)	1,260,016	533	474,080	163	995,499	313	7,683,000	3270

Source Calculated by the authors

District—Holstein, Yaroslavskaya, and Charolais breeds; in the Southern Federal District—Simmental, Brown Swiss, Charolais, and Kazakh Whiteheaded breeds; in the Volga Federal District—Gorbatovskaya Red, Ayrshire, Limousin, and Kalmyk breeds; un the Ural Federal District—Simmental, Black-and-white, and Aberdeen Angus breeds; in the Siberian Federal District—Simmental, Ayrshire, Hereford, and Kazakh Whiteheaded breeds; in the Far Eastern Federal District—Simmental, Holstein, and Galloway; in the North Caucasus Federal District—Montbéliarde, Holstein, and Aberdeen Angus breeds. Genetic diversity and different adaptability of different breeds are the unconditional competitive advantage of Russian cattle breeding. Therefore, their optimal ratio will increase the industry's stability under any transformations and changes in market conditions.

AO HCR kept 32.2% of the nationally required breeder bull herd and produced 26.9% of semen of the potential market. The share of AO HCR in the Russian market for bull semen reached 35.1%.

To increase the share of AO HCR in the market of breeding products, it is necessary to allocate subsidies for the purchase of bull semen from Russian agricultural producers for 65 rubles per dose. This measure will accelerate the solution of problems of import substitution in the Russian market of breeding products. Moreover, based on the use of breeding cattle of Russian selection, it will improve the quality of selection and breeding work in Russian cattle breeding.

Acknowledgements The work was prepared within the framework of the state assignment (121052600377-6) of the Ministry of Education and Science of Russia.

References

1. Abdelmanova AS, Mishina AI, Volkova VV, Chinarov RY, Sermyagin AA, Dotsev AV, Boronet-skaya OI, Petrikeeva LV, Kostyunina OV, Brem G, Zinovieva NA (2019) Comparative study of different methods of DNA extraction from cattle bones specimens maintained in a craniological collection. Agric Biol 54(6):1110–1121. https://doi.org/10.15389/agrobiology.2019.6.1110rus
2. Chinarov V (2019) Ways to improve the competitiveness of dairy cattle in the Russian Federation. In IOP conference series: earth and environmental science, vol 274, p 012044.https://doi.org/10.1088/1755-1315/274/1/012044
3. Chinarov VI (2019) Milk and meat cattle breeding of Russia: problems and prospects. Econ Agric Proc Enterp 2:8–11
4. Chinarov VI, Strekozov NI, Chinarov AV (2017) Organizational and economic solutions for dairy and beef cattle expanded reproduction. J Dairy Beef Cattle Breed 7:16–19
5. Dunin IM, Butusov DV, Shichkin GI, Safina GF, Chernov VV, Lastochkina OV, Mukhin AE et al (2020) Yearbook on breeding work in dairy cattle farms of the Russian Federation, 2019. Federal State Budgetary Scientific Institution "All-Russian Research Institute of Animal Breeding.", Moscow, Russia
6. Dunin IM, Butusov DV, Shichkin GI, Safina GF, Chernov VV, Lastochkina OV, Mukhin AE et al (2020) Yearbook on breeding work in beef cattle farms of the Russian Federation, 2019. Federal State Budgetary Scientific Institution "All-Russian Research Institute of Animal Breeding.", Moscow, Russia

7. Federal Customs Service of the Russian Federation (n.d.) Data analysis. http://stat.customs. gov.ru/analysis
8. Federal State Statistics Service of the Russian Federation (n.d.) Agriculture, hunting, and forestry. https://rosstat.gov.ru/enterprise_economy
9. Sermyagin AA, Gladyr EA, Zinoveva NA (2016) Genome-wide associations study for somatic cell score in Russian Holstein cattle population. J Dairy Sci 99(1):154
10. Strekozov NI, Sivkin NV, Chinarov VI, Bautina OV (2017) Evaluation of dairy breeds by reproductive and adaptive abilities. Zootechniya 7:2–6

Benefits of Circular Agriculture for the Environment: International Experience of Using Digitalization and Higher Education Development

Svetlana E. Karpushova⬭, Aliia M. Bazieva⬭, Natalia M. Fomenko⬭, and Elena S. Akopova⬭

Abstract The purpose of the work is to substantiate the need and develop recommendations for the most complete disclosure of the potential for the development of circular practices in agriculture based on higher education and digital technologies in the interests of sustainable agricultural development. The method of regression analysis is used in order to determine the consequences and prove the advantages of circular agriculture for the environment based on the study of international experience using a representative sample, which includes countries with developed agricultural economies—leaders in the World Bank ranking in terms of the share of agriculture farms in the structure of gross value added in 2020. Additionally, the authors study the international experience of using the capabilities of the digital economy and the development of higher education for implementing a circular model of agriculture. A critical necessity for smart technologies, digital personnel, and skilled employees in agriculture to implement its circular model is proved. It is substantiated that circular practices contribute to the development of agriculture but the potential for the development of these practices based on higher education and digital technologies is far from being fully realized, which hinders the sustainable development of agriculture. Recommendations for the national economic policy for the regulation of the process of transition to circular agriculture or its development based on stimulation of the dissemination of smart technologies and development of higher education in the

S. E. Karpushova
Volgograd State Technical University (Sebryakov Branch), Mikhaylovka, Russia
e-mail: sfkse@yandex.ru

A. M. Bazieva (✉)
Kyzyl-Kiya Institute of Technology, Economy and Law of Batken State University, Kyzyl-Kiya, Kyrgyzstan

N. M. Fomenko
Plekhanov Russian University of Economics, Moscow, Russia
e-mail: fnata77@mail.ru

E. S. Akopova
Rostov State University of Economics, Rostov-on-Don, Russia
e-mail: Akopova_rsue@icloud.com

interests of environmental protection in the aspect of production waste reduction and fighting climate change are developed.

Keywords Circular agriculture · Smart technologies · Environment · Waste reduction · Digitalization · Higher education development · Digital personnel · Fight against climate change

JEL Codes D91 · F64 · H52 · I26 · O13

1 Introduction

Historically, farming in many parts of the world was plagued by high sickness rates, insufficient manure, and the constant threat of a horrific calamity. Circular agriculture is not a strategy that suffocates growing businesses with rigid ideologies, market requirements, and unofficial rules [2]. It is a concept that refers to a concerted effort by all required delegates, including ranchers, to strike the optimal balance of environmental standards and contemporary innovation, new organizations, and beneficial business models. It places a premium on high yields and efficient resource and energy consumption, as well as the critical importance of squeezing the climate to the maximum extent possible [5, 27, 35].

It is a concept that views residues from agriculture biomass and food handling in the context of the food system as limitless resources [7]. By making more efficient use of scarce resources and wasting less biomass, we can reduce our reliance on imported chemical composts and distant sources of domesticated animal feed [20]. This means that the availability of alternative assets will determine the maximum production capacity and subsequent use of alternative assets.

This chapter hypothesizes that circular practices contribute to the development of agriculture but the potential for the development of these practices based on higher education and digital technologies is far from being fully realized, which hinders sustainable agricultural development [24, 25, 28–31]. The purpose of the work is to substantiate the need and develop recommendations for the most complete disclosure of the potential for the development of circular practices in agriculture based on higher education and digital technologies in the interests of sustainable agricultural development.

2 Materials and Method

Circular agriculture, which places a premium on a healthy harvest and government assistance for animals, recognizes a precise extended process. This process begins with robust microscopic organisms that are used to select plants and animals that are more resistant to diseases and irritations, as well as the effects of climate change.

Integrating agrobiodiversity into, on, and around fields would increase productivity by acting as a natural fertilizer and harvest security [26]. This can be accomplished, for example, by planting blossoms along field margins, inland squares, and in insect banks, which serve as hiding places for wild honey bees and other pollinators, as well as regular predators of various vermin species. While regular cycles benefit horticulture, they also contribute to a tremendously normal cultivating environment [34]. Agroecological "nature-inclusive agribusiness", which places a premium on biological system management, including preserving and utilizing nature and biodiversity on and around the homestead in a cultivating scene, is a more advanced form of circular horticulture [17].

The central premise of circular agriculture is to utilize agricultural biomass as frequently and successfully as possible. It entails avoiding the regular decomposition of excess biomass (crop residues, compost) and the subsequent production of carbon dioxide, nitrous oxide, and methane [8]. Additionally, it implies that less manure is required for agriculture as a whole to emit less CO_2.

Additionally, superior manure (excrement, soil, and fertilizer) promotes soil carbon retention, which is a systematic strategy for mitigating climate change. Thus, circular agriculture offers significantly more opportunities for reducing agribusiness's ozone-depleting chemical emissions than initiatives primarily focused on making traditional farming cycles more environmentally friendly. Horticulture, precisely because of this combination, has the potential to provide significant environmental benefits [3].

The central premise of circular agriculture is to utilize agricultural biomass as frequently and successfully as possible. It entails avoiding the regular decomposition of excess biomass (crop residues, compost) and the subsequent production of carbon dioxide, nitrous oxide, and methane [25]. Additionally, it implies that less manure is required for agriculture as a whole to emit less CO_2.

Additionally, superior manure (excrement, soil, and fertilizer) promotes soil carbon retention, which is a systematic strategy for mitigating climate change [33]. Thus, round horticulture offers far more opportunities for mitigating agribusiness's ozone-depleting chemical emissions than initiatives primarily focused on ecologically friendly farming cycles. Horticulture can provide significant environmental benefits precisely because of this combination [11].

To test the hypothesis put forward, a sample of 10 countries with developed agricultural economies—leaders of the World Bank ranking [33] in terms of the share of agriculture in the structure of gross value added in 2020 was formed. Circular agriculture, digitalization, and higher education development statistics in countries with developed economies in 2020 are given in Table 1

Based on the data from Table 1 by the method of regression analysis, the following economic and mathematical models were obtained:

$$Agr = 1.33 + 0.11\,ESR + 0.004\,NCP \qquad (1)$$

According to the obtained model (1), circular practices contribute to the development of agriculture. With an increase in the Efficient and Sustainable Resource Use

Table 1 Statistics of circular agriculture, digitalization, and higher education development in advanced agricultural economies in 2020

Country	Agriculture, forestry, and fishing, value added (% of GDP)	Efficient and sustainable resource use, points 1–100	Natural capital protection, points 1–100	Higher education factor (knowledge), points 1–100	Digital technologies factor (future readiness), points 1–100
	Agr	ESR	NCP	edu	dtc
Indonesia	13.7	62.88	64.3	41.26	46.695
Philippines	10.2	63.68	74.54	42.557	44.789
Thailand	8.6	59.43	74.73	54.193	49.936
Malaysia	8.2	55.8	71.07	73.636	64.048
Columbia	7.7	65.1	71.1	43.754	46.015
China	7.7	48.66	64.6	85.105	80.004
Peru	6.7	64.94	72.08	46.924	43.198
Brazil	5.9	65.5	71.03	44.349	51.618
New Zealand	5.7	58.11	69.64	66.603	75.023
Kazakhstan	5.3	45.95	43.21	62.942	63.839

Source Compiled by the authors based on materials from [12, 14, 33]

by 1 point, the share of Agriculture, forestry, and fishing, value-added increases by 0.11%. With an increase in the activity of the Natural Capital Protection by 1 point, the share of Agriculture, forestry, and fishing, value-added increases by 0.004%.

$$ESR = 80.93 - 0.31\,edu - 0.08\,dtc \tag{2}$$

$$NCP = 82.34 + 0.20\,edu - 0.46\,dtc \tag{3}$$

According to the obtained models (2) and (3), the factors of Higher Education and Digital Technologies make little contribution to the sustainable and circular development of agriculture. This is evidenced by negative regression coefficients, only one of which turned out to be positive. Thus, with the growth in the popularity of Higher Education by 1 point, the activity of Natural Capital Protection increases by 0.46 points. This confirms the hypothesis put forward.

3 Results

Circular economics advocates for the establishment of essential habitats such as soil, air, and water bodies. These biological systems perform a variety of functions, including cleaning, productive agriculture, fertilization, and water purification. In a direct economy, these administrations eventually run out of resources as a result of frequent product withdrawals or become overburdened as a result of toxic waste offloading [9]. When these items are used circularly and hazardous substances are avoided, the land, air, and water bodies remain healthy and beneficial.

One of the most fundamental challenges humanity will face in the coming years is ensuring an adequate supply of safe and nutritious food without dramatically expanding the planet's borders. In circular agriculture, waste is used as a raw material to create new valuable commodities such as crops, food, and feed. Another aspect of the concept is the requirement to reduce asset use and pollution [2]. Natural resources extraction and waste disposal have a detrimental effect on nature's reserves. These natural areas are critical for environmental administrations' preservation, as well as for the conservation of natural and cultural heritage [21]. Numerous governments and organizations are primarily concerned with preserving nature from crude material exploitation and waste disposal at the moment. This extraction and unloading process should be halted entirely to protect the ecosystem. It is accomplished through the application of a circular economy framework [19].

At the time, college adoption of advancements coincided with a paradigm shift in which innovation was viewed as a perplexing and interconnected environment conducive to computerized learning. As a result, regardless of the learning experiences enabled, the emphasis is shifted away from true innovation and towards the understudy. Digitization is critical for higher education institutions (HEIs) to recruit more and better students, improve the student experience, display resources, and manage the entire preparation cycle in this unique situation. Additionally, it enables observers to identify potential roadblocks in the preparation process and reduces the likelihood of students dropping out of school. Whatever the reason, the reluctance to perceive and seize opportunities for development towards a technological society persists [15].

As with prudent management, HEIs will invest in incorporating and advancing clean technologies into their operations and ensuring their widespread adoption in their current impact scenario. In analytical writing, clean innovation is also referred to as ecological, green, or natural sound innovation. It is an interaction or administration that mitigates adverse biological effects through significant increases in energy productivity, sustainable asset utilization, or natural security exercises [22]. Fundamentally, these cycles are more hygienic, make better use of assets, recycle more trash, and manage waste more effectively. Additionally, clean technology development is contingent on the advancement of data and communication technologies (ICT). HEIs embrace cloud-based media communications administrations, which eliminate the need for additional physical devices and equipment [32].

DT is an interaction that coordinates advanced innovation from multiple perspectives and necessitates changes to the innovation space, culture, and tasks, among other things. Organizations must self-assess and adjust their cycles in order to capitalize on developing technology and its rapid expansion into human activities. Thus, for DT to thrive, a shift in focus is required, as is an increase in innovation and a change in institutional culture. The DT is considered the fourth modern upheaval because it is mechanical and embraces new human capabilities despite organizational re-evaluation [18]. However, the third stage of computerized growth reception is examined as well, following computerized competence and computerized use.

Similarly, digital education expands the capacity for usage and application. DT is an interaction that occurs within the instructive topic and necessitates teaching evolution and adaptation to the understudy's new adapting needs. As a result, this interaction becomes more effective, allowing for community-based work to be accomplished.

As a result of technological advancements, it hybridizes, integrating traditional and virtual settings, online and offline, and displaying do-it-yourself patterns (Do It Yourself). Big Data and Artificial Intelligence (AI) are being used as instructional assets in novel learning environments, reinforcing the importance of difficult subjects in higher education. As a result of Big Data, students may discover patterns that correspond to novel exhibiting strategies, such as versatile realization, which tailors instruction to individual students based on their age, customs, or behaviour. This equipment includes low-cost instruction that strengthens clients' abilities and establishes an individualized profile of understudies. It will increase the visibility of areas of difficulty, allowing for the creation of an engaging course through the use of e-learning architecture. On the other hand, HEIs use AI to personalize the admissions process and determine which students are most likely to succeed in their certificates and positions of authority.

Additionally, this innovation enables instructors to monitor an understudy's progress or to take control of the displaying system if they notice a gap in comprehension. How we live, work, and collaborate is changing as a result of digital technology, computerization, and other forms of technological learning. As a result, instructional foundations face the challenge of developing a framework for continuous and vivid learning that is on par with sophisticated technologies and programming.

DT promotes rational and innovative training by incorporating new educational methods for both students and instructors, such as flipped classrooms, digital cooperative learning (DCL), gamification, augmented reality, virtual reality, or mixed reality. By focusing on innovation and business, the DT approach to education promotes learning methodologies that emphasize personalized preparation, personalization of information, and ability development through social learning [36]. The computerized age necessitates an adaptive education that enables the acquisition of new skills, outdoing oneself, and inventing in an era of constant change such as the current one. As a result, computerized education is defined as a method of in-person and remote instruction that makes use of contemporary technologies and aims to secure instructors' and students' skills and capacities for acquisition through a progressive preparation process.

Education catalyzes a global emphasis on personal fulfilment and achievable improvement. Additionally, access to a comprehensive and equitable education can help equip the populace with the necessary tools for problem-solving [22]. Thus, by linking quality education to innovation and promoting DT, undergraduates can provide information, abilities, and inspiration to assist youth in understanding the SDGs, mobilize youth, provide educational or professional training to assist in implementing SDG arrangements and expand opportunities for limited collaboration between undergraduates and agricultural experts to address SDG [37]. According to the concept of sustainable administration, this is a collection of human, moral, and natural characteristics that enable social orders to rationally advance organizations, foundations, and networks, thereby ensuring the global monetary and social texture's intensity and reinforcement [4].

In terms of HE, it should ensure that DT is administered effectively in order to achieve the desired model of a transparent, progressive, inventive, and organized foundation. As a result, astute administration should establish management frameworks based on sound principles, ensuring that associations achieve increased visibility and advancement [16]. Partners must be candid about the manageability of instructional organizations' actions in this regard. The pursuit of novel and improved deduction strategies in HEIs is one of the most difficult issues of steady progress. As a result, effectively managing instructional organizations is a prerequisite for developing an entirely sophisticated instructional approach. Regardless of its advantages, DT is detrimental to HE. In this sense, a lack of self-control on the part of the understudy may disrupt the educational flow of events. Additionally, the most widely used method of education and learning is not so much human as it is generic; it is insufficiently comprehensive because not everyone uses electronic equipment. On the other hand, it has the potential to invalidate certain abilities and basic capacities.

4 Conclusion

Concerns about the environment and public health play a role in the decision to initiate a new wave of horticulture development. Having a strong market strategy, innovative capabilities, access to a globally competent organization, and the ability to secure venture funding are critical components of successful execution and business development [13]. Furthermore, scaling up has begun, primarily through the expansion of projects undertaken by the organizations or associations involved. The food system as a whole has not yet made the transition to circularity (yet).

At both the production and framework levels, the difficulties and risks associated with establishing circular agriculture are identified. There are lengthy and prohibitively expensive enlistment procedures for new goods at the manufacturing level, a lack of communication about new products among potential clients, and extended cycles for perfecting the new roundabout model. Stopping a relatively insignificant process results in few monetary or ecological gains at the framework level [23]. Other linear processes continue to operate as a result of their waste streams, for example, because valuing food within the current food framework does not generate externalities [1]. Another issue raised by the use of natural waste is the introduction of hazardous materials or germs into the food chain. Additionally, if circularity is promoted solely for specialized and financial reasons, such as supplement reuse and business case development, it may have negative social consequences for vulnerable groups [6].

Existing strategies could also be examined to reduce sponsorships in agriculture, energy, and transportation that obstruct manageable asset utilization. Endowments that encourage water, energy, and manure waste could also be eliminated or reduced, with reserve funds directed towards agrarian research, enhanced water and land-use executives, compensatory pay support for small ranchers, and designated small sponsorships to the board to achieve explicit roundabout rural practices. For instance, sponsorships could be reimbursed if ranchers agree to the CEOs' highly visible soil ripening practices, which trap a significant amount of carbon [10].

Sustainable resource management policies should incentivize smallholder ranchers to pursue breakthroughs in precision farming and harvest efficiency. It is inextricably linked to the imperative of achieving net-zero energy costs for water reuse, which will require a rethinking of critical public policies.

References

1. Abad-Segura E, González-Zamar MD, Belmonte-Ureña LJ (2020) Effects of circular economy policies on the environment and sustainable growth: worldwide research. Sustainability 12(14):5792
2. Barros MV, Salvador R, de Francisco AC, Piekarski CM (2020) Mapping of research lines on circular economy practices in agriculture: from waste to energy. Renew Sustain Energy Rev 131:109958
3. Berndtsson M (2015) Circular economy and sustainable development
4. Boev VU, Ermolenko OD, Bogdanova RM, Mironova OA, Yaroshenko SG (2019, April) Digitalization of agro-industrial complex as a basis for building organizational-economic mechanism of sustainable development: foreign experience and perspectives in Russia. In Institute of scientific communications conference. Springer, Cham, pp 960–968
5. Bogoviz AV, Sergi BS (2018) Will the circular economy be the future of russia's growth model? In Exploring the future of russia's economy and markets. Emerald Publishing Limited
6. Camana D, Manzardo A, Toniolo S, Gallo F, Scipioni A (2021) Assessing environmental sustainability of local waste management policies in Italy from a circular economy perspective: an overview of existing tools. Sustain Produ Consum
7. Dagevos H, Lauwere CD (2021) Circular business models and circular agriculture: perceptions and practices of dutch farmers. Sustainability 13(3):1282
8. Duque-Acevedo M, Belmonte-Ureña LJ, Yakovleva N, Camacho-Ferre F (2020) Analysis of the circular economic production models and their approach in agriculture and agricultural waste biomass management. Int J Environ Res Public Health 17(24):9549
9. Fourie A (2006) Municipal solid waste management as a luxury item. Waste Manag (New York, NY) 26(8):801–802
10. Friant MC, Vermeulen WJ, Salomone R (2021) Analyzing European Union circular economy policies: words versus actions. Sustain Produ Consum 27:337–353
11. Gao W, Chen Y, Dong W (2010) Circular agriculture as an important way to a low-carbon economy. Zhongguo Shengtai Nongye Xuebao/Chin J Eco-Agric 18(5):1106–1109
12. Global Green Growth Institute (2021) Global Green Growth index 2020. https://greengrowthindex.gggi.org/#download-reports-popup
13. Heshmati A (2017) A review of the circular economy and its implementation. Int J Green Econ 11(3–4):251–288
14. IMD (2021) World Digital Competitiveness Report 2020. https://www.imd.org/centers/world-competitiveness-center/rankings/

15. Johansson N, Henriksson M (2020) Circular economy running in circles? A discourse analysis of shifts in ideas of circularity in Swedish environmental policy. Sustain Prod Consum 23:148–156
16. Khabarov V, Volegzhanina I (2019, December) The knowledge management system of an industry-specific research and education complex. In IOP conference series: earth and environmental science, vol 403, no 1. IOP Publishing, p 012197
17. Kleppmann F, Stengler E (2013) Current developments in European waste-to-energy. In Interessengemeinschaft der thermischen abfallbehandlungsanlagen deutschland eV: Düsseldorf, Germany.
18. Lajoie-O'Malley A, Bronson K, van der Burg S, Klerkx L (2020) The future (s) of digital agriculture and sustainable food systems: an analysis of high-level policy documents. Ecosyst Ser 45:101183
19. Liang L, Chen YQ, Gao WS (2010) Integrated evaluation of circular agriculture system: a life cycle perspective. Huan jing ke xue= Huanjing kexue 31(11):2795–2803
20. Mehmood A, Ahmed S, Viza E, Bogush A, Ayyub RM (2021) Drivers and barriers towards circular economy in agri-food supply chain: a review. Bus Strategy Dev
21. Numbeo (2021) Quality of life index. https://www.numbeo.com/quality-of-life/rankings_current.jsp
22. Peterson M, Epler R (2020) Sustainability developments in cities of the world. Cont Broaden Market Concept
23. Polverini D (2021) Regulating the circular economy within the ecodesign directive: progress so far, methodological challenges and outlook. Sustain Prod Consum 27:1113–1123
24. Popkova EG, Popova EV, Sergi BS (2018) Clusters and innovational networks toward sustainable growth. In Exploring the future of russia's economy and markets. Emerald Publishing Limited
25. Popkova EG, Sergi BS (2018) Will industry 4.0 and other innovations impact Russia's development? In Sergi BS (ed. Exploring the future of Russia's economy and markets: towards sustainable economic development. Emerald Publishing Limited, Bingley, UK, pp 51–68
26. Rosemarin A, Macura B, Carolus J, Barquet K, Ek F, Järnberg L, Okruszko T (2020) Circular nutrient solutions for agriculture and wastewater–a review of technologies and practices. Current Opin Environ Sustain 45:78–91
27. Schroeder P, Anggraeni K, Weber U (2019) The relevance of circular economy practices to sustainable development goals. J Ind Ecol 23(1):77–95
28. Sergi BS, Popkova EG, Bogoviz AV, Litvinova TN (2019). Understanding industry 4.0: AI, the internet of things, and the future of work. Emerald Publishing Limited, Bingley, UK
29. Sergi BS, Popkova EG, Bogoviz AV, Ragulina YV (2019) The agro-industrial complex: tendencies, scenarios, and policies. In Sergi BS (ed) Modeling economic growth in contemporary Russia. Emerald Publishing, Bingley, UK
30. Sergi BS, Popkova EG, Bogoviz AV, Ragulina YV (2019). Costs and profits of technological growth in Russia. In Sergi BS (ed) Tech, smart cities, and regional development in contemporary Russia. Emerald Publishing Limited, Bingley, UK, pp. 41–54
31. Sergi BS, Popkova EG, Bogoviz AV, Ragulina YV (2019) The agro-industrial complex: tendencies, scenarios, and regulation. In Sergi BS (ed) Modeling economic growth in contemporary Russia. Emerald Publishing Limited, Bingley, UK, pp 233–247
32. Ugur NG (2020) Digitalization in higher education: a qualitative approach. Int J Technol Edu Sci 4(1):18–25
33. Vollaro M, Galioto F, Viaggi D (2016) The circular economy and agriculture: new opportunities for reusing Phosphorus as fertilizer. Biobased Appl Econ 5(3):267–285
34. World Bank (2021) Agriculture, forestry, and fishing 2020, value added (% of GDP). https://data.worldbank.org/indicator/NV.AGR.TOTL.ZS?most_recent_value_desc=true&view=chart
35. Yi-xiang WBQW, Jin-gui LEI (2009) On circular economy development and low-carbon agriculture construction. J Poyang Lake 3
36. Zhenjian L, Jiahua L, Yunbao X (2021) Research on the path of agriculture sustainable development based on the concept of circular economy and big data. Acta Agric Scand Sec B—Soil Plant Sci, 1–12

37. Zucchella A, Previtali P (2019) Circular business models for sustainable development: a "waste is food" restorative ecosystem. Bus Strateg Environ 28(2):274–285

Ecologization of Cultivating Honey Plants in the Region

Viktoria V. Vorobyova⬤, Daria V. Rozhkova⬤, and Pavel T. Avkopashvili

Abstract The authors reveal the peculiarities of the territorial location of beekeeping by regions of Russia and natural-climatic zones of the Altay Territory. Additionally, the authors indicate disproportions in the location of the honey plants and the concentration of beekeepers. The paper compares the efficiency of the production of sunflower oilseeds using intensive technology (which involves the application of mineral fertilizers) and environmentally friendly technology (which involves bee pollination during sunflower growing).

Keywords Beekeeping · Production location · Placement disproportions · State of apiaries · Sunflower · Crop pollination · Economic efficiency

JEL Codes Q53 · Q55 · Q150

1 Introduction

Several studies were conducted to assess the efficiency of bee pollination of sunflower crops in the Altay Territory. Beekeeping is a branch of agricultural production that does not absorb but rather multiplies natural resources during its activity. The social, economic, and ecological importance of beekeeping is determined by the high value of the primary, side, and associated products. Moreover, bees are used in the pollination of honey-bearing, berry, and fruit crops, which increases the yields and quality of the resulting products.

V. V. Vorobyova (✉) · P. T. Avkopashvili
Altai State University, Barnaul, Russia

P. T. Avkopashvili
e-mail: regcenoe@bk.ru

D. V. Rozhkova
Altai Branch of the Russian Presidential Academy of National Economy and Public Administration, Barnaul, Russia
e-mail: danya2510@yandex.ru

Between 1992 and 2019, honey production in all countries increased by 62.83%, with a tendency for bee colonies to decrease while honey production increased. In 2019, the leading honey producers were China (24.13% of global gross honey production), Turkey (5.90%), Brazil (4.26%), and Canada (4.34%). These countries accounted for 38.63% of all honey produced worldwide.

Among the countries with developed beekeeping, the highest growth rate of gross production for 1992–2019 was observed in Brazil, Canada, and China—the volume of production increased by 2.44–2.65 times. The decline in production was observed in Mexico (2.97%) and the USA (29.22%). Also, there were structural shifts in the structure of honey production by country over the period 1992–2019. The combined share of the USA, Mexico, and Argentina decreased by 8.37%. In turn, the combined share of China, Brazil, and Canada increased by 10.53% (Table 1).

In recent years, there has been a decline in the number of bee colonies in the USA and Europe due to their high mortality and low profitability of bee production. In 1961, there were 5.51 million and 21.10 million bee colonies in the USA and Europe. In turn, in 2019, there were 2.81 million (a decrease of 1.96 times) bee colonies in the USA and 16.22 million bee colonies in Europe (a decrease of 1.30 times) (Fig. 1).

One of the reasons for the decrease in the number of bee colonies is their death due to the use of chemical agents to protect honey crops from pests and diseases (studies

Table 1 Gross honey production in the world, thousand tons

Countries	1992		2000	2010	2019		
	Thousand tons	% of total			Thousand tons	% of total	% of 1992
China	183,175	16.10	251,839	409,149	447,007	24.13	244.03
Turkey	60,318	5.30	61,091	81,115	109,330	5.90	181.26
Canada	30,330	2.67	31,860	81,672	80,345	4.34	264.90
Brazil	61,000	5.36	93,000	59,000	78,927	4.26	129.39
USA	100,560	8.84	99,945	80,042	71,179	3.84	70.78
Ukraine	57,111	5.02	52,439	70,873	69,937	3.78	122.46
India	51,000	4.48	52,000	60,000	67,141	3.62	131.65
Russia	49,556	4.36	54,248	51,535	63,526	3.43	128.19
Mexico	63,886	5.62	58,935	55,684	61,986	3.35	97.03
Argentina	18,841	1.66	21,865	38,073	45,981	2.48	244.05
Hungary	10,742	0.94	15,165	16,500	20,000	1.08	186.19
Central African Republic	9500	0.83	13,000	15,000	16,206	0.87	170.59
Greece	12,898	1.13	14,356	16,237	n.d	x	x
Spain	23,958	2.11	28,860	34,550	n.d	x	x
World	1,137,749	100.00	1,260,063	1,588,061	1,852,598	100.00	162.83

Source Compiled by the authors based on [3]

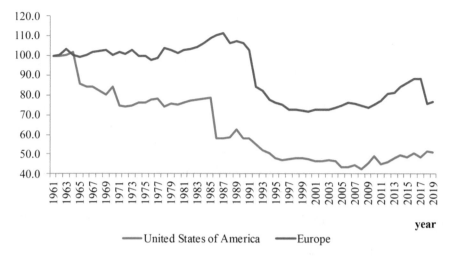

Fig. 1 Baseline rates of bee population decline in the USA and Europe (1961 = 100%), %. *Source* Compiled by the authors based on [3]

of the chemical composition of dead bees, honey, and bee pollen in hives show high pesticide content [7]). Another reason is the spread of genetically modified crops [1, 2]. It is noteworthy that, after the death of bees, no insects or other bees fly into hives full of honey, which indicates the presence of toxins or repellents in the hives and products [10].

In 2010–2019, Russia accounted for no more than 3.25–3.45% of global honey production, occupying the following positions in the top ten world honey producers:

- Fourth place (after the EU, China, and Turkey) in the number of bee colonies;
- Eighth place (after China, Argentina, the EU, the USA, Mexico, Turkey, and India) in terms of honey production.

V. I. Lebedev and A. S. Ponomarev also note the following rating positions of Russia [8, 11]:

- Fifth place (after Ethiopia, Ukraine, China, and India) in the number of beekeepers (about 137 thousand people);
- Ninth place in the average number of colonies in an apiary;
- Tenth place in the productivity of bee colonies and honey export volume.

Russia and Ukraine have similar requirements for agricultural producers engaged in cultivating honey crops on the need to comply with preventive measures of bee mortality in any chemical treatments of crops. However, practice shows widespread violations of these regulations, including sunflower crops as the main honey crop in many areas [5, 9].

2 Methodology

The information on the number of bee colonies by regions of Russia and munici-palities of the Altay Territory and entomophilic crops were taken from the Unified Interagency Information and Statistical System of the Federal State Statistics Service (Rosstat). The information on the volume of honey production and the number of bee colonies by country was obtained from the UN Food and Agriculture Organi-zation (FAO) database. The honey supply was calculated by multiplying the area of entomophilic crops, nectar productivity of lands, and the coefficient of avail-ability of these supplies to bees (equals 0.33). The efficiency of bee pollination of sunflower crops was calculated in regulatory-technological maps, taking into account the most common technology of sunflower growing in the steppe regions of the Altay Territory. When calculating the normative-technological map, we considered the additional costs associated with the transportation of additional products received, transportation of bees, and payments to beekeepers for living near apiaries.

3 Results

The Altay Territory is located in seven natural and climatic zones—from the arid western steppe regions to the foothill regions with excessive moisture. On average, the number of bee colonies per 100 ha of agricultural land in the region has increased from 1.21 to 1.71. However, it remains three times lower than the minimum number of bee colonies needed to maintain ecological balance (the norm is at least 5–7 beehives per 100 ha of land [12]) (Fig. 2).

In most municipalities of the Altay Territory, the number of bee colonies is within 20.0% of the norm; there were 32 (52.46%) of such municipalities in 2010 and 30 (49.18%) in 2020. These municipalities are located mostly in the western part

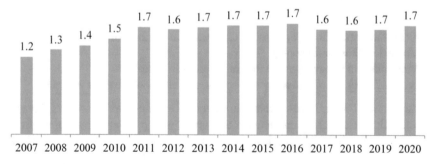

Fig. 2 Number of beehives per 100 ha of agricultural land in the Altay Territory in 2007–2020, pcs. *Source* Compiled by the authors based on [4]

Table 2 Distribution of municipalities in the Altay Territory by the number of beehives per 100 ha of agricultural land

Number of bee colonies per 100 ha of agricultural land	Number of municipalities, pcs				Number of bee colonies per 100 ha of agricultural land, on average			
	2010		2020		2010		2020	
	Total	% of total	Total	% of total	Total	% of total	Total	% of total
Less than 0.5 pieces	23	37.70	20	32.79	0.26	5.2	0.25	5.1
From 0.5 to 1.0 pieces	9	14.75	10	16.39	0.73	14.6	0.72	14.4
From 1.0 to 1.5 pieces	7	11.48	7	11.48	1.25	24.9	1.31	26.3
From 1.5 to 2.0 pieces	8	13.11	6	9.84	1.79	35.7	1.73	34.7
Above 2.0 pieces	14	22.95	18	29.51	4.58	91.5	4.71	94.2
Average	61	100.00	61	100.00	1.50	29.9	1.71	34.2

Source Compiled by the authors based on [4]

of the region with arid conditions. At the same time, these areas had the highest concentration of sunflower crops—from 57.0 to 65.3% in 2010–2020. The number of bee colonies close to optimal was observed only in 14 districts of the region (22.95% of the total number) in 2010 and 18 districts (29.51% of the total number) in 2020 (Table 2).

To estimate the honey supply throughout the Altay Territory, we took the areas of cultivated entomophilous crops in the municipalities in the region and the normative information of the average values of nectar productivity. Estimated honey supply (available to bees, i.e., about 33.0% of the potential honey supply), excluding the natural honey stock for the whole region in 1995–2020, ranged from 84.4 thousand tons to 139.6 thousand tons. At the same time, there is a steady dynamic of its decrease—for the analyzed period, the average annual rate of its decrease was more than 2.00% (Fig. 3).

Decrease of honey supply in the Altay Territory is connected not so much with a general reduction of area under honey crops as with the change of their inner structure (structure influence is significant enough since nectar productivity of different field honey plants varies from 40 to 70 kg/ha in sunflower and buckwheat to 240 kg/ha in annual and perennial grasses). If in 1995, about 93.0% of the total honey supply was produced by forage crops (perennial and annual grasses), then in 2020, their share decreased to 74.0% (by 19.0%). Under these conditions, the buckwheat and sunflower share in the honey reserve formation increased from 3.2% to 13.5% and from 2.8% to 11.0%, respectively (Fig. 4).

With the significant potential to increase the gross honey harvest in the region, beekeeping faces several problems associated primarily with the widespread use of

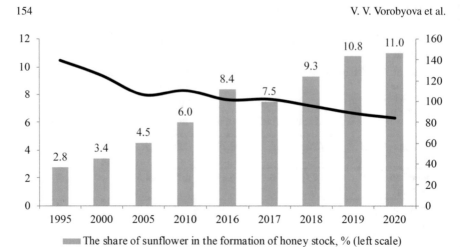

■ The share of sunflower in the formation of honey stock, % (left scale)

■ Honey supply of honey-bearing plants, thousand tons (right scale)

Fig. 3 Estimated honey supply in the Altay Territory (excluding the natural honey stock) and the share of sunflower in its formation. *Source* Compiled by the authors based on [4]

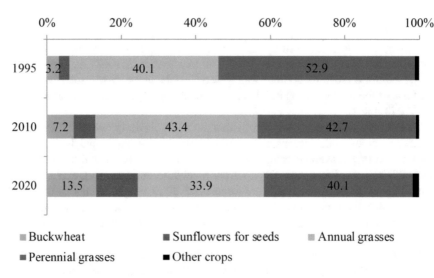

Fig. 4 Distribution of honey supply available for bees by main entomophilic crops in the Altay Territory in 1995–2020, %. *Source* Compiled by the authors based on [4]

chemicals in the cultivation of sunflower and other crops within the flight range of bees, which leads to their deaths.

The standard costs (determined by us on technological cards) with the bee pollination amounted to 726.1 thousand rubles per 100 ha, which is 28.4% lower than in the variant without bee pollination. In terms of 1 ton of oilseeds, it is lower by 386%. In absolute terms, bee pollination per 100 ha of sunflower resulted in a 5.48% increase

Table 3 Standard costs of growing sunflowers for grains with different variants of bee pollination, RUB/100 ha

Costs	Costs without bee pollination		Costs with bee pollination		
	Total	% of total	Total	% of total	% of the actual technology
Remuneration with deductions	164,613	16.23	171,986	23.69	104.48
Material costs - total	632,941	62.42	222,713	30.67	35.19
including fuel and petroleum products	164,565	16.23	156,085	21.50	94.85
seeds	57,000	5.62	57,000	7.85	100.00
chemical protection	403,123	39.75	–	–	0.00
electricity	8253	0.81	9628	1.33	116.67
Depreciation and repairs	97,165	9.58	93,469	12.87	96.20
Transportation	9206	0.91	10,952	1.51	118.98
Payment for bee pollination services	–	0.00	150,000	20.66	x
Total costs	1,014,024	100.00	726,096	100.00	71.61
including per 1 hectare	10,140	x	7261	x	71.61
per one ton of oilseeds	11,267	x	6915	x	6138

Source Compiled by the authors

in labor costs with deductions, electricity—by 16.67%, transportation costs—by 18.98%. Seed costs remained unchanged, and the reduction of costs on fuel and oil products was from 164.6 thousand rubles to 156.1 thousand rubles (5.15%) (Table 3). Costs of bee pollination services were calculated based on the norm of placement of bee colonies in sunflower (1 bee colony per 1 ha of crops), the area of crops, and the contract price of placing bee colonies near the fields.

Bee pollination increases the profitability of sunflower oilseed production from 47.47% to 85.48% (by 38.01%), which is 42.65% higher than the profitability of oilseed production on average in the Altay Territory (Fig. 5).

4 Conclusion

The research allowed us to estimate the uneven distribution of honey resources of the Altay Territory and bee colonies by municipalities. The authors revealed the concentration of sunflower crops in the steppe areas of the region with a fairly low supply

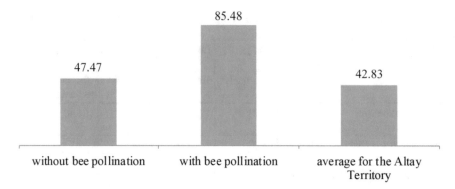

Fig. 5 Profitability of the production of sunflower oilseeds with alternative bee pollination, %. *Source* Compiled by the authors

of bee colonies. To increase the volume of oilseeds production, it is recommended to pay attention to the reserve associated with bee pollination of crops [6], which can increase the yield of sunflowers, reduce the cost of production, and significantly increase the production profitability. Gradual reduction of the chemicalization in the future will allow agricultural producers in the Altay Territory to enter the markets of organic products, including vegetable oil. The high margin of sunflower cultivation will diversify the production of medium-sized enterprises specializing in the production of oilseeds, which will create conditions for a more even use of production resources during the year and the equal receipt of funds.

References

1. Barron AB (2015) Death of the bee hive: understanding the failure of an insect society. Current Opin Insect Sci 10:45–50. https://doi.org/10.1016/j.cois.2015.04.004
2. Delanoë A, Galam S (2014) Modeling a controversy in the press: the case of abnormal bee deaths. Physica A 402:93–103. https://doi.org/10.1016/j.physa.2014.01.054
3. FAOSTAT (n.d.) Production. http://www.fao.org/faostat/en/#data. Accessed 8 June 2021
4. Federal State Statistics Service (2021) Unified interdepartmental information and statistical system. https://fedstat.ru/organizations/. Accessed 8 June 2021
5. Fudina YV (2018) Analytical review of production and consumption of honey in Russia. Sursky Vestnik 4:72–74
6. Holland JM, Sutter L, Matthias A, Philippe J, Pfister SC, Schirmel J, Cresswell JE et al (2020) Moderate pollination limitation in some entomophilous crops of Europe. Agric Ecosyst Environ 302:107002.https://doi.org/10.1016/j.agee.2020.107002
7. Kasiotis KM, Anagnostopoulos C, Anastasiadou P, Machera K (2014) Pesticide residues in honeybees, honey and bee pollen by LC–MS/MS screening: reported death incidents in honeybees. Sci Total Environ 485–486:633–642. https://doi.org/10.1016/j.scitotenv.2014.03.042
8. Lebedev VI, Prokofieva LV (2007) Aspects of the formation of the market of beekeeping products in Russia. Beekeeping 1:3–5
9. Mykhailova LI, Hrytsenko VL (2018) Organizational and economic principles of functioning of the beekeeping market. Econ Agric 8:35–43

10. Ponomarev AS (2007) Bee mortality in North America and Europe. Beekeeping 5:32–33
11. Ponomarev AS (2021) Russian beekeeping in 2020: challenges, results, and trends. Beekeeping 1:8–11
12. Vorobyova VV, Bugai YA (2021) Developing personal subsidiary farms in the food supply system of the Altai Krai. In IOP conference series: earth and environmental science, vol 670, p 012001https://doi.org/10.1088/1755-1315/670/1/012001

Growth Reserves of Agricultural Production in the Ryazan Region

Olga A. Bodryagina⬤, Svetlana G. Vezlomtseva⬤, Olesya A. Zarubina⬤, Evgeny E. Maslennikov⬤, and Galina Z. Cibulskya⬤

Abstract The paper focuses on the problems of perspective development of the agricultural sector of the economy of the Ryazan Region. The authors present the position of the industry in the overall results of the regional economy. Additionally, the authors evaluate the positions of collective agricultural organizations in production for 2015–2019. Finally, the authors substantiate the opportunities of these farms for the next three years in the expansion of sown areas, the growth of production, and maintaining the annual amount of investment in fixed assets received by them for 2019.

Keywords Ryazan Region · Agricultural sector of the economy · Collective agricultural organizations · Agriculturally used areas · Cattle stock · Investments in fixed assets · Optimizing the size of small farms

JEL Codes D13 · Q24 · R11

O. A. Bodryagina (✉) · S. G. Vezlomtseva · O. A. Zarubina · G. Z. Cibulskya
Academy of Law and Management of the Federal Penitentiary Service, Ryazan, Russia
e-mail: olgabodryagina@mail.ru

S. G. Vezlomtseva
e-mail: dsveta@yandex.ru

O. A. Zarubina
e-mail: ole-lisa@yandex.ru

G. Z. Cibulskya
e-mail: 89155933846@mail.ru

E. E. Maslennikov
Ryazan Branch of the Kikot Moscow University of the Ministry of Internal Affairs of Russia, Ryazan, Russia
e-mail: maslennikov78@yandex.ru

1 Introduction

Agriculture is one of the most important industries and activities of the Ryazan Region. Its efficiency depends on the natural and climatic areas, the quality of agricultural land, the availability of the necessary amounts of fixed and working capital, and the position and interest of economic entities of the industry.

2 Materials and Methods

The sources of information are materials from the collection of the Rosstat regional office of Ryazan Region "Agriculture, hunting, and forestry of the Ryazan region" [7] published in 2020, and the All-Russian agricultural census of 2016 [4].

The research methods are the analysis and evaluation of the dynamics of agricultural development in the region as a whole and three categories of economic entities for four years.

3 Results

Assessing the agricultural production in the economy of the Ryazan region, the authors specified that collective agricultural organizations are the main users of land and producers of agricultural products. Further, the authors justified the available reserves for increasing production volumes by using the arable land available to these farms for crops, provided that the volume of investments in fixed production means received in 2019 is maintained over the next three years.

The authors propose to increase the sowing of forage crops, which will increase the number of cattle and, accordingly, milk production necessary to meet the needs of the population in the region fully.

It is also recommended to optimize the size of agricultural organizations with small areas under crops.

4 Discussion

During the research, the authors analyzed the results of four categories of economic entities in the dynamics over several years to find ways of improving the efficiency of agricultural production in the Ryazan Region. According to the authors, one of the options for solving the issue is to optimize the size of agricultural organizations and farms.

Agriculture is part of the most important branches and activities of social production in Russia. Agriculture produces raw materials for food, employs many farmers, and forms the conditions for the successful functioning of several other industries and activities. On the other hand, agriculture depends on the natural and climatic conditions, manufacturers of the necessary machinery, equipment, mineral fertilizers, chemicals, and other fixed and working capital. Agricultural land is the most important and practically irreplaceable production means for agricultural production [2].

The Ryazan Region is located in the forest-steppe zone of the Central Federal District of Russia. According to the Rosstat regional office of the Ryazan Region, the total area of agricultural land of the region amounted to 2504.6 thousand hectares in 2019 [7]. By the end of the year, the area of lands at the disposal of agricultural producers was 2492.7 thousand ha, including 2328.5 thousand ha of arable lands, of which 1470.2 thousand ha (63.1%) of agricultural fields, 813.8 thousand ha (34.9%) of forage lands (hayfields and pastures), and the remaining 2% are perennial plantations [7]. Of the total arable area, 31.0% are chernozems, 28.8% are sodpodzolic and light gray, 24.5% are dark gray and podzolic chernozems, 13.4% are sod-podzolic, and the remaining 2.3% are soils of other types [6]. According to the average long-term agroclimatic data in the Ryazan Region, the duration of frost-free days is 132 days, with 157 days with an average daily temperature of +10 °C (i.e., the growing season). The amount of precipitation for the year is 582 mm [6]. These data indicate favorable conditions for agricultural production in the Ryazan Region.

Table 1 shows the data that reflect the dynamics of the share of agriculture in the economy of the Ryazan Region. Despite a declining trend, the share of the industry in gross value added (GVA) is quite significant. If we consider the results of only commercial sectors (industry, construction, and electricity and water supply) and agriculture itself, its share is quite significant [1]. Simultaneously, the share of agriculture in the value of fixed assets, including agricultural land, is lower than its share in GVA for the entire economy. This means that agriculture has a higher rate of

Table 1 The position of agriculture in the economy of the Ryazan Region in 2015–2019 (in % of the results for the Region)

No	Indicators	2015	2017	2018	2019
1	Rural population at the beginning of the year	28.6	28.3	28.1	27.9
2	Specific weight in gross value added (GVA):				
	– from the results of the entire economy	9.6	7.5	7.7	–
	– from the results of the commodity-producing industries	20.0	15.7	16.8	–
3	From the value of fixed assets:				
	– of all industries and activities	4.8	5.6	5.7	–
	– of productive industries	14.8	17.4	17.6	–
4	Investments from the amount in fixed assets	11.3	11.4	17.8	18.5

Source Compiled by the authors

Table 2 Dynamics of the share of collective agricultural organizations in the production of agricultural products in the Ryazan Region in 2015–2019

Types of products	2015	2017	2018	2019
Natural volumes				
Grain	85.7	83.5	84.3	84.7
Sugar beet	87.6	84.5	88.4	89.6
Potato	31.2	35.5	40.9	37.5
Vegetables	14.3	19.5	17.7	16.9
Cattle and poultry for slaughter	75.0	81.2	83.4	86.6
Milk	85.0	86.3	87.3	89.0
Eggs	91.5	91.9	93.3	94.2
Cost in actual prices				
Gross output, including:	62.3	65.2	66.4	71.4
crop production	56.9	58.3	59.3	66.0
livestock farming	70.0	74.0	75.8	80.0

Source Compiled by the authors

return on assets—respectively, two times in 2015 (9.8:4.8), 34% in 2017 (7.5:5.6), and 35% in 2018 (7.7:5.7). The importance of the AIC for the region is also indicated by a gradual increase in investment in fixed capital, despite the reduction in the share of the rural population in the total population of the region.

Like almost all regions of the Central Federal District of Russia, the primary producers of agricultural products in the Ryazan Region are collective agricultural organizations and private households of farmers and rural population.

Table 2 shows the share of collective farms in the natural volume of production of the main types of agricultural products in the Ryazan Region.

These data show that the share of collective agricultural organizations practically increases every year in physical volumes and actual current prices of products. The share of agricultural organizations is significantly lower only in the production of potatoes and vegetables; moreover, there are downward trends. The share of the production of grain, sugar beet, and all types of livestock products is many times higher than that of households and farmers, and there is a gradual increase [5]. In value terms, the results of agricultural organizations are growing continuously in crop and livestock production.

However, it should be noted that a significant increase in the share of agricultural organizations in the total sectoral output results, to a certain extent, from a sharp deterioration in the results of personal subsidiary plots of the rural population and the low volume of indicators of farms.

First, it should be noted that significant areas of agricultural land in the region still remain unused. Thus, for 2017–2019, the total arable area of the three categories of economic entities in the region decreased insignificantly, from 1475 thousand hectares in 2017 to 1470 ha in 2019. At the same time, the areas of crops and clean fallows were 1064 thousand hectares and 1147 thousand hectares, respectively. For this reason, 411 thousand hectares remained unused in 2017 and 323 thousand hectares in 2019. Households reduced their sowing areas only by 2 thousand hectares. Sowing areas on farms increased by 8 thousand hectares. Agricultural organizations increased their sowing areas by 52 thousand hectares [3]. Thus, only agricultural organizations had the opportunity to increase their cultivated areas to a greater extent, bearing in mind that over the years, above all, they have significantly increased investment. Over three years, the renewal rate of tractors in these farms ranged from 3.5 to 5.4%, with the elimination rate 3.4–4.4%. The renewal rate of grain harvesters ranged from 6.8 to 9.1%, with the elimination rate 2.6–4.4%.

It is evident that to increase the sown areas, agricultural organizations need additional equipment—tractors, seeders, combine harvesters, trucks, etc. From the authors' point of view, they have the financial capacity to acquire them. Table 1 presents data on the amount of investment in agriculture. For 2019, they accounted for 18.5% (8.5 billion rubles) of total investment in the region's economy. If the same amount of investment will be kept for the next three years annually (i.e., in 2020–2022), the farms, having at their disposal 27.0 billion rubles, will be able to buy necessary additional technical means to expand cultivated areas by 323 thousand ha.

In the authors' opinion, it is advisable to use these additional cultivated areas as follows:

- 120 thousand hectares for grain crops; the gross yield will increase by 366 thousand tons (at an average annual yield over five years (2015–2019) equal to 3.05 tons per ha);
- 5 thousand hectares for potato crops; the gross yield will increase by 75 thousand tons (at a yield of 15 tons per ha);
- 10 thousand hectares for sugar beef; the gross yield will increase by 450 thousand tons (at a yield of 45 tons per ha);
- 30 thousand hectares for sunflower; the gross yield will increase by 48 thousand tons (at a yield of 1.6 tons per ha);
- 128 thousand hectares for forage crops; the gross yield will increase by 320 thousand tons (at a yield of 2.5 tons per ha);
- 30 thousand hectares for pure fallow.

In 2015–2019, sown areas of fodder crops (corn for green mass, annual and perennial grasses, etc.) decreased from 167 to 158 thousand hectares. In 2019, their share of the total sown area of all three categories of economic entities was only 16%.

It is known that roughage and succulent fodder are mainly intended for cattle, sheep, and goats. The decrease in these crops was accompanied by a decrease in the number of these animals. In five years, the number of cattle has decreased by almost 10 thousand (5.9%), sheep and goats by 11 thousand (18.6%). The increase in cow productivity from 5522 kg in 2015 to 7072 kg in 2019 allowed to increase gross milk production from 375,000 tons in 2015 to 459,000 tons in 2019 (22.4%). Thus, the region produced 413 kg of milk per capita in 2019. However, the actual consumption was 230 kg since the region had contractual obligations to supply dairy products to other Russian regions. While milk production increased by 31% per capita in 2010–2019, its consumption decreased by 11%. These circumstances indicate the need to increase milk production by increasing the productivity of cows and their herd. The reduction of forage crops and, consequently, the volume of roughage and succulent forage production in the future will not allow increasing the number of cows. The share of concentrated feed in milk production has already reached 42% of total consumption. Concentrated fodder (mainly purchased) is more expensive than roughage and succulents of own production. For this reason, despite the significant increase in cow productivity for 2015–2019, the cost of milk increased from 17,140 to 20,450 rubles per ton (19.3%), and the growth of cattle from 137,530 to 170,450 rubles per ton (23.9%). This process reduces the economic efficiency of product manufacturers. In this regard, the authors propose to significantly increase the sown area of forage crops by using all the arable land of the Ryazan Region. The yield of pastures (about 480 thousand hectares), which are used less effectively due to reducing the number of cattle, sheep, and goats, will also increase. The proposed variant of expansion of sowing areas will allow increasing the number of cattle of agricultural organizations for the next three years by at least 70 thousand animals, including 30 thousand cows. Accordingly, gross milk production will also increase.

One of the reserves for increasing the efficiency of agricultural production in the region is the optimization of the size of agricultural organizations and farms. According to the agricultural census of the Russian Federation for 2016, in the Ryazan Region, 96.5% of agricultural organizations have land (an average of 3150 hectares per farm), including:

- 7.8% of farms had up to 100 ha (an average of 39 ha);
- 4.1% of farms had 100.1–200 ha (an average of 157 ha);
- 14.4% of farms had 200.1–500 ha (an average of 327 ha) [4].

Of all farms, 86.7% had sown areas (an average of 2687 ha per farm), including:

- 6% of farms had up to 100 ha (an average of 56 ha);
- 4.5% of farms had 100.1–200 ha (an average of 157 ha);
- 14.2% of farms had 200.1–500 ha (an average of 341 ha) [4].

Only 43% of farms had cattle (an average of 1077 animals per farm), of which 14.3% had up to 100 animals (an average of 39 animals, including 16 cows) [4].

According to the research results, the authors found that the share of management personnel reaches up to 23% in agricultural organizations with the area of crops up to 500 ha due to the need to maintain certain groups of managers, specialists, and employees. On farms with up to 100 heads of cattle, there are only 8–10 animals per farm worker. On farms with 500 or more animals, there are 15–21 animals per farm worker.

5   Conclusion

Thus, small farms are marked with lower labor productivity and less effective implementation of technical means. This problem can largely be solved through developing production cooperation between agricultural organizations and agricultural organizations and farms of the administrative district.

References

1. Arkhipova LS, Gorokhova IV (2021) Regional features of agricultural development in Russia. In Bogoviz AV (ed) The challenge of sustainability in agricultural systems.Springer, Cham, Switzerland, pp 55–63. https://doi.org/10.1007/978-3-030-73097-0_8
2. Bodryagina OA (2015) Rational land use as a condition for the growth of agricultural production in the region (Dissertation of Candidate of Economic Science). Federal State Budgetary Scientific Institution VNIOPTUSKh, Moscow, Russia
3. Bodryagina OA (2017) Analysis of production status of economic entities in agriculture of the Ryazan Region. J Econ Entrep 2–2(79):325–330
4. Federal State Statistics Service of the Russian Federation (Rosstat) (2016) All-Russian agricultural census 2016. Preliminary results: statistical bulletin, vol 2. Information & Publishing Centre "Statistics of Russia", Moscow, Russia
5. Pechenevsky VF, Zakshevsky GV (2019) Prospects for the spatial development of agricultural production in the region. In IOP conference series: earth and environmental science, vol 274, p 012008. https://doi.org/10.1088/1755-1315/274/1/012008
6. Romanenko GA, Tyutyunnikov AI, Shutkov AA, Makarov IP (1995) Agro-industrial complex of Russia: resources, products, and economy, vol 1. Russian Academy of Agricultural Sciences, Novosibirsk, Russia
7. Rosstat Regional Office of Ryazan Region (Ryazanstat) (2020) Agriculture, hunting, and forestry in the Ryazan region. Ryazanstat, Ryazan, Russia

Assessment of Export Prospects of Russian Agricultural Producers

Alexander A. Dubovitski⊙**, Elena A. Yakovleva**⊙**, Olga Yu. Smyslova**⊙**,**
Gayane A. Kochyan⊙**, and Elena V. Zelenkina**⊙

Abstract The purpose of this study is to substantiate the prospects for the export of Russian agricultural producers in the context of the globalization of markets. The study is conducted according to the data of the Russian Federation for the period 2011–2020. The study carries out a quantitative assessment of the parameters that determine the possibilities of expanding exports, based on statistical indicators by analyzing the dynamics of production and the elasticity of exports. The authors substantiate the expediency of expanding the export of agricultural products by Russian regional producers; localize the list of commodity groups that are most relevant in terms of the formation of the commodity structure. The calculations made allow estimating the unused export potential at the level of 4,011.1 million USD. The largest potential for export from Russia is Animal or vegetable fats and oils in the amount of 1,254 million USD and Cereals—1,077.6 million USD. Oilseeds and Meat and edible meat offal are also promising with export growth potential—786.6 and 415.0 million USD, respectively. The share of the Russian Federation in world exports is likely to increase from 1.76 to 2.01% if it is implemented. It is proved

A. A. Dubovitski (✉)
Michurinsk State Agrarian University, Michurinsk, Russia
e-mail: economresearch@mail.ru; Daa1-408@yandex.ru

E. A. Yakovleva
Voronezh State Forestry Engineering University Named After G.F. Morozov, Voronezh, Russia
e-mail: elena-12-27@mail.ru

O. Yu. Smyslova
Lipetsk branch of the Financial University Under the Government of the Russian Federation, Lipetsk, Russia
e-mail: savenkova-olga@mail.ru

G. A. Kochyan
Kuban State Technological University, Krasnodar, Russia
e-mail: gayanek@mail.ru

E. V. Zelenkina
Moscow University for Industry and Finance "Synergy", Moscow, Russia
e-mail: zelenkina_e@mail.ru

© The Author(s), under exclusive license to Springer Nature Singapore Pte Ltd. 2022   167
E. G. Popkova and B. S. Sergi (eds.), *Sustainable Agriculture*,
Environmental Footprints and Eco-design of Products and Processes,
https://doi.org/10.1007/978-981-16-8731-0_17

that the expansion of exports will have a positive impact on the economic results of regional agricultural producers and the country's trade balance.

Keywords Agriculture · Efficiency · Production placement · Agri-food market · Development prospects · Export · Diversification · Concentration

JEL Codes Q13 · Q17

1 Introduction

Modern processes of global economic integration against the background of active development of transport infrastructure provide favorable conditions for the expansion of world trade and create new opportunities for regional producers, including agricultural products.

This statement is consistent with the existing research results. The higher the degree of integration between countries and the lower the trade barriers, the higher the market volume and the level of trade flows [1, 2]. The researchers note that trade liberalization is a key factor in promoting agricultural trade between countries participating in regional trade agreements [3–5]. The world market becomes a catalyst for the economic growth of regional agricultural enterprises [6, 7], this, in turn, creates prerequisites for increasing the volume of production and export of agricultural products by various countries shortly [8, 9]. The availability of the agri-food market at the national and regional levels determines the development of export-oriented industries [10], although it is accompanied by certain risks in the conditions of instability and negative dynamics of energy prices [11], and also due to some natural features of agricultural production [12].

Measures to support the promotion of agricultural products to foreign markets and ensure their compliance with international requirements are designed to minimize risks [13, 14]. At the same time, many authors emphasize the importance of the structural transformation of the agricultural sector [3] based on strategic options for entrepreneurial activity in agriculture [15].

2 The Theoretical Basis of the Study

The possibility of reducing risks and ensuring the growth of production and export of agricultural products is solved by researchers in different ways. One author focuses on diversification [16], believing that more specialized farms are more at risk of yield and income [17]. In this case, the priority of the diversification strategy in the development of agriculture in conditions of instability is explained by the innovative nature of diversification [18]. Other researchers associate wide access to markets with the development of specialization, the ability for producers to distribute their land for more valuable crops [3]. As Jongwanich [19] notes, specialization can contribute to an increase in intensive margins, stimulating economic growth, especially in the processing industries.

One of the directions of deepening the specialization of agricultural enterprises to develop exports is the formation of regional export-oriented production clusters [20, 21]. An attempt to substantiate its parameters was made by the author's team of Lukyanova et al. [6] the researchers developed extensive, intensive, and extensive-intensive scenarios for the development of agribusiness in the Republic of Bashkortostan. This approach can contribute to the realization of the export potential of the regional agricultural economy. However, it should be noted that agriculture is characterized by a significant heterogeneity of territories, soil and climatic conditions, and institutional, technological, social, and environmental factors of production, which makes it necessary to consider development opportunities in the context of the regional aspect [22–26]. Thus, when managing the development of regional agricultural farms, it is necessary to pay special attention to the formation of the production structure following the market potential.

3 Methodology

The scientific hypothesis of the study is based on the assumption that in the conditions of the globalization of the agricultural products market, the export prospects of Russian agricultural producers are determined by the potential capacity of the world market and the ability of regional producers to meet these needs.

The purpose of this study is to substantiate the export prospects of Russian agricultural producers in the context of market globalization.

Research objectives: (1) economic assessment of the regional development of agricultural sectors; (2) analysis of the prospects for expanding the production and export of the main types of agricultural products in the region; (3) substantiation of the directions of realization of potential opportunities for the development of agricultural enterprises in the conditions of globalization of industry markets.

When conducting the research, general scientific methods were used (logical and comparative analysis using a review of information and statistical data). The authors used some absolute and structural indicators with the calculation of export diversification and concentration indices. The export diversification index was determined by the authors based on the absolute deviation of the share of individual food products in the country's exports from their share in world exports:

$$S_j = \frac{\sum_{i=1}^{n} |h_{ij} - h_i|}{2} \tag{1}$$

where: S_j–export diversification index of j–country;

h_{ij}–the share of product i in the total exports of j–country;

h_i–the share of product i in total world exports.

The export concentration index was determined according to the Hirschman index:

$$H_j = \sqrt{\left[\sum_{i=1}^{I}\left(\frac{e_i}{x}\right)^2 - \sqrt{1/I}\right]/1 - \sqrt{1/I}} \tag{2}$$

where: H_j–export concentration index of j–country;

I–number of types of products by classification;

i–product index (from 1 to I);

e_i–the cost of exporting i goods by j–country;

e–the total export value of j–country:

$$e = \sum_{i=1}^{I} e_i \tag{3}$$

The quantitative assessment of the parameters determining the possibilities of expanding food exports was carried out based on statistical indicators by analyzing the dynamics and structure of agricultural production.

The basis for determining promising export growth indicators was the potential estimates of the capacity and availability of world food markets. The authors determined the export growth potential of individual goods taking into account the export and production elasticity indices. The coefficient of elasticity of exports by production (R) is calculated by the percentage ratio of the change in the quantity of exported products (ΔE) to the percentage change in the production of products within the country (ΔP):

$$R = \Delta E/\Delta P \tag{4}$$

where:

$$\Delta E = (E^2 - E^1)/E^1 \tag{5}$$

$$\Delta P = (P^2 - P^1)/P^1 \tag{6}$$

E^1; E^2–the number of exported products in the 1st and 2nd year;
P^1; P^2 the volume of production in the 1st and 2nd year.

$$R = \left[(E^2 - E^1)/E^1\right] - \left[(P^2 - P^1)/P^1\right] \tag{7}$$

The elasticity indices were determined as an average value for the analyzed number of years for the relevant types of product groups.

The export growth potential was estimated taking into account the elasticity of exports and the forecast of growth in the production of main types of agricultural products:

$$E^{i+1} = R * P^{i+1} \tag{8}$$

The study was conducted according to the data of the Russian Federation for the period 2010–2020, export indicators were analyzed using world trade statistics for 2016–2020.

4 Results

The global market of agricultural and processed food products has been showing constant growth in recent years. According to the International Trade Center, since 2016, the volume of world trade in this sector has grown by an average of 3.6% per year and in 2020 amounted to 1.6 trillion USD [27]. Russia is not among the leaders in the world in terms of total exports of agricultural products and food. However, in recent years, the country has demonstrated a steady increase in trade indicators in this area (Table 1).

In 2016–2020, the value of exports of agricultural products and products of its processing increased by 1.6 times, by 11.2 billion USD. The average annual growth rate was a significant 13.2%. At the same time, the share of the Russian Federation in world exports increased by 0.5 percentage points and amounted to 1.7602% in 2020. The structural deviation of Russian exports in this sector from the global average values grew during 2016–2018. During this time, the diversification index increased by 0.0724 mainly due to the growth of exports for three main groups

Table 1 Absolute and structural indicators of exports of agricultural and processed food products from the Russian Federation, 2016–2020

Indicator	2016	2017	2018	2019	2020
Export value (thousand USD)	17,044,501	20,705,591	24,884,904	24,753,303	28,287,096
The share of the Russian Federation in world exports (%)	1.2126	1.3611	1.5679	1.5658	1.7602
Agricultural export Diversification index (S)	0.1575	0.1814	0.2299	0.1573	0.1565
Agricultural export Concentration index (H)	0.4477	0.5225	0.4913	0.4465	0.4261

Source Compiled by the authors based on [27]

of goods: cereals, fish, and oils, which together accounted for almost 70% of the export structure in 2018. In the future, the growth of many other positions, primarily meat, sugar, and oilseeds, allowed the structure to be somewhat smoothed, and the diversification index by 2020 decreased to 0.1565. These conclusions are confirmed by the dynamics of the export concentration index, which in 2017 and 2018 reached the maximum values, and in subsequent years decreased to a minimum. All this indicates a serious structural transformation of the country's agricultural economy, accompanied by significant changes in the composition and structure of agricultural exports (Table 2).

The best dynamics indicators in 2016–2020 were demonstrated by several product groups. In absolute terms, these are cereals, fish, and seafood, animal or vegetable fats, and oils, as well as oilseeds. In relative terms, the largest increase was shown by exports of meat (almost three times), oilseeds, and sugar (about two times).

The group of cereals products with a value of more than 9 billion USD remains the main one in agricultural exports from the Russian Federation. According to this indicator, the country ranks third in the world with a specific weight of 7.85%. The main markets are Turkey and Egypt, which account for more than a third of all exports. Turkey and Egypt mainly supply wheat of 1.7 and 1.8 billion USD, respectively. In each of these markets, Russia occupies more than 20%.

The second most important type of products exported from the Russian Federation is fish and seafood, their export volume in 2016–2020 increased by one and a half times (on average 10.8% per year) and amounted to 4.6 billion USD. According to this indicator, the country ranks sixth in the world with a specific weight of 4.16%. The main markets are the Republic of Korea, China, and the Netherlands. They collectively account for 85% of exports. Mainly crustaceans (776 million USD in 2020) and frozen fish (666 million USD) are supplied to Korea. Russia closes about 60% of world exports to this country. Frozen fish (1,158 million USD), crustaceans (339 million USD), and shellfish (43 million USD) are supplied to China. Crustaceans (446 million USD), frozen fish (200 million USD), and fish fillets (159 million USD) are supplied to the Netherlands.

An important place in the export structure is occupied by the group of goods animal or vegetable fats and oils worth 3.9 billion USD in 2020. According to this position, the Russian Federation ranks seventh in the world with a specific weight of 3.84. For 2016–2020, the average annual increase in this group of goods was 15% for a total amount of 1.7 billion USD. The main markets are China, Turkey, and India. Russia occupies a significant share of world exports to China for sunflower oil—22% (564 million USD), rapeseed—43% (251 million USD), soy—55% (242 million USD). Mainly sunflower oil is supplied to Turkey of 456 million USD and a share of 18.5%.

The oilseeds Group with a value of 1.6 billion USD ranks 11th in the world with a market share of 1.49%. The cumulative increase for this group in 2016–2020 was 214.43%. The main markets are China and Turkey. The main types of exports of goods of this group to China are soybeans (265.9 million USD in 2020), rapeseed (169.1 million USD), and flax seeds (106.7 million USD). Moreover, Russia's share in China's total exports of these goods is very high—64.0, 57.8, and 46.3%, respectively.

Table 2 Indicators of the dynamics of exports of agricultural and processed food products from the Russian Federation, 2016–2020 (top10)

Product code	Name of product groups	The cost of exports from Russia in 2020 (million USD)	Rating of the Russian Federation in world exports in 2020	Export value growth, 2016–2020		The share of the Russian Federation in world exports in 2020 (%)	Growth of the share of the Russian Federation in world exports from 2016 to 2020 (%)
				absolute (million USD)	relative (%)		
10	Cereals	9,340.5	3	3,734.4	66.61	7.85	2.03
3	Fish and crustaceans, molluscs, and other aquatic invertebrates	4,639.6	6	1,624.3	53.87	4.16	1.40
15	Animal or vegetable fats and oils and their cleavage products; prepared edible fats; animal.	3,890.5	7	1,681.9	76.15	3.84	1.32
12	Oilseeds and oleaginous fruits; miscellaneous grains, seeds, and fruit; industrial or medicinal.	1,634.1	11	1,114.4	214.43	1.49	0.90
23	Residues and waste from the food industries; prepared animal fodder	1,430.6	5	485.4	51.35	1.79	0.39
2	Meat and edible meat offal	862.7	25	644.1	294.65	0.64	0.45
21	Miscellaneous edible preparations	819.9	31	314.3	62.16	1.00	0.22

(continued)

Table 2 (continued)

Product code	Name of product groups	The cost of exports from Russia in 2020 (million USD)	Rating of the Russian Federation in world exports in 2020	Export value growth, 2016–2020		The share of the Russian Federation in world exports in 2020 (%)	Growth of the share of the Russian Federation in world exports from 2016 to 2020 (%)
				absolute (million USD)	relative (%)		
19	Preparations of cereals, flour, starch or milk; pastrycooks' products	754.9	29	236.3	45.56	0.95	0.16
18	Cocoa and cocoa preparations	740.9	18	257.1	53.14	1.50	0.44
17	Sugars and sugar confectionery	731.6	13	486.7	198.73	1.73	1.19

Source Compiled by the authors based on [27]

The main export to Turkey is sunflower seeds. In 2020, they were sold in the amount of 234.6 million USD, and the share of Russia at the same time was 41.7%. There is an intensive growth in exports of meat and edible meat offal and sugars and sugar confectionery. In terms of meat exports, the Russian Federation ranks only 25th in the world and has a share of 0.64%. However, in 2016–2020, the increase in the value of exports amounted to 294.65%, which indicates a significant potential for this group of goods in the future. The main meat supplies are carried out in China and Vietnam. The main types of exports of goods of this group to China are meat and edible offal of poultry (262.9 million USD in 2020) and cattle meat (47.6 million USD). Russia's share in Chinese imports is also quite large—61.5 and 64.0%, respectively. Pork is the main commodity export product to Vietnam. The volume of its deliveries in 2020 amounted to 119.4 million USD, and the share in total exports to this country is 45.1%. The increase in the export value of Sugars and sugar confectionery for 2016–2020 amounted to 198.73%. At the same time, the share in world exports has grown quite strongly—from 0.54 to 1.73%. The main sugar markets are Kazakhstan (128.8 million USD in 2020), Uzbekistan (129.9 million USD), and Azerbaijan (44.5 million USD). In the markets of these countries, Russia occupies 27.8, 28, and 10%, respectively.

The considered trends of export changes from the Russian Federation form its modern commodity structure (Fig. 1).

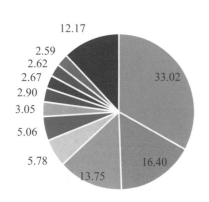

Cereals

Fish and crustaceans, molluscs…

Animal or vegetable fats and oils. . .

Oil seeds and oleaginous fruits. . .

Residues and waste from the food industries…

Meat and edible meat offal

Miscellaneous edible preparations

Preparations of cereals, flour, starch or milk…

Cocoa and cocoa preparations

Sugars and sugar confectionery

Other product groups

Fig. 1 Value structure of exports of agricultural and processed food products from the Russian Federation, %, 2020. *Source* Compiled by the authors based on [27]

The Russian Federation is characterized by a rather narrow specialization in the world market. Despite the wide range of goods supplied to the world market of goods, two main commodity groups—Cereals and Fish and crustaceans, molluscs, and other aquatic invertebrates-account for half of the value of all agricultural exports from the country. And together with Animal or vegetable fats, these three groups occupy 63.21%. This confirms the conclusions that Russia's exports deviate from the structure of world exports, that is, the country exports a limited group of goods. In this respect, the commodity structure of Russian exports is identical to the corresponding structure of a "classic" developing country.

When assessing the potential for export expansion, an important conclusion made by Hidalgo et al. [28] should be taken into account. The researcher points to the principle of forming the commodity structure of different countries, according to which the range of manufactured and exported products cannot change dramatically. It was proved that the commodity composition of countries changes gradually and in a certain "product space". Under this principle, the "traditional" groups of goods that demonstrate positive dynamics are promising for expanding exports from the Russian Federation. We are talking about expanding the export of cereals, fish and seafood, meat, sunflower seeds and vegetable oil, sugar.

The export potential of the country is laid by the development of the production of relevant types of products within the country. Without increasing production volumes, it is impossible to increase exports. In the Russian Federation, the agricultural sector plays one of the main roles in the economy. The development of agriculture in Russia in recent decades has been accompanied by an increase in labor productivity and structural optimization of agribusiness [16, 29]. In 2011–2020, the country achieved a significant increase in agricultural production (Fig. 2, Table 3).

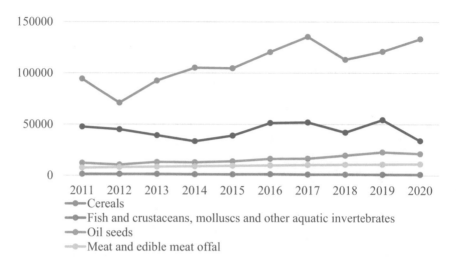

Fig. 2 Dynamics of agricultural production in the Russian Federation, thousand tons. *Source* Compiled by the authors based on [6]

Table 3 Indicators of the dynamics of production and export of agricultural and processed food products from the Russian Federation, 2011–2020

Name of product groups	Equation of the trend line of production volume in the Russian Federation	The equation of the trend line of the size of exports from the Russian Federation	Export and production elasticity coefficient
Cereals	$y = -299.23x^2 + 8{,}842.4x + 72{,}056$ $R^2 = 0.7233$	$y = 28{,}194x^2 + 20{,}9466x + 5E + 06$ $R^2 = 0.6533$	1.1183
Fish and crustaceans…	$y = 0.433x^2 - 55.878x + 1{,}495.1$ $R^2 = 0.7451$	$y = 27{,}437x^2 - 29{,}925x + 2E + 06$ $R^2 = 0.9395$	1.1279
Oilseeds	$y = 99.544x^2 + 187.98x + 11{,}034$ $R^2 = 0.9266$	$y = 23{,}073x^2 - 131{,}623x + 46{,}3347$ $R^2 = 0.9317$	1.2516
Meat and edible meat offal	$y = -16.526x^2 + 591.56x + 6{,}938$ $R^2 = 0.9992$	$y = 13{,}322x^2 - 63{,}992x + 118{,}555$ $R^2 = 0.9833$	1.4521
Sugar beet / sugar	$y = -41.193x^2 + 595.57x + 42{,}107$ $R^2 = 0.0051$	$y = 9{,}581.2x^2 - 57{,}321x + 303{,}684$ $R^2 = 0.8513$	1.2972

Source Compiled by the authors based on [10, 27]

The most stable growth trend has been formed in the production of grain crops and sunflower products. However, in some years there are serious deviations in productivity associated with the variability of weather conditions. The constructed equations are characterized by sufficient approximation reliability (0.7233 and 0.9266), which allows predicting further growth of these indicators with a high degree of probability. Sugar beet production is characterized by maximum variability, which is explained by the strong dependence of the industry on external factors, primarily weather. The constructed trend line is characterized by a small increase. Despite the positive dynamics, the reliability of the trend is very low ($R^2 = 0.0051$). The most stable growth in volumes is demonstrated by meat production. Moreover, the reliability of the trend is confirmed at the level of $R^2 = 0.9992$, which is explained by the absence of a significant impact of weather conditions on the livestock industries of intensive, industrial enterprises. The volume of fish and seafood production is characterized by fluctuation. Stable trends in the growth of world prices for these goods contribute to ensuring a constant increase in the value of exports of this commodity group. In general, for all calculated items of goods, the coefficient of elasticity of export and production is greater than one. This allows us to predict the outstripping growth rates of the export value over the growth of production volumes by the corresponding amount of elasticity.

In terms of the production of some goods, the Russian Federation occupies higher places in the world ranking than in terms of the value of exports. For example, the country ranks 1st in the production of sugar beet, and only 13th in the export of sugar. Russia is in 4th place in meat production and only 25th place in export. The additional potential is laid at the stage of optimizing the movement of goods due to the logistics component and improving the safety of products. The prospective parameters for expanding the export of agricultural and processed food products from the Russian Federation are presented in Table 4.

The total unused export potential of the exporter of the Russian Federation is 4,011.1 million USD. The largest value increase can be achieved by realizing the export potential of Animal or vegetable fats and oils in the amount of 1,254 million USD and Cereals in the amount of 1,077.6 million USD. Lower export growth rates, but better in relative terms, can be provided for the products of Oilseeds and Meat and edible meat offal. According to these groups, the export growth potential is 786.6 and 415.0 million USD, respectively. The share of these groups of goods in the export structure will increase. As a result, the concentration index of agricultural exports from the Russian Federation will increase by 0.551 relative to the level of 2020 and will amount to 0.4812. The share in world exports is likely to increase from 1.76 to 2.01%.

Table 4 Prospective parameters of export of agricultural and processed food products from the Russian Federation

CBI sector	Product description	Export in USD million (2020)	Product's export potential value			Prospective export level in USD million
			absolute value in USD million	relative value (%)	Top 3 markets	
10	Cereals	9,340.5	1,077.6	11.54	Bangladesh, Indonesia, Algeria	10,418.1
3	Fish and crustaceans, molluscs, and other aquatic invertebrates	4,639.6	182.1	3.92	China, Korea Republic and Japan	4,821.7
15	Animal or vegetable fats and oils and their cleavage products; prepared edible fats; animal.	3,890.5	1,254.0	32.23	China, India, and the Islamic Republic of Iran	5,144.5
12	Oilseeds and oleaginous fruits; miscellaneous grains, seeds, and fruit; industrial or medicinal.	1,634.1	786.6	48.14	China, Belgium, and Germany	2,420.7
2	Meat and edible meat offal	862.7	415.0	48.11	Kazakhstan, Kyrgyzstan, and China	1,277.7
17	Sugars and sugar confectionery	731.6	295.7	40.42	Kazakhstan, Uzbekistan, and Turkey	1,027.3

Source Authors calculations

5 Conclusion

The global market of agricultural and processed food products has shown constant growth in recent years, which will continue in the foreseeable future. The economic state of the agriculture and processing industries in the Russian Federation is characterized by the dynamics of increasing the efficiency of the industry and expanding exports. Several groups of goods demonstrated consistently high indicators: in absolute terms—cereals, fish, and seafood, animal or vegetable fats and oils, oilseeds. In relative terms, the increase is in the export of meat, oilseeds, and sugar. Russia ranks higher in the world production volume than in the export volume. In the current conditions, the expediency of expanding the export of manufactured products is obvious to Russian regional producers. The realization of the underutilized export potential is possible due to the preservation of trends in the growth of production volumes, optimization of a commodity movement, due to logistics components, and increasing the safety of products. The total unused export potential of the exporter of the Russian Federation can be estimated at 4,011.1 million USD. If it is implemented, the concentration index of agricultural exports from the Russian Federation will increase by 0.551 relative to the level of 2020 and will amount to 0.4812. The share in world exports is likely to increase from 1.76 to 2.01%. As a result, the expansion of exports will have a positive impact on the economic results of regional producers of agricultural products and the country's trade balance.

References

1. Agirbov Yu, Mukhametzyanov R, Britik E (2020) Russia in the world production and market of potatoes and fruit and vegetable products. Econ Agric Process Enterprises 9:74–83
2. Ishchukova N, Smutka L (2014) The formation of Russian Agrarian trade structure: inter-industry vs. intra-industry trade activities. Acta Univ Agric Silvicult Mendelianae Brunensis 62:1293–1299
3. Divanbeigi R, Paustian N, Loayza N (2016) Structural transformation of the agricultural sector: a primer. World Bank Research and Policy Briefs. http://documents.worldbank.org/cur ated/en/561951467993197265/pdf/Structural-transformation-of-the-agricultural-sector-a-pri mer.pdf. Accessed 20 July 2021
4. Serrano R, Pinilla V (2010) Causes of world trade growth in agricultural and food products, 1951–2000: a demand function approach. Appl Econ 42(27):3503–3518
5. Sugiharti L, Purwono R, Esquivias M (2020) Analysis of determinants of Indonesian agricultural exports. Entrepren Sustain Issues 7(4):2676–2695
6. Lukyanova M, Zalilova Z, Kovshov V, Farrakhova F (2021) Export potential in rural areas. Manag Issues. https://doi.org/10.1007/978-3-030-73097-0_94
7. Tilov A, Poliakova V, Sumbatyan S (2021) The agricultural market of Russia: trends and development priorities. In: Bogoviz A (ed) The challenge of sustainability in agricultural systems. Springer, Cham, pp 3–9
8. Esteban Á (2021) The determinants of world wheat trade, 1963–2010: a gravity equation approach. Hist Agraria 83:165–190
9. Sergi B, Popkova E, Bogoviz A, Ragulina Y (2019) The agro-industrial complex: tendencies, scenarios, and regulation. In: Sergi, B (ed) Emerald Group Publishing Limited
10. Otinova M, Salnikova E, Tyutyunikov A (2021) Factors for the development of an export-oriented agri-food market. In: Bogoviz, A (ed) The challenge of sustainability in agricultural systems. pp 505–513
11. Butakova M, Borisova O, Goryaninskaya O (2021) Exports of vegetable oils to Asian markets: Opportunities, risks, and prospects. In: Proceedings of the IOP conference series: earth and environmental science vol. 670(1). https://doi.org/10.1088/1755-1315/670/1/012045
12. Altukhov A (2018) Prevention of risks and threats to food security is a necessary condition for the spatial development of agriculture. Bull Kursk State Agric Acad 7:150–158
13. Bugai Y, Minenko A, Khorunzhin M (2019). State and problems of exporting the products of the agro-industrial complex in the Altai region. In: Proceedings of the IOP conference series: earth and environmental science vol. 395(1). https://doi.org/10.1088/1755-1315/395/1/012105

14. Graskemper V, Yu X, Feil J (2021) Analyzing strategic entrepreneurial choices in agriculture–empirical evidence from Germany. Agribusiness. https://doi.org/10.1002/agr.21691
15. Polyakov D (2013) The modern development of the world market of agricultural products. Innov Invest 7:157–161
16. Babushkin V, Dubovitski A, Klimentova E, Bazarova T, Melekhova N (2021) Rural unemployment in Russia: reasons and regulation mechanism. Turismo Estudos Práticas (UERN) 1:1–10
17. Bozzola M, Smale M (2020) The welfare effects of crop biodiversity as an adaptation to climate shocks in Kenya. World Dev 135 https://doi.org/10.1016/j.worlddev.2020.105065
18. Chemirbayeva M, Malgarayeva Z, Azamatova A (2020) The economic strategy of diversification of enterprise activities under conditions of globalization. Entrepren Sustain Issues 8(2):1083–1102
19. Jongwanich J (2020) Export diversification, margins and economic growth at the industrial level: evidence from Thailand. World Econ 43(10):2674–2722
20. Otsuka K, Ali M (2020) Strategy for the development of agro-based clusters. World Dev Perspect 20 https://doi.org/10.1016/j.wdp.2020.100257
21. Rastvortseva S (2017) Agglomeration economics in regions: the case in the Russian industry. Reg Sci Inq 9(2):45–54
22. Dubovitski A, Klimentova E, Nikitin A, Babushkin V, Goncharova N (2020) Ecological and economic aspects of efficiency of the use of land resources. In: Proceedings of the E3S Web of Conferences vol. 210, p 11004. https://doi.org/10.1051/e3sconf/202021011004
23. Dubovitski A, Karpunina E, Klimentova E, Cheremisina N (2019) Ecological and economic foundations of effective land use in agriculture: The implementation prospects of food security. In: Proceedings of the 33rd IBIMA Conference: Education Excellence and Innovation Management through Vision 2020. pp 2687–2693
24. Karpunina E, Petrov I, Klimentova E, Sozaeva J, Korkishko I (2020a) Mechanisms of self-development of subsidized regions. In: Proceedings of the 35th IBIMA conference: education excellence and innovation management: a 2025 vision to sustain economic development during global challenges. pp 2282–2293
25. Karpunina E, Lapushinskaya G, Arutyunova A, Lupacheva S, Dubovitski A (2020b) Dialectics of sustainable development of digital economy ecosystem. Lect Notes Netw Syst 129:486–496
26. Nikitin A, Klimentova E, Dubovitski A (2020) Impact of small business innovation activity on regional economic growth in Russia. Rev Inclusion 7:309–321
27. ITC (2020) Trade Map. https://www.trademap.org/Index.aspx Accessed 20 July 2021
28. Hidalgo Cю, Klinger B, Barabasi A-L, Hausmann R (2007) The product space conditions the development of nations. Science 317(5837):482–487
29. Klimentova E, Dubovitski A, Yurina E, Bayanduryan G, Agabekyan R (2021) Regional features of rural unemployment in Russia. Ekonomika Poljoprivreda-Econ Agric 68(2):357–374
30. Gupta D, Davidson B, Hill M, McCutcheon A, Pandher M, MacDonald D, Mekala G (2021) Vegetable cultivation as a diversification option for fruit farmers in the Goulburn Valley, Australia. Int J Agric Sustain https://doi.org/10.1080/14735903.2021.1923286
31. Rosstat (2021) Official statistics. https://rosstat.gov.ru/ Accessed July 20 2021

The Current State of the Organic Market in Russia

Irina V. Chernyaeva(ID)**, Larisa V. Shirshova**(ID)**,
and Natalia V. Lashchinskaya**(ID)

Abstract The regulatory framework for the production and identification of organic agricultural products in Russia is revealed in this article. The directions, volumes of production, and processing of organic products in Russia, the formation of food resources of organic products (demand, volume of production, exports, imports of products) are indicated, the assortment and price proposals for organic products are considered, a SWOT analysis of the organic products market is also carried out.

Keywords Organic production · Agriculture · Activity licensing · Product market · Food quality · Market resources · SWOT analysis

JEL Codes Q53 · Q55 · Q150

1 Introduction

In recent years, the global food and beverage market has seen a significant increase in sales of organic products. Over the period 2018–2019 alone, the global organic market increased from USD 97.0 billion to USD 129.0 billion, or by 33.0% (with a 9.48-fold market growth over the period 1999–2019, the average annual market growth rate accounted for about 11.3%). More than 1.5% of agricultural land in all countries of the world was involved in organic production of organic products in 2019 (72.29 million hectares, which is 4.8 times higher than the level in 2000).

I. V. Chernyaeva (✉)
Altai State University, Barnaul, Russia
e-mail: gurkina-22@mail.ru

L. V. Shirshova
Peoples' Friendship, University of Russia, Moscow, Russia
e-mail: larisa2030@list.ru

N. V. Lashchinskaya
MIREA - Russian Technological University, Moscow, Russia

Table 1 Structure of organic agricultural land by regions of the world, %

Years		Africa	Asia	Europe	Latin America	Northern America	Oceania	World
2000	Thousand ha	52.7	60.5	4,581.1	3,917.6	1,059.0	5,310.2	14,981.0
	% to total	0.4	0.4	30.6	26.2	7.1	35.4	100.0
2005	Thousand ha	490.4	2,678.7	6,988.4	5,055.1	2,219.6	11,813.8	29,246.1
	% to total	1.7	9.2	23.9	17.3	7.6	40.4	100.0
2010	Thousand ha	1,072.1	2,457.9	10,028.8	7,539.6	2,472.7	12,145.1	35,713.9
	% to total	3.0	6.9	28.1	21.1	6.9	34.0	100.0
2015	Thousand ha	1,686.2	3,846.7	12,663.9	6,941.2	2,973.9	22,257.0	50,365.1
	% to total	3.3	7.6	25.1	13.8	5.9	44.2	100.0
2019	Thousand ha	2,030.8	5,911.6	16,528.7	8,292.1	3,647.6	35,881.1	72,285.7
	% to total	2.8	8.2	22.9	11.5	5.0	49.6	100.0
	% to 2000	3,855.4	9,766.1	360.8	211.7	344.5	675.7	482.5

Source Compiled by the authors based on [5]

The main areas of organic farming are located in Oceania—35.9 million hectares (including 35.7 million hectares in Australia) or 49.6% of the total land in all countries of the world, Europe (16.5 million hectares or 22.9%), and South America (8.3 million hectares or 11.5%). For the period 2000–2019, the largest increase in organic land was observed in Asia and Africa—97.7 times and 38.5 times as compared to 2.1–3.6 times growth in Europe, North, and South America (Table 1).

The segment of food markets, which is associated with organic products, is constantly growing in value and kind for all product groups. At the same time, the structure of the assortment of organic products is changing—new types of products are being constantly included in it, i.e., there is a commodity differentiation of the market, which is associated with a change in supply and a change in demand.

2 Materials and Methods

The research was carried out to identify the features of the development of the market for organic products in Russia, the formation of an assortment of organic food and beverages. Sources of information on the area of certified and currently converted agricultural land in countries of the world, including the EU countries, Russia, Kazakhstan, and Ukraine, and on the number of enterprises certified in international certification centers, include statistical database Eurostat, European, and global organic farming statistics (FiBL Statistics). The statistical data were processed using standard MS Excel functions.

3 Results

In 2019, the area of ecological agricultural production in the Russian Federation amounted to 674.4 hectares (about 0.1% of the total area of agricultural land in the country); the turnover in the retail trade of organic products and beverages exceeded 160 million euros, which is 2.5 times higher than the level of 2010. It should be also noted here that the statistical data on the sales of organic products of Russian origin are clearly underestimated, because they are collected according to the data of international certification bodies and don't include information on the sales of those enterprises that have passed certification, according to Russian standards (State All-union Standard). The difference in the methodology for collecting statistical data has been repeatedly mentioned in the international press [11].

Despite the underestimated indicators of the turnover of organic products in Russia, recorded by international databases, it should be recognized that the market for organic food and beverages in Russia is just emerging. The largest markets have been in North America (USA, Canada), Europe (Germany, France, Great Britain, Switzerland), and Japan for many years. Each large market is regulated by its standards in the area of organic products circulation (Japan–JAS, USA–USDA/NOP, EC–Regulation 834/2007) and it covers all stages of food and beverage production from the manufacture of agricultural products directly, their storage, processing, transportation, and labeling. This takes into account the existing regional and national conditions [12].

Among the countries located in the post-Soviet space, Russia is the leader in terms of land area with organic production, leaving behind Ukraine (468.0 thousand hectares) and Kazakhstan (294.3 thousand hectares), but being surpassed by Ukraine in terms of the number of certified producers. If the area of fully transformed and converted agricultural land will be taken into account, then in the structure of sown areas for organic production in Russia, the leaders were cereals (45.76% of the total area), sunflower (12.49%), root crops (without sugar beet and potatoes—6.97%), linen (2.49%) in 2019. For Kazakhstan, the list of agricultural crops is almost the same, but the structure of organic crops is different: cereals (66.23% of the total area), linen (16.09%), soybeans (6.64%), legumes (6.48%), and root crops (without sugar beets and potatoes—2.73%) were the leaders. The broadest assortment among the analyzed countries was observed in Ukraine, which is understandable given the proximity of European consumption markets (Ukraine in 2018–2019 exported organic products to the EU countries in the amount of 266.0–337.9 million euros, ranking second in the rating of the largest suppliers [1] and having more fertile land than the average for Russia and Kazakhstan (Table 2).

Among grain crops that are grown using organic technology, wheat and grain maize were in the lead in Russia and Ukraine, the share of crops of which accounted

Table 2 Fully transformed and converted organic production area in Kazakhstan, Russia, and Ukraine

Indicators		Russia		Kazakhstan		Ukraine	
		2010	2019	2010	2019	2010	2019
Cereals	Total	4,522	239,963	89,070	145,360	117,738	87,383
	% to total	81.77	45.76	54.57	66.23	65.60	53.14
Linen	Total	–	13,076	17,105	35,311	1150	924
	% to total	–	2.49	10.48	16.09	0.64	0.56
Legumes	Total	–	61,946	13,697	14,230	5,080	2,748
	% to total	–	11.81	8.39	6.48	2.83	1.67
Root crops (no sugar beets and potatoes)	Total	–	36,554	26,359	5,986	15,250	3,094
	% to total	–	6.97	16.15	2.73	8.50	1.88
Sunflower	Total	724	65,524	10,030	3,427	20,500	15,766
	% to total	13.09	12.49	6.15	1.56	11.42	9.59
Soy	Total	–	100,735	6,528	14,574	1800	42,662
	% to total	–	19.21	4.00	6.64	1.00	25.94
Fruits			1	20	–	469	1,930
Potato			11	5,984	–	410	12
Berry			–	–	–	435	1,077
Sugar beet			–	–	–	6,100	129
Other			272	603	589	10,539	8,724

Source Compiled by the authors based on [5]

for more than 89.72 and 88.86% across countries, and in Kazakhstan, wheat accounted for more than 91.80% (Table 3).

Despite many factors (psychological unpreparedness of most of the commodity producers; high risks in organic farming of a decrease in crop yields, an increase in

Table 3 Fully transformed and converted grain harvest area (organic production) in Kazakhstan, Russia, and Ukraine in 2019

Cereals	Russia		Kazakhstan		Ukraine	
	ha	% to total	ha	% to total	ha	% to total
Barley	12,959	5.40	9,860	6.78	2,341	2.68
Buckwheat	1,568	0.65	200	0.14	1,417	1.62
Oats	5,253	2.19	1,689	1.16	3,291	3.77
Wheat	125,346	52.24	133,440	91.80	16,820	19.25
Maize	89,927	37.48	0	0.00	60,823	69.61
Rice, rye, triticale	4,910	2.05	171	0.12	2,691	3.08

Source Compiled by the authors based on [5]

clogging of crops, diseases of crops and their infestation with pests; need to undergo lengthy certification procedures and their high cost; low shelf life of perishable products with a total ban on the use of chemicals), preventing the development of organic farming in Russia, the production of environmentally friendly and organic products is steadily increasing [1, 2, 7, 9]. At the same time, organic farming technologies are used by many agricultural producers quite consciously rather than spontaneously. However, it should be noted that the bulk of the produced environmentally friendly products in Russia is formed on uncertified lands by small forms of farming—households and peasant (farmer) households with low production volumes [10]. This is because the additional profit from the sale of certified organic products with small sales volumes will not cover the financial costs that are aimed at obtaining a certificate and functioning in the "organic" status (depending on the number of types of certified products, up to 500–800 thousand rubles). In the EU countries and the United States, the costs of farmers for certification and obtaining certificates of organic products are completely covered by subsidies from the budget, but such a practice isn't provided for in federal regulations in Russia. Compensation of individual costs for organizing the production of organic products is provided by the legislation of only 7 territories: Voronezh, Ulyanovsk, Saratov, Belgorod, Tyumen, and Krasnodar regions, as well as the Republic of Tatarstan [3].

If the advantages for the main participants in the organic food and beverage market will be considered, then commodity producers gain access to international markets with higher prices, while extracting additional foreign exchange income, reducing the cost of chemicalizing agricultural production and its gradual biologization, which leads to an increase in profitability of the activity. The benefits for consumers are associated primarily with the consumption of products that don't lead to deterioration in health, with a higher taste. With the development of agricultural production, including on an organic basis, the state receives an expansion of the taxable base for basic taxes (value-added tax, personal income tax, etc.), more complete use of the production resources of the territories and the food industry enterprises located on them, efficient use of production, engineering and road infrastructure, which also indirectly affects tax revenues to the budgets of all levels, as well as insurance payments [6].

However, it should be also taken into account that consumers of organic products in Russia are faced with a rather high cost, especially for livestock products (the cost of eggs in eco-shops is 7.4 times higher than prices in retail chains, chicken meat—5.2 times, pork and beef—2.7 times), vegetables, and potatoes. As for the online stores, the purchase price for an amount of at least a certain limit should be added to the cost of delivery of goods in many outlets [4, 8].

In this regard, while considering the concentration of the market for organic food and beverages, it was found that the market saturation is higher in Moscow and the Moscow Region in Russia because the purchase of food at inflated prices doesn't radically affect the family budget with higher incomes of their residents. In the regions of Russia, the market for certified organic products is more spontaneous and decentralized.

While considering the price factor of the development of the market for organic food and beverages, high prices are formed under the influence of two factors: the absence of real competition in the market and sale at monopolistically inflated prices (sale of products of households and small peasant (farmer) households at markets and fairs by many consumers isn't presented as an alternative to certified organic products, due to there is no guarantee that no chemicals were used in their production, storage, processing or transportation); declaring organic products as unique and beneficial to health significantly reduces the price elasticity of demand.

At the same time, the development of the market with a low level of the population of a country with high incomes will inevitably lead to the fact that sellers of organic products will have to reorient their sales toward the population with average incomes, which implies a gradual reduction in selling prices. However, the loss in price can be offset by increased sales and lower costs, due to the effect of positive economies of scale.

4 Conclusion

Summarizing the main strengths, weaknesses, opportunities, and threats of the organic food and beverage market in Russia, the strengths include the low level of chemicalization of agricultural production (except for industries with intensive production: poultry, pig, vegetable growing); the presence of a large array of agricultural land, including for various reasons withdrawn from economic circulation; the presence of associations of producers of organic products that contribute to the informational promotion of healthy nutrition, the promotion of products in foreign and domestic markets; availability and successful implementation of scientific research developments of scientific research institutes, as well as universities in the area of organic farming, processing, and storage of products into business practice.

The weaknesses of the organic food and beverage market in Russia include soil contamination with radionuclides in certain parts of the country; underdevelopment of agricultural technologies and technologies for processing organic products in small forms of management; high costs for storage and transportation of organic products, including perishable products; low interest of the majority of agricultural producers in independently entering foreign markets for their products; insufficiently high incomes of the majority of the country's population; insufficient development of information and consulting centers in the regions, as well as municipalities of the country.

The opportunities for the development of the market for organic food and beverages in Russia include low competition in the domestic market and the unworthiness of the supply of organic products by local producers and importers; the use of mass media resources, while conducting information campaigns for healthy nutrition, product promotion; development of consumer cooperation of small businesses in the production of grain, milk, meat, and vegetables; use of foreign and accumulated domestic experience in organizing the production of organic products; proximity to concentrated organic markets and strong demand for organic products in the EU, China, and Japan; the possibility of attracting cheap borrowed resources from commercial banks within the framework of the program for the development of agriculture and agri-food markets, support for the export of agricultural products, including processed ones with a higher added value.

The threats to the development of the organic food and beverage market in Russia include insufficiently developed legislation at the federal level, for which reason individual regions adopt their own regulatory and legal acts. However, not all regions can provide full support products due to the peculiarities of the formation of financial resources; the arrival on the Russian market of international-level Internet platforms selling food products, including organic products; low funding for agricultural science, including research in the area of organic production; instability of economic relations with partners in the sale of organic products during the period of toughened sanctions against Russia on the part of many countries, and the adoption of counter-sanctions; strengthening of protectionism in the USA and EU in the area of agriculture.

References

1. Alekhin VT (2019) Problems connected with the transition to organic farming. Plant Protect Quarant 3:10–11
2. Avarsky ND, Astrakhantseva EY (2017) Methodological aspects of the development of organic agriculture in Russia. AIC Econ Manag 8:38–56. https://doi.org/10.33305/178-38
3. Bryzgalina MA (2020) Certification subsidizing and organic product development. Sci Rev Theory Pract 10:580–593
4. Egorov AY, Pechenkina VV (2012) The market of organic agricultural products. Econ Agric Russia 8:50–59
5. FiBL Statistics—European and global organic farming statistics (2021) The World of Organic Agriculture. https://www.organic-world.net/yearbook/yearbook-2021.html Accessed 10 June 2021
6. Korshunov SA, Lyubovedskaya AA, Asaturova AM, Ismailov VY, Konovalenko LY (2019) Organic agriculture: innovative technologies, experience, prospects. Rosinformagrotech
7. Kruchinina FS (2017) State regulation of organic products market in Russia. Bull Voronezh State Univ Eng Technol 79, 296–305. https://doi.org/10.20914/2310-1202-2017-2-296-305
8. Rozhkova DV (2019) Organic production as a priority direction of the development of the "green" economy. Bull Knyaginsky Univ 2:59–68
9. Vorobyova VV, Vorobyov SP, Shlegel SV (2021) Economic Viability of the Specialized Agricultural Micro-enterprises in the Altai region. IOP Conf Ser Earth Environ Sci 670 https://doi.org/10.1088/1755-1315/670/1/012051
10. Vorobyova VV, Vorobyov SP, Shmakov AA (2019) The Factors Determining Profitability of Grain Production in a Region. Environment, Technology, Resources. In: Proceedings of the 12th international scientific and practical conference. https://doi.org/10.17770/etr2019vol1.4036.
11. Willer H, Kilcher L (2011) The world of organic agriculture. statistics and emerging trends 2011. FiBL-IFOAM Report. IFOAM, Bonn and FiBL, Frick
12. Willer H, Kilcher L (2012) The world of organic agriculture. Statistics and emerging trends 2012. Frick: FiBL—Forschungsinst

Farm Size in Organic Agriculture: Analysis of European Countries and Russia

Natalia Yu. Nesterenko⑩**, Alexander V. Kolyshkin**⑩**, and Tamara V. Iakovleva**⑩

Abstract Organic agriculture is one of the possible technological directions of sustainable agriculture based on the principles of agroecology. Globally and in Russia, interest in organic agriculture is determined by its role in sustainable development (including economic, social, and environmental aspects of agricultural production) and its potential to form sustainable consumption and healthy lifestyles of the population. Researchers are interested in forms of organization of organic production, which is directly related to the type of products produced and the features of the economy in a particular country. Some countries are dominated by small-scale production, mostly family farms, while in other countries and economies, larger agricultural organizations, producing, among other things, organic products, are more common. The world has many examples of a successful combination of the production strategy and conventional and organic technologies (e.g., Nestle). The paper aims to determine the prevailing size of organic farms in Russia and some European countries to identify product and organizational strategies for developing this segment of the agricultural sector in Russia.

Keywords Organic farming · Sustainable agriculture · Agroecology · Correlation and regression analysis · Family farms · Agroholding · Farm size

JEL Codes Q10 · Q12 · Q13

N. Yu. Nesterenko (✉) · A. V. Kolyshkin · T. V. Iakovleva
The Herzen State Pedagogical University of Russia, Saint-Petersburg, Russia
e-mail: natkrav@mail.ru

A. V. Kolyshkin
e-mail: alexvk75@mail.ru

T. V. Iakovleva
e-mail: tamara80@yandex.ru

1 Introduction

The topic of sustainable agriculture began its scientific development with the study of the environmental aspects of agriculture. In the 1930s, there emerged the concept of agroecology. Agroecology was initially defined as part of the science of ecology, concerned with the study of ecological aspects of agricultural production, mainly crop production [1, 2]. The interdisciplinary nature of the research was due to the mutual influence of soil dynamics and agricultural practices. The concept of agroecology was further developed by expanding the range of issues addressed and including social, cultural, and economic aspects as a response to the concept of sustainable development formulated at that time. At the beginning of the twenty-first century, political issues related to the government's role in achieving sustainable development were incorporated into this concept. As a result, the International Symposium "Agroecology for Food Security and Nutrition," organized by the FAO, defined agroecology as a concept that combines the environmental, economic, social, cultural, and political aspects of sustainable agriculture [3].

As a set of agricultural practices, agroecology seeks to find ways to improve agricultural systems by using natural processes, creating beneficial biological interactions and synergies among components of agroecosystems, minimizing synthetic and toxic externalities, and using ecological processes and ecosystem services to implement and develop agricultural techniques [4]. As a result of summarizing the stages of evolution of this concept, A. Wezel formulated the basic principles of agroecology [5]. These principles are as follows:

- Waste recycling;
- Reduced inputs;
- Ensuring soil health;
- Ensuring the health of animals;
- Provision of biodiversity;
- Provision of synergy;
- Economic diversification;
- Co-creation of knowledge;
- Preservation of social values and diet;
- Equity provision;
- Interconnectedness of the actors of the agrifood system;
- Encouraging community organizations and greater participation in decision-making and management of the agrifood system.

Russian scholars are studying the prospects and conditions for the transition of the agricultural and food system to the principles of sustainable development [6–9]. Peculiarities of pricing in the market of organic products of Germany and Russia are investigated in the work of Nesterenko and Shagalkina [10].

Another critical issue in achieving sustainable agriculture through implementing agroecology principles is the question of farm size. Many researchers estimate the size of a farm through the area of agricultural land used for production. Small and medium-sized farms have successfully addressed the challenges of food security, biodiversity enhancement, and rural communities' social development. The report of the FAO High-Level Panel of Experts [11] pays much attention to the impact of farm size on the implementation of sustainable development goals and the formation of sustainable agriculture. Based on a synthesis of many studies, it is shown that small and medium-sized farms (up to 50 hectares) account for 51%–77% of global production of virtually all agricultural commodities and nutrients studied (including vegetables, sugar crops, root and tuber crops, legumes, oilseeds, livestock products, fruits, fiber, and grains) [12–14]. Additionally, small and medium-sized farms use available land more efficiently and obtain higher agricultural land productivity, including lower harvest and post-harvest losses.

The farm size dramatically influences the prevailing supply chains. Large farms can realize long supply chains due to large production volumes, often using foreign trade models. Small and medium-sized farms are more focused on local demand. Short supply chains (often direct sales to the population) provide a significant advantage over large farms due to closer contact with consumers, greater flexibility to changes in demand, and a wider range of products. The concept of agroecology focuses on local agricultural producers, showing their potential in shaping sustainable agriculture. In this study, we understand local agricultural producers as producers of agricultural products grown or produced, processed, and sold to the end consumer within a small geographical distance. The most critical parameter of local agricultural production is the formed short supply chain from production to the end consumer.

Of the above principles of agroecology, local small and medium-sized producers of organic products (including organic) can fulfill a large part of them, in particular:

- Waste recycling (use of waste in animal feed and soil fertilization);
- Reduction of inputs (increased self-sufficiency in fodder and fertilizers to improve economic results);
- Ensuring soil's health through the preferential use of organic fertilizers;
- Ensuring the health and welfare of animals (free grazing);
- Ensuring biodiversity by reducing (rejecting) the use of mineral fertilizers;
- Ensuring synergy through the complementarity of elements of agroecosystems (crop and livestock);
- Preservation of social values and culture through the reproduction of local culture and food traditions;
- Producer–consumer interconnectivity through close contact, including through social networks and the use of short supply chains.

Increasing the sustainability of agriculture while maintaining high production volumes and developing small farms remains highly urgent and is expressed in finding a balance between mass industrial production and environmentally friendly technologies. Of particular interest are the factors contributing to the development of small-scale production in the segment of organic agriculture and the barriers preventing this.

2 Materials and Methods

The growth of certified organic farmland is a key indicator of the development of organic agriculture in the country. According to the annual reports of FiBL & IFOAM on the development of organic agriculture in the world [15–23], between 2010 and 2018, the area of organic land in Russia increased by more than 13 times, which is the strongest growth among European countries.

Another aspect of the analysis of organic farms is the area of organic land. Russian and foreign scientists study the impact of farm size on its efficiency, including in organic farming [24, 25]. Factors for increasing the efficiency of agricultural production by increasing the size of the farm are investigated in detail. The most important factors are fixed production costs and opportunities for more rational use of production capacity. At the same time, the development of production technology and the digitalization of agriculture is leading to an increasing prevalence of production-sharing models, which eliminates the traditional disadvantages of small-scale production associated with the high capital production intensity.

There is no established criterion for classifying farms as small, medium, or large in the world literature and practice due to the features of the economy of different countries, the historical experience of the organization of the economy, and other reasons. Nevertheless, several authors give the following gradation:

- Small-sized farms—up to 2 hectares [26];
- Small-sized farms—up to 20 hectares [13];
- Medium-sized farms—up to 50 hectares [13].

For the purposes of our research, we choose the following size limits of farms:

- Small-sized farms—up to 30 hectares;
- Medium-sized farms—from 30 to 60 up to 30 hectares;
- Large-sized farms—over 60 hectares.

To analyze the size of the organic farm in some European countries, we used a correlation-regression analysis of the dependence of the increase in the area of certified organic land from the increase in the number of registered agricultural producers in the selected European countries. This research does not consider the sectoral structure of organic agriculture for several reasons. First, it is not easy to assess the impact of the production structure on the agricultural land area of one farm. Second, as a result of the correlation analysis, the statistical sample included only the European countries with a strong correlation between the studied parameters. The basis of statistical data is the report "The World of Organic Agriculture. Statistics and Emerging Trends," which annually publishes information on the area of organic agricultural land and the number of certified producers.

The first stage of the research involved a correlation analysis of the relationship between the area of certified agricultural land and the number of certified producers. The sample does not include organizations engaged exclusively in the processing of organic raw materials. The period from 2010 to 2018 was chosen as a period

Fig. 1 The dynamics of organic land in Europe. *Source* Calculated by the authors based on [15–23]

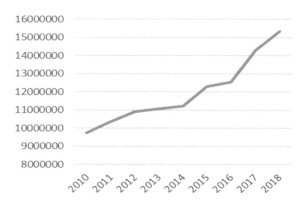

Fig. 2 The dynamics of producers of organic products in Europe. *Source* Calculated by the authors based on [15–23]

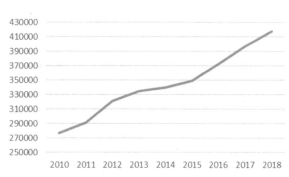

of statistical research as the period of the most dynamic development of organic agriculture in Europe. During the analyzed period, the area of organic land increased by 57% (Fig. 1), while the number of producers of organic products increased by 50% (Fig. 2).

Correlation analysis showed the strength of the relationship between the dynamics of the area of organic agricultural land and the dynamics of the number of certified producers. Out of the total number of European countries, we selected nine countries whose correlation coefficient was above 0.97: Belgium, Croatia, Estonia, France, Germany, Hungary, Norway, Slovenia, and Great Britain (Table 1). Thus, we identified the countries in which the dynamics of the number of certified producers and the dynamics of the area of organic land are closely related to each other over the period 2010–2018.

In the second stage of the research, we analyzed the impact of the increase in the number of certified producers on the growth of organic land in each country. The resulting regression coefficient was interpreted as the average size of one organic farm (Table 1).

The analysis (Table 2) indicates that the above sample includes countries with a predominance of small organic farms (up to 50 hectares), medium organic farms (over 50 hectares), and large organic farms (over 100 hectares).

Table 1 Parameters of the relationship between the area of organic land and the number of certified organic producers

Country	Correlation coefficient	Regression coefficient
Belgium	0.99	33.08
Germany	0.99	53.15
Slovenia	0.99	10.77
France	0.99	58.10
Norway	0.99	13.73
Hungary	0.98	33.97
Croatia	0.98	23.89
Estonia	0.98	148.21
United Kingdom	0.97	132.32

Source Calculated by the authors based on [15–23]

Table 2 Distribution of countries in Europe relative to the average size of the organic farm

Country	Farm size category	The average size of an organic farm
Slovenia	small organic farms	10.77
Norway		13.73
Croatia		23.89
Belgium	medium organic farms	33.08
Hungary		33.97
Germany		53.15
France		58.10
Great Britain	large organic farms	132.32
Estonia		148.21

Source Calculated by the authors based on [15–23]

The prevailing farm size in each particular country is related to many factors: economic conditions, natural conditions, historical factors of entrepreneurial activity in rural areas, and many others. Since there is little experience in developing organic agriculture in many countries, we conclude that there are currently no specific factors affecting the average size of the organic farm in a particular country. In other words, the practice of organizing organic farms is not different from the practice of organizing any other farm.

A more important result of the correlation and regression analysis is the characteristics of the effect of organic farm size on the formation of sustainable agriculture. International Federation of Organic Agriculture Movements has formulated four principles of organic agriculture [27]:

- Health (soils, plants, animals, people, and planet as one and indivisible concept);

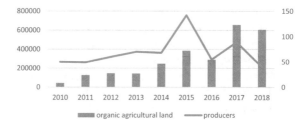

Fig. 3 Dynamics of organic land and certified producers in Russia. *Source* Calculated by the authors based on [28]

- Ecology (reliance on ecological systems and cycles of living organisms, working with them, reproducing them, and helping to ensure their sustainability);
- Fairness to the environment and opportunities to improve life;
- Careful consideration (prudent and responsible rational use) to protect the health and well-being of present and future generations and the environment.

The advantages of small and medium-sized farms in achieving sustainable agriculture in the segment of organic agriculture are greatly enhanced by using environmentally friendly technologies. Therefore, it can be argued that small and medium-sized farms producing organic products contribute to the formation of sustainable agriculture more significantly compared to large organic farms.

Russia is the fastest growing country in Europe in terms of the growth of certified organic agricultural land. Between 2010 and 2018, according to the FiBL & IFOAM report [15–23], the area of organic land increased more than 13-fold from 44,017 ha in 2010 to 606,975 ha in 2018. However, the number of producers during this period decreased from 50 organizations in 2010 to 40 organizations in 2018 (by 20%). It is important to note that during the studied period, the number of producers varied significantly. The maximum number was 142 producers in 2015 (Fig. 3).

The chaotic nature of the development of organic agriculture indicates that the country has not worked out a common product and regional development strategy for this segment of the agrifood system. Legislative and institutional support will form a clearer positioning of organic farms.

3   Results

The research results showed that the movement of organic agriculture in terms of farm size is not homogeneous. The examples of several countries show that this segment can be dominated by small (Slovenia, Norway, Croatia), medium (Germany, France, Belgium, Hungary), and large farms (UK, Estonia). This is related not only to the historical experience of the organization of agriculture but also to the type of products produced. Thus, the production of organic raw materials for further processing (e.g., grain or fiber) involves the production of a standardized product in large quantities. Large organic farms are mainly engaged in the production of raw materials and mass production. The large size of the organic farm allows optimizing costs and using production resources rationally. Large organic farms have greater access to innovation through capital opportunities. Economies of scale can reduce the cost of production, which makes certified organic products more affordable to consumers.

Small-scale production of organic products with minimal processing (dairy, meat products, fresh vegetables, fruits, and berries) is more common among small and medium-sized organic producers. Due to the economic and environmental advantages of polyculture production, small and medium-sized organic farms provide economic and environmental benefits by combining crop and livestock production and economic and environmental advantages of using waste and organic fertilizers in crop production. Proximity to consumers and the ability to reproduce local culture contribute to the social goals of sustainable agriculture.

4   Discussion

A study of the average size of organic farms of different countries allows us to understand that organic agriculture in different countries has different product and organizational specifics. To identify strategic directions for developing organic agriculture in Russia, it is necessary to identify the potential for the effective development of different products. Currently, this segment of the agrifood system is represented by three product lines:

- Production of organic raw materials for further deep processing (wheat, soybeans, and buckwheat);
- Production of fresh organic products (vegetables, fruits, and berries) and products with minimal processing (dairy and meat products);
- Collection of wild plants (mushrooms, berries, and medicinal herbs).

Analysis of the efficiency potential of organic farms of different sizes will allow to correlate the product areas with different organizational forms and sizes of farms and develop effective positioning of the products produced. In Russia, organic agriculture is currently represented by sixty organizations, with a small proportion of small and medium-sized farms. It is important to note that dairy and meat production farms are mostly located in the European part of Russia near Moscow. They focus on middle- and upper-middle-income consumers, produce a wide range of products, and are in direct contact with consumers through social networks. Large producers of organic products are engaged mainly in the production of grain. Interestingly, the large organic producers also include organizations that produce alcoholic beverages. Thus, we can say that Russia, in general, has formed a product niche of organic agriculture. The large size of the country and the potential for conversion of agricultural land to organic production allow us to predict the successful development of this segment in several product areas at once.

5 Conclusion

Organic farms of different sizes (small, medium, and large) have particular potential in shaping sustainable agriculture. A variety of product and organizational forms allow for diversification of risks in developing this segment of the agrifood system. The export potential of organic agriculture can be realized by large and medium-sized producers making large amounts of organic products. In this regard, the most important task is to increase the added value of organic products through deeper processing (i.e., from the production of organic raw materials to the production of finished products with a long shelf life).

Small organic farms solve other problems of sustainable agriculture: the formation of sustainable consumption, solving social problems of rural areas, and the preservation and reproduction of national culture, culinary traditions. The government implements programs in this direction to support national food brands. Government support for the creation of organic farms includes free certification for small and medium-sized businesses and reimbursement of costs for certification of exported products. Other government support measures are provided on an equal basis with other agricultural organizations without regard to production technologies.

The difficulties of developing small organic farms lie in the need to build a complete chain of organic production from raw materials to finished products. Another difficulty is a strong dependence on the population's income. The COVID-19 pandemic has had a significant impact on public demand for environmentally friendly products. Concerns for personal health and the lack of opportunities for gastronomic tourism in other countries led to an increase in sales of organic farm products (certified and non-certified). Additionally, the active participation of delivery services and marketplaces in the formation of supply chains of farm products to the doors of consumers can solve the problems of sales and delivery in small and medium-sized farmers.

References

1. Bensin BM (1930) Possibilities for international cooperation in agroecological investigations. Int Rev Agric Part I Monthly Bull Agric Sci Pract 21(8):279–280
2. Tischler W (1950) Ergebnisse und probleme der agraroekologie. Schrift Landwirtschaft Fakultaet Kiel 3:71–82
3. FAO (2015) Agroecology for food security and nutrition. In: Proceedings of the FAO International Symposium. Rome, Italy. http://www.fao.org/3/a-i4729e.pdf
4. Wezel A, Casagrande M, Celette F, Vian JF, Ferrer A, Peigné J (2014) Agroecological practices for sustainable agriculture. Rev Agron Sustain Dev 34(1):1–20. https://doi.org/10.1007/s13 593-013-0180-7
5. Wezel A, Herren BG, Kerr RB, Barrios E, Gonçalves ALR, Sinclair F (2020) Agroecological principles and elements and their implications for transitioning to sustainable food systems. A Rev Agron Sustain Dev 40:40. https://doi.org/10.1007/s13593-020-00646-z

6. Buzdalov IN (2015) Agriculture in Russia: a view through the prism of sustainable development. AIC: Economics. Management 8:3–16
7. Nesterenko NY, Kolyshkin AV, Iakovleva TV (2021) Sustainable agriculture in Russia: the role of eco-friendly and organic technologies. In Bogoviz AV (ed) The challenges of sustainability in agricultural systems. Cham, Switzerland: Springer, pp 41–48. https://doi.org/10.1007/978-3-030-73097-0_6
8. Nesterenko N, Artemova D (2018) Prospects for the development of sustainable organic food supply chains in Russia. Econ Agric Russia 7:2–16
9. Nesterenko N, Pakhomova N (2016) Organic agriculture in Russia: conditions for transition to a sustainable development path. Econ Agric Russia 12:34–41
10. Nesterenko N, Shagalkina M (2019) Comparative characteristics of the organic food market in Russia and Germany. IOP Conf Ser Earth Environ Sci 274:012059. https://doi.org/10.1088/1755-1315/274/1/012059
11. HLPE (2019) Agroecological and other innovative approaches for sustainable agriculture and food systems that enhance food security and nutrition. A report by the high-level panel of experts on food security and nutrition of the committee on world food security. http://www.fao.org/3/ca5602en/ca5602en.pdf
12. Graeub BE, Chappell MJ, Wittman H, Ledermann S, Bezner Kerr R, Gemmill-Herren B (2016) The state of family farms in the world. World Dev 87:1–15. https://doi.org/10.1016/j.worlddev.2015.05.012
13. Herrero M, Thornton PK, Power B, Bogard JR, Remans R, Fritz S, Havlík P (2017) Farming and the geography of nutrient production for human use: a transdisciplinary analysis. Lancet Planet Health 1(1):33–42. https://doi.org/10.1016/S2542-5196(17)30007-4
14. Ricciardi V, Ramankutty N, Mehrabi Z, Jarvis L, Chookolingo B (2018) How much of the world's food do smallholders produce? Glob Food Sec 17:64–72. https://doi.org/10.1016/j.gfs.2018.05.002
15. Willer H, Lernoud J (eds) (2012) The world of organic agriculture: Statistics and emerging trends, 2012. http://www.organic-world.net/yearbook/yearbook-2012.html
16. Willer H, Lernoud J (eds) (2013) The world of organic agriculture: Statistics and emerging trends, 2013. http://www.organic-world.net/yearbook/yearbook-2013.html
17. Willer H, Lernoud J (eds) (2014) The world of organic agriculture: statistics and emerging trends, 2014. http://www.organic-world.net/yearbook/yearbook-2014.html
18. Willer H, Lernoud J (eds) (2015) The world of organic agriculture: Statistics and emerging trends, 2015. http://www.organic-world.net/yearbook/yearbook-2015.html
19. Willer H, Lernoud J (eds) (2016) The world of organic agriculture: Statistics and emerging trends, 2016. http://www.organic-world.net/yearbook/yearbook-2016.html
20. Willer H, Lernoud J (eds) (2017) The world of organic agriculture: Statistics and emerging trends, 2017. http://www.organic-world.net/yearbook/yearbook-2017.html
21. Willer H, Lernoud J (eds) (2018) The world of organic agriculture: Statistics and emerging trends, 2018. http://www.organic-world.net/yearbook/yearbook-2018.html
22. Willer H, Lernoud J (eds) (2019) The world of organic agriculture: Statistics and emerging trends, 2019. http://www.organic-world.net/yearbook/yearbook-2019.html
23. Willer H, Lernoud J (eds) (2020) The World of Organic Agriculture: statistics and emerging trends 2020. http://www.organic-world.net/yearbook/yearbook-2020.html
24. Heinrichs J, Kuhn T, Pahmeyer C, Britz W (2021) Economic effects of plot sizes and farm-plot distances in organic and conventional farming systems: A farm-level analysis for Germany. Agric Syst 187:102992. https://doi.org/10.1016/j.agsy.2020.102992
25. Ladvenicová J, Miklovičová S (2015) The relationship between farm size and productivity in Slovakia. Visegrad J Bioecon Sustain Dev 4(2):46–50. https://doi.org/10.1515/vjbsd-2015-0011

26. Garibaldi LA, Carvalheiro LG, Vaissière BE, Gemmill-Herren B, Hipólito J, Freitas BM, Zhang H (2016) Mutually beneficial pollinator diversity and crop yield outcomes in small and large farms. Science 351(6271):388–391
27. IFOAM (2020) Principles of organic agriculture. https://www.ifoam.bio/sites/default/files/2020-03/poa_english_web.pdf
28. FIBLstatistics (n.d.) Key indicators on organic agriculture worldwide. https://statistics.fibl.org/world/key-indicators.html

On the Issue of Food Security of the EAEU Countries During a Pandemic

Olga B. Digilina⬛, Andrey O. Zlobin, and Andrey A. Chekushov

Abstract The main purpose of this article is to determine the impact of the pandemic on the food security of the EAEU, which have different levels of agricultural development and experience various problems with the logistics of delivering food to the population. The authors emphasize that at present the efforts of all countries should be aimed not only at maintaining the stable functioning of the internal agri-food market, but also at determining the further prospects for its development. It became necessary to develop an international treaty on the creation of a single market for organic agricultural products, which will allow organizing the unhindered circulation of such products in the domestic market, as well as starting negotiations on the access of producers of the EAEU member states to the markets of third countries. Joint efforts of the EAEU countries can create technologies for the complex processing of food raw materials, new methods of growing, storing, and transporting agricultural products.

Keywords Food security · Pandemic · Food market · Regional economic integration

JEL Codes O18 · F52

O. B. Digilina (✉)
RUDN University, Moscow, Russia

A. O. Zlobin
Vladimir State University Named After Alexander and Nikolai Stoletov, Vladimir, Russia

A. A. Chekushov
Financial University Under the Government of the Russian Federation, Moscow, Russia
e-mail: aachekushov@fa.ru

1 Introduction

In 2020, all countries of the world to one degree or another have experienced the impact of coronavirus infection and the ensuing coronavirus crisis caused by the introduction of quarantine measures by the governments of many countries. The pandemic has become a serious challenge not only for individual countries but also for the integration associations of the modern world.

One of such international organizations for regional economic integration is the Eurasian Economic Union (EAEU), the purpose of which was to ensure the freedom of movement of goods, capital, and labor between the member states of the Union, to pursue a coordinated policy in various spheres of economic activity. The coordinated actions of the member countries of integration are aimed at ensuring sustainable economic growth by combining common efforts to effectively use the available resource potential and strengthen their competitive advantages in the world economy.

Agriculture is one of the spheres of cooperation between the member states of the Union—the Republic of Belarus, the Republic of Kazakhstan, the Republic of Armenia, the Kyrgyz Republic, and the Russian Federation. His condition affects such an important indicator as food security. Food security of a country or a group of countries presupposes a state of the economy in which food independence and self-sufficiency of each country or group of countries is ensured, the economic accessibility of the population to food and high-quality drinking water is guaranteed following the required physiological nutritional standards [1].

2 Materials and Method

The theoretical basis of this article is the fundamental theories and concepts presented in the classical and modern works of domestic and foreign scientists in the field of studying economic integration and food security.

The methodological basis of the study was a structural analysis of the state of food security of the Eurasian Economic Union, which was based on the use of dialectical principles of cognition and a system-functional approach.

The informational basis of the article is the documents of the EAEU countries, statistical databases of these countries, and content analysis of publications on the Internet.

3 Results

In 2020, all countries of the world to one degree or another have experienced the impact of coronavirus infection and the ensuing coronavirus crisis caused by the introduction of quarantine measures by the governments of many countries. The pandemic has become a serious challenge not only for individual countries but also for the integration associations of the modern world.

Assessment of the level of ensuring food security in the EAEU is based on indicators of the physical availability of agricultural products and food; economic affordability of food; the level of nutrition of the population (energy value of the food ration of the population). The availability of agricultural products and food is determined by their most important types—grain, meat; milk; sugar; vegetable oil; potatoes; eggs; vegetables, melons, and gourds; fruits and berries; fish, as well as in terms of the index of the sufficiency of production. As rational norms for the consumption of basic food products in each of the EAEU member states, the average per capita consumption norms established at the legislative level are used.

To assess food security in terms of the level of nutrition of the population, data on the actual level of caloric intake per capita in each of the EAEU member states are used [1].

Given that maintaining food security requires significant efforts, the introduction of quarantine measures in 2020 has raised questions about maintaining food security and ensuring food adequacy.

The member states of the Union were forced to urgently take measures to maintain food security.

In Belarus, despite the deployment of a pandemic, it was decided to engage in agricultural work in full and not to abandon the import of agricultural products.

According to the statements of the country's officials, external challenges had practically no effect on the agriculture of Belarus: since the end of 2019, the production of the agro-industrial complex has grown by 4.5–5.5% monthly [2].

According to available information, the export of agricultural products of the country in the first quarter of 2020 increased by 9.8% and amounted to more than $ 1.4 billion. The increase in supplies was observed in almost all regions: to Russia, they increased by 4.4%, to countries The CIS excluding the Russian Federation—by 24.7%, and in the PRC—2.4 times. Belarus became the leader in terms of agricultural production growth in the EAEU (5.5%). In Armenia, the growth was 4.1%, Kazakhstan—2.5%, Kyrgyzstan—0.5%, Russia—3%.

Although it was not without negative consequences: in the spring of 2020, the country experienced problems with some types of products (buckwheat, rice), prices for meat products increased not only due to the corona crisis but also due to the bird flu epidemic in Europe (for turkey meat—by 22.3%, for chicken meat—by 8.1%).

In general, the situation in the agro-industrial complex of Belarus is characterized as satisfactory [3].

As for the food security of the Republic of Kazakhstan, at present, the country through its production can almost completely provide itself with basic food products. Export potential is growing. So, in 2019, the export of agricultural products grew by 6.5% and amounted to 3.3 billion US dollars. Supplies to China (by 50.5%), the EAEU countries (by 8.2%), and Central Asia (by 7.4%) increased significantly [4].

A fairly high level of food security in Kazakhstan is due to some factors: a large volume of agricultural production, its active export to the near and far abroad, the low cost of quality products.

Nevertheless, the corona crisis played a certain role, prompting agricultural producers to make decisions on the need to increase investment in agricultural processing, improve the storage system, logistics, and effective marketing of products, provide agriculture with qualified personnel, and create an effective information system [4].

In Armenia, the situation with food security is somewhat worse.

According to officials, Armenia is fully self-sufficient in vegetables and fruits, lamb and beef. The demand for poultry meat and pork is covered only by imports (domestic production of poultry meat provides only 1/4 of the total demand, pork—2/3). The missing volumes of products are imported mainly from Russia [5]. The country depends on the external market for such products as wheat and vegetable oil.

In 2020, the situation on the international market became more complicated due to an increase in demand for agricultural products compared to supply. In addition, wheat plantings have decreased in the country: if in 2015 there were 108 thousand hectares of wheat sown, then in 2019—only 60 thousand. In 2019, the lowest yield was recorded—19–20 centners per hectare instead of the usual 20–30. As a result of the existing problems, in just a week in May 2020, grain prices increased by $ 25 per ton [6].

In general, after an increase in production in the first quarter of 2020 by 3.9%, in the second quarter, the country recorded a sharp decline—by 13.7% compared to the same period in 2019 activity in Armenia in comparison with the same period in 2019 decreased by 10.2% [7].

Despite these results, experts believe that the pace of Armenia's integration development within the framework of the Eurasian Economic Union has been preserved [6].

As for the Kyrgyz Republic, its economy has always been characterized by instability, and in 2020 the situation was further aggravated by a pandemic and quarantine. Almost 15 thousand companies closed in March and April. The volume of Kyrgyzstan's GDP in the first half of 2020 decreased by 5.3% compared to the same period in 2019, budget revenues decreased by almost 10 billion rubles.

In recent years, the country has received support in the form of humanitarian aid and financing of various projects within the EAEU in the amount of about $ 700 million, as well as assistance in the form of direct investments in the country's economy on preferential terms.

The Russian-Kyrgyz Development Fund financed 70 large joint projects and through partner banks—more than 2,700 small and medium-sized business projects. The country does not count on financial support from other EAEU countries.

Despite the imposition of severe restrictions in March 2020, Russia remained one of the few countries that continued to assist the republic: closed borders did not interfere with the delivery of humanitarian supplies to rural regions of Kyrgyzstan. Russia allocated $ 8 million for the purchase of food (mainly fortified wheat flour and vegetable oil) [8].

In Kyrgyzstan, since the beginning of the pandemic, there has been a rapid increase in food prices, especially meat: in the first four months of 2020 alone, this figure exceeded 10%. This is even though for 10 months of 2020, the production of livestock and poultry in the republic increased by 1.5% compared to the same period in 2019 [3].

According to officials, the rise in meat prices was associated with the purchase of local cattle by entrepreneurs from Uzbekistan and Tajikistan at an inflated price. This was due to the introduction of restrictive measures by Kazakhstan on the export of meat in live weight and the growth of the dollar.

In the fall of 2020, the Ministry of Agriculture of Kyrgyzstan decided to ban the export of livestock and some other food products (barley, corn, rice, wheat flour, vegetable oil, sugar, eggs, and iodized salt) and feed (hay, straw, cereals) for six months [2].

In general, the Kyrgyz Republic, despite the status of an agrarian country, provides itself with only three types of vital food products: milk and dairy products, potatoes, vegetables, and melons. The rest of the products are imported. The provision of the state of the country's domestic food market, taking into account imports, is: bread products—101.5%, potatoes—105.8%, milk and dairy products—133.2%, meat—61.6%, vegetables and melons—92.9%, berries—27.1%, poultry eggs—78.8%, sugar—86.8% [3].

The bulk of the republic's population lives in rural areas and is engaged in farming. And the share of agriculture in total GDP is almost a quarter. At the same time, agriculture has always been perceived as a weak, subsidized sector of the country.

Kyrgyzstan is located on a transit route (the Great Silk Road) and could well become a country on whose territory there is a hub for the collection, fattening of cattle from the CIS countries, processing, and export of meat to Arab countries.

However, judging by the decrease in the share of the agricultural sector in GDP (its share decreased from 23% in 2008 to 12% in 2019), there is extremely slow growth in domestic food production [9].

For many years, the country's economic infrastructure was used ineffectively and was poorly suited for the rapid deployment of investment projects. The lack of clear priorities of the state economic policy accumulated risks of the country's food security.

While domestic food production has not been hit hard by the coronavirus crisis, the restrictions have posed a serious threat to food security. Kazakhstan's restrictions on wheat exports (removed from June 1, 2020) have created problems for the Kyrgyz Republic, which imports high-gluten wheat and its hard varieties from Kazakhstan.

Replacing the export ban with export quotas in April 2020 (22,000 tons of flour and 30,000 tons of grain) partially improved the food situation in Kyrgyzstan. However, access to food is hindered by declining incomes and rising food prices.

The prices for flour, potatoes, and vegetable oil have grown the most in the republic. According to the World Bank, a 5% increase in consumer prices could raise the national poverty rate by 3.6 percentage points [9].

Since the spread of the coronavirus in the Kyrgyz Republic, measures have been taken to strengthen food security and support agriculture: stocks of state food reserves have been increased, food prices have been controlled, restrictions on the transportation of agricultural products and the movement of workers in the industry have been lifted.

According to the Minister of Agriculture of the Kyrgyz Republic, the agricultural industry needs fundamental changes in the direction of consolidation of farms, digitalization of the industry, and the development of the processing industry [3].

As for the food security of the Russian Federation, as of July 2020, Russia is provided with such products as grain by 155%, sugar by 125%, and meat by 97%. But for several products, for example, milk and potatoes, the country has not reached the level of self-sufficiency [10]

In January 2020, the Russian Federation adopted a new Food Security Doctrine (with a horizon until 2030).

According to it, food independence is defined as the level of self-sufficiency in percent, calculated as the ratio of the volume of domestic production of agricultural products, raw materials, and foodstuffs to the volume of their domestic consumption [11].

In the new doctrine, some indicators have been changed. So the self-sufficiency threshold for sugar and vegetable oil was increased from 80 to 90%, for fish products—from 80 to 85%. The norms of self-sufficiency have not changed for grain (95%), meat (85%), milk and dairy products (90%), potatoes (95%). New product groups have appeared—vegetables and melons. Russia should be provided with fruits and vegetables by 90 and 60%, respectively [11].

Based on the available data, the food sector in Russia is successfully surviving the coronavirus and food shortages are not expected.

According to Rosstat, the volume of meat and offal in Russia in January—May 2020 amounted to 3.5 million tons, which is 6.5% more than in 2019. There was also an increase in the production of meat products. Milk yield, sugar, and salt production increased. Production of processed and canned fish, crustaceans, and shellfish increased by 8% compared to 2019.

However, in the context of the spread of COVID-19 in the country, a ban was introduced on the export of some goods; at the end of March 2020, a non-tariff quota for grain crops was introduced. Until the end of the first half of 2020, only 7 million tons of grain could be exported abroad. This limitation did not apply to the EAEU countries [11].

According to the Global Food Security Index (excludes 59 different indicators), prepared by analysts of The Economist Intelligence Unit with the support of Corteva Agriscience, Russia in terms of food security at the end of 2020 ranked 24th among 113 countries (73.7 points out of 100 possible). In 2019, the country ranked only 42nd (69.7 points).

In terms of food security, the country came close to Belarus, which was in 23rd place in the ranking (73.8 points). In 2020, Russia managed to rise in the ranking, since in recent years the country has limited imports and developed its production (Russia jumped on food, n/a).

4 Conclusion

In general, summing up the state of food security in the EAEU, it should be said that the coronavirus pandemic has become a serious global challenge for the world community and the EAEU countries. The issues of stable saturation of the market with foodstuffs have become especially acute.

Existing threats and risks are pushing the EAEU states to join forces to saturate the market with food and protect against low-quality products. Each country has certain advantages: the production and processing of fruits and vegetables are developing in Armenia and Kyrgyzstan, the production of dairy products in Belarus, the production of animal proteins (poultry, beef, pork) is growing in Russia, Kazakhstan has advantages in the production of wheat (Food security in the EAEU: how has the pandemic affected the common market? n/a).

All countries have learned certain lessons from the coronavirus crisis. And the most important is the importance of interaction between countries and their beneficial cooperation.

The efforts of all countries should be aimed not only at maintaining the stable functioning of the domestic agri-food market, but also at determining prospects.

In any case, now there is a need to develop an international treaty to create a single market for organic agricultural products in the EAEU. It will allow organizing the unhindered circulation of such products on the domestic market, as well as starting negotiations on the access of manufacturers of the EAEU member states to the markets of third countries.

Joint efforts of the EAEU countries can create technologies for the complex processing of food raw materials, new methods of growing, storing, and transporting agricultural products [8].

Joint work to ensure food security will help turn global challenges into opportunities for further deepening Eurasian integration in all areas of interaction will unlock the potential of each state, and, in general, strengthen the Union.

References

1. Vartanova ML (2020) Assessment of the level of food supply for the population of the states of the Eurasian Economic Union during the COVID-19 pandemic. https://cyberleninka.ru/art icle/n/otsenka-urovnya-prodovolstvennogo-obespecheniya-naseleniya-gosudarstv-evraziysk ogo-ekonomicheskogo-soyuza-v-period-pandemii-covid
2. Eurasian integration https://www.ritmeurasia.org/news--2020-12-09--kak-minselhoz-kirgizii-planiruet-spasat-prodovolstvennuju-bezopasnost-52275
3. Eurasian integration https://www.ritmeurasia.org/news-2020-05-14--prodovolstvennaja-bez opasnost-vo-vremja-pandemii-vyhodit-na-pervyj-plan-49015
4. How COVID-19 affects food security in Kazakhstan. https://cabar.asia/ru/kak-covid-19-vli yaet-na-prodovolstvennuyu-bezopasnost-kazahstana
5. Armenia has no problems with ensuring food security http://analitikaua.net/2020/u-armenii-net-problem-s-obespecheniem-prodovolstvennoj-bezopasnosti/

6. Will Armenia face a grain deficit, or what is there with food security? https://ru.armeni asputnik.am/economy/20200628/23546573/Stolknetsya-li-Armeniya-s-defitsitom-zerna-ili-Chto-tam-s-prodovolstvennoy-bezopasnostyu.html
7. How the "Velvet Revolution" and the coronavirus crisis changed the Armenian economy. It is reported by Rambler. https://news.rambler.ru/caucasus/44760243/?utm_content=news_m edia&utm_medium=read_more&utm_source=copylink
8. Provision of the population of the states of the Eurasian Economic Union during the COVID-19 pandemic (2020) Natural and humanitarian research, 31(5). http://academiyadt.ru/wp-content/uploads/egi/egi-31.pdf
9. COVID-19 in the Kyrgyz Republic: socioeconomic impact and vulnerability assessment and policy responses https://kyrgyzstan.un.org/sites/default/files/2020-08/UNDP-ADB/252 0SEIA_11/2520August/25202020/2520Rus.pdf
10. Russia jumped on food. http://www.rbc.ru
11. Food security in Russia 2020 in the context of the COVID-19 pandemic http://id-marketing. ru/goods/prodovolstvennaja_bezopasnost_rossii_2020_v_uslovijah_pandemii_covid_19/
12. Food security in the EAEU: how has the pandemic affected the common market? https://fin ance.rambler.ru/markets/45661063-prodovolstvennaya-bezopasnost-v-eaes-kak-pandemiya-povliyala-na-obschiy-rynok/
13. Products for labour. https://rg.ru/2020/12/09/rossiia-prodolzhit-pomogat-maloobespech ennym-semiam-kirgizii.html

Self-Sufficiency in a Highly Productive Seed and Breeding Base as a Factor in the Sustainability of the Food Security of the Russian Federation in the Context of the Transformation of the World Food System

Vera A. Tikhomirova ⓘ

Abstract The destabilization of world economic ties due to the global epidemic of coronavirus has revealed the colossal dependence of the overwhelming majority of countries in the world on access to resource support for agro-industrial production. Interruptions in the supply of agricultural raw materials and food products, as well as more frequent epizootics of farm animals and poultry, stimulate economies with developed food production to develop independent highly productive seed and breeding funds. The article is devoted to the study of the degree of dependence of the Russian food system on the import of genetic material from crops, breeding animals, and birds. In the course of the study, the author revealed an increase in the dependence of Russian agro-industrial production on the supply of sunflower and corn seeds, the import of hatching eggs of poultry, breeding pigs, and cattle for breeding. The almost total dependence on the import of these resources not only creates risks in foreign economic activity but also threatens the stability of the country's national food system. The study showed the need to revise the Russian state policy in the field of ensuring national food security by including in the Food Security Doctrine of the Russian Federation indicators and indicators for monitoring the dynamics of self-sufficiency in the genetic base of the most significant agricultural plants, animals, and birds.

Keywords Russian Federation · Food security · Export · Import · Self-sufficiency · Genetics of agricultural plants · Animals and birds · Seed base · Breeding base

JEL Codes F10 · F15 · F52 · F53 · O13 · O53

V. A. Tikhomirova (✉)
Private Educational Institution of Higher Education "Moscow International Academy", Moscow, Russia
e-mail: vera-t@myrambler.ru

FSBI "Agroexport", Moscow, Russia

© The Author(s), under exclusive license to Springer Nature Singapore Pte Ltd. 2022   209
E. G. Popkova and B. S. Sergi (eds.), *Sustainable Agriculture*,
Environmental Footprints and Eco-design of Products and Processes,
https://doi.org/10.1007/978-981-16-8731-0_21

1 Introduction

Ensuring food security is one of the basic functions of any modern state, the effectiveness of which depends on its further development, the stability of the national economy, and the stability of the current political course. At the same time, it is important to understand that the set of measures taken to achieve it is not static and transforms over time. Accordingly, the study of current trends in the development of the world food system allows you to expand and deepen your understanding of the essence of this phenomenon.

The challenges and contradictions of our current geopolitical situation initiate the evolution of the existing approaches to achieving food security at the national level and require the state to create and maintain the maximum possible volumes of reserves of strategic food categories, as well as develop food production from internal resources in volumes sufficient to ensure threshold values indicators of food security.

In addition, at present, the objective necessity of organizing large-scale scientific and applied research aimed at increasing the productivity of the domestic breeding and seed base is becoming obvious. The destabilization of global food production chains caused by the coronavirus pandemic makes this agenda more relevant every day, initiating states around the world to search for possible solutions to the issue of ensuring uninterrupted access to the highly productive genetic material of agricultural animals and plants [5]. At the same time, an increase in the degree of global political contradictions puts the above problem far beyond the economic plane and negatively affects the implementation of the principles of globalization.

Within the framework of this study, the author sets himself the goal of identifying strategically important segments of the Russian national food system, analyzing the degree of their dependence on the supply of imported genetic resources, and also suggesting possible directions for the development of the concept of national food security of the Russian Federation.

2 Methodology

The methodology of this study is based on a comprehensive functional analysis of the current state of the world food system. Within the framework of this approach, the use of the economic-statistical method, as well as the method of analogies, makes it possible to assess the level of food security of the Russian Federation, to identify key disproportions in the field of organizing access to the resource supply of the agro-industrial complex (AIC) and to determine the most promising directions for improving the Russian food supply system in the most balanced way. Visualization of the material through computational-constructive and graphic methods of scientific knowledge allows you to visually illustrate the results obtained.

3 Results

The crisis in the system of the global division of labor has become one of the drivers of the launch of centrifugal and de-globalization trends in the world economy. Currently, the socio-economic agenda of the political leadership of countries around the world is mainly focused on maintaining a stable supply of food to the population and curbing agflation. The almost unprecedented rise in world prices for agricultural raw materials and food products is accompanied by a decrease in the purchasing power of the population around the world [2].

Countries that managed to organize competitive self-sufficiency in agricultural products were less affected by the disruption of global supply chains. However, the boom in food demand that followed the lockdown caused a significant increase in world prices in the segment and provoked an outflow of agricultural raw materials and food products to the foreign market, which, in turn, was accompanied by an increase in the cost of these products at the national level.

Against the background of a reduction in the economic access of national food systems to the import of resource provision for the agro-industrial complex, a tendency is formed in which every day an increasing number of subjects of international relations are showing interest in achieving sustainable self-sufficiency in a highly productive seed and breeding base.

The cases of epizootics of farm animals and birds that have become more frequent in recent years are destabilizing the national agri-food systems that do not have an independent gene pool. At the time of this writing, the African swine fever (ASF) epizootic has spread to China, Vietnam, and the Philippines [4], and highly pathogenic avian influenza has caused colossal damage to the national food systems of Japan, South Korea, the Netherlands, France, Poland, and the Czech Republic [3]. If we take into account the fact that the standard set of measures to combat the spread of the above diseases, as a rule, includes the culling of the herd, it becomes obvious that the resumption of production requires large-scale replenishment of the livestock and is directly correlated with the objective possibility of access to the breeding base.

Often, the indicators of the productivity of genetics, which the vast majority of national food systems have, do not allow achieving the required level of profitability, which prompts agricultural enterprises to search for suppliers in the external market [9]. At the same time, the total dependence on imports of the breeding and seed base undermines the national, economic, and therefore food security of the state.

Solving the problem of physical and economic access to the genetic material of highly productive agricultural plants, animals and birds is an urgent need for many modern national food systems, which have determined food security as a development priority [9].

Guided by the purpose of the study, the author proposes to proceed to the consideration of the peculiarities of the resource provision of the Russian food system. Over the past twenty years, agriculture has evolved from one of the most backward sectors of the Russian economy into a large-scale agro-industrial complex capable of providing not only domestic consumption but also a significant export potential [8].

It is widely known that to maintain a high level of food self-sufficiency at the national level, countries need not only to develop crop production but also to have sustainable access to their planting material. The Food Security Doctrine of the Russian Federation, updated in early 2020, for the first time in the history of our state, identified self-sufficiency in seeds of major crops as one of the indicators of the state of the national food security system [1]. At the same time, it should be noted that in the text of this regulatory legal act, the assessment of the provision of seeds for domestic breeding crops is presented only in general form without breaking down into separate categories. This approach significantly reduces the quality characteristics of monitoring carried out for this indicator, since does not allow identifying the subsectors of national crop production that are most dependent on the import of planting material.

The study of customs data in the field of import of agricultural seeds by the Russian Federation showed a significant degree of dependence of the domestic agro-industrial complex on the import of planting material of sunflower and corn, the processing products of which are widely used in the production of fodder for the needs of animal husbandry [6].

At present, Russia is the fourth largest world importer of sunflower seeds and is in sixth place in terms of purchases of planting material for corn seeds. Figure 1 demonstrates that the cost of Russian imports of sunflower planting material from year to year shows a positive trend, corn—keeps at a fairly high level. The basis of supplies in the segment of sunflower seeds is made up of products from the USA and Turkey, corn—from Hungary and France [10].

A high degree of dependence in these segments indicates a low level of competitiveness of seeds of domestic selection. This circumstance, given the significant volumes of domestic consumption of sunflower oil, sunflower meal, and corn, as

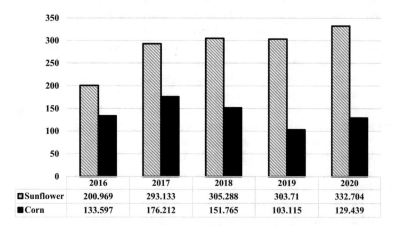

Fig. 1 Import of corn and sunflower seeds for sowing by the Russian Federation (USD million)
Source Compiled by the author based on [10]

well as the export of sunflower, corn, and their processed products to the foreign market, may further negatively affect the national food security of Russia.

According to forecasts of authoritative analytical agencies, in the next forty years, global climatic changes will lead to an increase in the number of Russian territories included in the agricultural turnover in Siberia and the Far East [7]. The growth of acreage will require expanding our country's access to the seed base of crop production. Accordingly, the relevance of the development by the Russian Federation of an independent genetic fund for agricultural oilseeds and forage crops will only increase. By historical standards, Russia does not have much time left to successfully solve this problem, avoid dependence on the changing situation on the global planting material market and take full advantage of its position.

With a further revision of the Doctrine of Food Security of the Russian Federation, according to the author, it is possible to single out oilseeds and forage crops as independent indicators of the country's food security, the level of self-sufficiency of which is maintained through state intervention.

In addition to self-sufficiency in seeds of the most cultivated forage and oilseeds, the Russian food system is characterized by an acute shortage of highly productive genetic material of its products for the needs of poultry and animal husbandry. The need to organize self-sufficiency with these resources is not reflected in the Food Security Doctrine of the Russian Federation [1]. Let's consider them in more detail.

Poultry meat is one of the largest Russian export categories, the volume of supplies of which to the external market is much higher than the export of other categories of meat products. Over the past five years, the value of poultry meat exports from the Russian Federation has grown by 72.5% and by the end of 2020 amounts to USD 427.3 million, pork—by 84.5% to USD 265.1 million, beef—by 89.0% to 87.78 million US dollars [10].

At the same time, when assessing the level of self-sufficiency in poultry meat, it is important to understand that successes in the development and optimization of domestic production of these products are accompanied by a catastrophic lack of self-sufficiency in hatching eggs of the poultry of highly productive breeds. At present, Russia is the world's largest net importer of hatching eggs, the import value of which was estimated at the US $ 208.6 million in 2020 (Fig. 2). The main suppliers of poultry hatching eggs to Russia are the Netherlands, Germany, and the Czech Republic, the poultry sectors of which at the beginning of 2021 were significantly affected by the epidemic of highly pathogenic avian influenza [4]. This circumstance had an extremely negative impact on the domestic poultry industry, which, in the face of increased demand, was forced to quickly diversify the sources of imports.

The situation is similar in the domestic pork industry, whose products are the second-largest export position of the Russian Federation in this category. According to the World Organization for Animal Health, the Russian Federation has not yet eradicated the ASF epizootic on its territory, which negatively affects the growth rate of exports in this direction [3]. The combination of these circumstances determined the country's dependence on the import of highly productive purebred breeding pigs for breeding. Although, according to the statements of representatives of the Russian

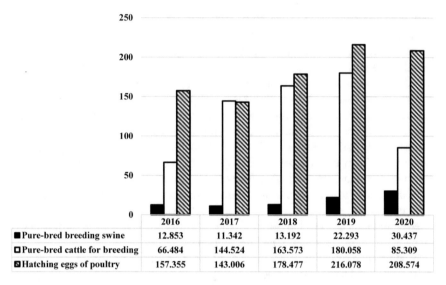

	2016	2017	2018	2019	2020
■ Pure-bred breeding swine	12.853	11.342	13.192	22.293	30.437
□ Pure-bred cattle for breeding	66.484	144.524	163.573	180.058	85.309
▨ Hatching eggs of poultry	157.355	143.006	178.477	216.078	208.574

Fig. 2 Imports by the Russian Federation of hatching eggs of poultry, pure-bred pigs, and cattle for breeding (USD million) *Source* Compiled by the author based on [10]

branch unions of industry enterprises, there is no shortage of breeding animals for breeding in the country, the quality of domestic breeding material in its characteristics is still significantly inferior to foreign analogues.

The above characteristic of the domestic pig industry is confirmed by customs statistics, according to which, by the end of 2020, Russia is the sixth-largest importer of purebred breeding pigs in the world in terms of supply value, behind China, Vietnam, and Poland in terms of this indicator. The bulk of purchases fall on products from Canada and Denmark [10].

The Russian sector of cattle is also critically dependent on the import of foreign genetics, largely due to the emergence of this sector of the domestic agro-industrial complex. Cattle production is the most difficult type of activity in comparison with poultry and pig breeding and requires the enterprises of the industry to organize a developed technological base. The Russian Federation ranks third among the importers of purebred cattle for breeding and is most dependent on the supply of genetic material of this species from Germany and Denmark. The cost of Russian imports for this product category at the end of 2020 amounted to 85.3 million US dollars [10].

Based on the foregoing, it can be concluded that the organization of sustainable import substitution by the breeding base of poultry, pigs, and cattle is an urgent direction in the development of the concept of the agro-industrial complex of the Russian Federation. The stability of the Russian food system will depend on the implementation of this approach in the future. The development of domestic genetics of highly productive poultry and pigs is a priority task. Import substitution with breeding cattle, due to the greater resource intensity and labor costs of this process, can be attributed to long-term goals.

In the author's opinion, the goals of self-sufficiency in poultry hatching eggs and pig breeding base should be included in the Food Security Doctrine of the Russian Federation.

4 Conclusion

Based on a comprehensive study of the degree of dependence of the Russian food system on the import of genetic material from agricultural plants, animals, and birds, the author concluded that to maintain the national food security of the Russian Federation in the future, it will be necessary to transform the system of access to the resource supply of the agro-industrial complex.

Against the background of the destruction of production chains, the transformation of commodity flows, and the rise in prices for logistics costs, the availability of economic access to a highly productive seed and breeding base for the needs of agricultural production only through imports is not a guarantee of stable provision of the national agro-industrial complex with these resources.

The Russian side needs to identify the benchmark points of the existing disproportions in the field of self-sufficiency in fodder crops, genetics of the main farm animals and birds, and also to legislate the threshold for self-sufficiency in these categories. Otherwise, the growth of world prices in the segments of the fodder base, hatching eggs, and purebred breeding animals for breeding may level out the achievements of the last twenty years, during which agriculture has become a strategically important branch of the Russian [8].

The further genesis of the domestic agro-industrial complex requires that such production resources as highly productive oilseeds and forage crops, as well as the genetics of farm animals and poultry, be supplied to the domestic market primarily through self-sufficiency, and not by import. Compliance with this condition will allow the Russian agro-industrial complex to reach a new level of development and lay a solid foundation for ensuring national food security.

References

1. Decree of the President of the Russian Federation dated January 21, 2020, No. 20 On the approval of the Doctrine of food security of the Russian Federation. Official Internet portal of the President of the Russian Federation. http://kremlin.ru/acts/bank/45106. Accessed 19 March 2021
2. FAO Statistical Yearbook-2020 Official Internet portal of FAO. http://www.fao.org/3/cb1329en/CB1329EN.pdf. Accessed 19 May 2021
3. Global control of African swine fever: A GF-TADs initiative (2020–2025) OIE Official Internet Portal. https://www.oie.int/fileadmin/Home/eng/Animal_Health_in_the_World/docs/pdf/ASF/ASF_GlobalInitiative_Web.pdf. Accessed 19 May 2021

4. OIE situation reports for avian influenza The OIE Official Internet Portal. https://www.oie.int/fileadmin/Home/eng/Animal_Health_in_the_World/docs/pdf/ASF/ASF_GlobalInitiative_Web.pdf. Accessed 19 May 2021
5. Opinion of the CPC Central Committee and the State Council of the People's Republic of China on comprehensive assistance to the revival of the countryside and accelerating the modernization of agriculture. Official Internet portal of the State Council of the People's Republic of China. http://www.gov.cn/xinwen/2021-03/06/content_5590842.htm. Accessed 27 Mar 2021
6. Seed market in Russia 2020: research and forecast until 2024. RIOF EXPERT. Official Internet portal of RBC. https://www.trademap.org/Index.aspx. Accessed 18 Mar 2021
7. The Demeter 2021. Produce and feed: the daily challenge of a confused world. Official Internet portal of The Demeter Club. https://www.clubdemeter.com/fr/le-demeter. Accessed 18 Mar 2021
8. Tikhomirova VA (2020) Implementation of the doctrine of food security in Russia: assessment and prospects: Bulletin of the peoples' friendship University of Russia. Book Ser Econ 28(4):751–764
9. Tikhomirova VA (2019) Ensuring food security: international and Russian experience. Ph.D. in Economics thesis. 08.00.14; RUDN. 62–99, p 190
10. Trade Statistic for International Business Development. Official Internet portal of the International Trade Center. https://www.trademap.org/Index.aspx. Accessed 18 Mar 2021

Biochemical Indicators of Grain Sorghum Varieties in the Rostov Region

Olesya A. Nekrasova⬤, Elena V. Ionova⬤, Nina S. Kravchenko⬤, Natalya G. Ignateva⬤, and Vladimir V. Kovtunov⬤

Abstract The purpose of the current study is to estimate the grain sorghum varieties according to the main biochemical indicators to use them in the further breeding process aimed at improving the grain quality. The objects of the study are 15-grain sorghum varieties. Based on the conducted biochemical analysis of grain sorghum, there were identified two-grain sorghum varieties 'ZSK 443/16' and 'Zernogradskoe 204/4'. The two-factor variance analysis determined that Factor B ('year of study') had the main effect on raw protein content in the kernel (47.2%), and Factor A ('genotype') had a slighter effect on it (15.9%). The interaction between the factors was 14.3%. The interaction of Factors A and B had the greatest effect on the formation of starch in the kernel (47.3%), while there were 26.2% of genotypic variability (Factor A) and 15.3% belonged to the year of study (Factor B). Factor A ('genotype') produced a greater effect on the tannin content in the kernel (91.0%), the interaction of weather conditions with genotype was 7.5%. Factor B ('year of study') had 0.6% of the effect on the total variation of the trait.

Keywords Grain · Sorghum · Protein · Starch · Oil · Tannin

JEL Codes G1 · G180

O. A. Nekrasova (✉) · E. V. Ionova · N. S. Kravchenko · N. G. Ignateva · V. V. Kovtunov
Agricultural Research Center "Donskoy", Zernograd, Russia
e-mail: nekrasova_olesya@rambler.ru

E. V. Ionova
e-mail: ionova-ev@yandex.ru

N. S. Kravchenko
e-mail: ninakravchenko77@mail.ru

N. G. Ignateva
e-mail: ninakravchenko78@mail.ru

V. V. Kovtunov
e-mail: kowtunov85@mail.ru

1 Introduction

Sorghum is a unique cereal plant, both in terms of its biological characteristics and economically valuable traits. The main advantages of the grain crop are high and stable annual productivity, resistance to abiotic and biotic factors, and versatility of utilization.

Sorghum is the most important fodder, technical, and food crop. Sorghum grain is considered fodder, that is, its main purpose is to produce feed for husbandry. In this regard, the appropriateness of its use depends on the quality indicators, digestibility, and nutritional value.

The feed quality greatly depends on the amount and ratio of various chemical elements that make up its dry matter. This includes raw protein (nitrogen substances), nitrogen-free extractive substances (oils, fiber, starch, etc.), minerals (phosphorus, calcium, potassium, iodine, etc.), and vitamins (A, B, C, D, etc.). The indicators of the chemical composition of feed-in modern conditions are the basis for estimating their nutritional value.

In the course of a breeding process aimed at developing sorghum varieties and hybrids that meet the requirements of the present agriculture, an important thing is to study thoroughly the ability of initial material to form highly productive grain of good quality under certain soil and weather conditions.

The purpose of the current study is to estimate the grain sorghum varieties according to the main biochemical indicators to use them in the further breeding process aimed at improving the grain quality.

2 Materials and Methods

The current paper has used the research works on the study of the characteristics and economically valuable traits of grain sorghum, indicators of grain quality of crops [1–15].

The objects of the study were 15-grain sorghum varieties. The study was carried out on the experimental plots of the laboratory grain sorghum breeding and seed production of the FSBSI Agricultural Research Center "Donskoy".

Grain quality indicators were estimated according to generally accepted methods: raw protein percentage in the kernel was estimated by the Kjeldahl method; starch content in the kernel was estimated by the Evers polarimetric method; tannin content in the kernel was estimated by the method, based on the reaction of polyphenols with vanillin at HCl; raw oil in the kernel was estimated by the method of S.V. Rushkovsky according to the amount of oil-free residue.

Mathematical and statistical data processing was carried out according to the method of B. A. Dospekhov.

The year 2018 was characterized by an insufficient amount of precipitation (-38.6 mm to the norm) in May, which harmed the field germination and strength of sorghum seed growth. June and August were also characterized by insufficient precipitation (−67.1 mm and − 40.4 mm to the norm, respectively), which together with the increased temperature regime in the summer months (+3.4 °C to the norm in June, + 2.8 °C to the norm in July, and + 2.7 °C to the norm in August) negatively affected panicle formation and flowering. Precipitation in July (+14 mm to the norm) contributed to better kernel formation and its filling.

In May 2019, good soil moisture supply (+12.6 mm to the norm) and high temperature (+2.5 °C to the norm) positively influenced the field germination and strength of sorghum seed growth. Insufficient precipitation in June (−60.5 mm to the norm) and in the first decade of July (4.2 mm of precipitation in the first decade) and an increased temperature regime in June (+4.7 °C to the norm) had a negative impact on panicle formation and flowering. Precipitation in the second and third decades of July (+14 mm to the norm) contributed to better kernel formation and its filling. The shortage of precipitation (−31.6 mm to the norm) and high temperature (+1.5 °C to the norm) in August negatively influenced kernel formation.

The moisture availability in soil (+28.6 mm to the norm) in May 2020 had a positive effect on the field germination of sorghum seeds, at the same time, the low air temperature (−1.1 °C to the norm) had a negative effect on the rate of initial growth. Shortage of precipitation in June (−32.5 mm to the norm), as well as the increased temperature regime (+2.6 °C to the norm), had a negative impact on the development of plants. Precipitation in July and August was at the average long-term level (57.7 mm and 45.2 mm, respectively), which positively influenced panicle and kernel formation.

3   Results and Discussion

One of the important issues of sorghum breeding is the development of varieties and hybrids with a high protein percentage. Protein percentage in the kernel is of decisive importance in characterizing the nutritional and feed qualities of the variety.

Through the years of study, protein percentage in kernel varied from 11.5% (the variety 'Luchistoe') to 12.9% (the variety 'Zernogradskoe 88') (Fig. 1).

All studied varieties were characterized by the mean protein percentage in the kernel (10.6–13.0%). There were identified the samples 'ZSK 443/16', 'Zernogradskoe 204/4', 'Lazurit 601/16' (12.5% each), and 'Zernogradskoe 88' (12.9%) which produced the largest amount of protein in the kernel.

To determine the proportion of the effect of 'genotype' and 'year of study' on protein percentage in the kernel there was conducted a two-factor variance analysis. There has been identified that the values of F_{fact} of the factors 'genotype', 'year of study, and their interaction exceeded the value of F_{theor}, which allows considering the obtained values in the trial as reliable. The effect of the factors was unequal. Factor B ('year of study') had the main effect on raw protein content in the kernel (47.2%), and Factor A ('genotype') had a slighter effect on it (15.9%). The interaction between the factors was 14.3% (Table 1).

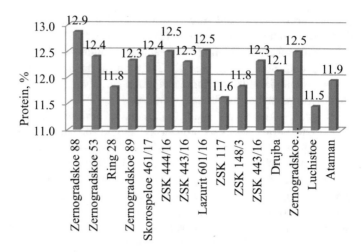

Fig. 1 Distribution of the grain sorghum varieties according to raw protein percentage in the kernel, 2018–2020. *Source* Developed and compiled by the authors

Table 1 Two-factor variance analysis of the grain sorghum varieties according to raw protein percentage in kernel

Source of variance	Sum of squares	Freedom degree	Variance analysis	F_{fact}	F_{tab095}	Effect, %
Factor A	11.2	14	0.8	4.4	1.9	15.9
Factor B	37.5	2	18.8	103.9	3.2	47.2
Interaction AxB	12.6	28	0.5	2.5	1.8	14.3

Source Developed and compiled by the authors

Most kernel endosperm is starch, it is the main biochemical indicator that characterizes grain quality of grain sorghum intended in the production of food starch.

The mean starch content through the years of study varied from 71.2% (the variety 'ZSK 444/16') to 73.5% (the variety 'Zernogradskoe 204/4') (Fig. 2). All sorghum varieties had high starch content in the kernel (71–75%).

The largest starch content was identified in the varieties 'ZSK 148/3' (73.1%), 'ZSK 443/16' (73.2%), and 'Zernogradskoye 204/4' (73.5%).

The results of two-way variance analysis according to starch content in the kernel, presented in the table, show that the variances of genotypic variability, the variability caused by the conditions of the year of study, and the interaction of these two factors are reliable (Table 2).

The interaction of the factors 'genotype' and 'year of study' had the greatest effect on the formation of starch in the kernel (47.3%), while there were 26.2% of genotypic variability (Factor A) and 15.3% belonged to the year of study (Factor B).

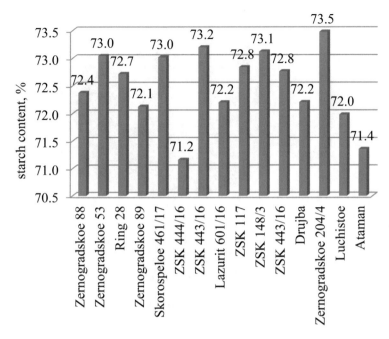

Fig. 2 Distribution of the grain sorghum varieties according to starch content in the kernel, 2018–2020. *Source* Developed and compiled by the authors

Table 2 Two-factor variance analysis of the grain sorghum varieties according to starch content in kernel

Source of variance	Sum of squares	Freedom degree	Variance analysis	F_{fact}	F_{tab095}	Effect, %
Factor A	35.6	14	2.5	48.5	1.9	26.2
Factor B	20.5	2	10.2	195.7	3.2	15.3
Interaction AxB	64.1	28	2.3	43.8	1.8	47.3

Source Developed and compiled by the authors

The sorghum seed coat contains tannins, which can reduce the quality of the starch during milling.

The variation of the indicator in the grain sorghum varieties ranged from 0.15% (the variety 'Ataman') to 7.57% (the variety 'Zernogradskoe 53').

Distribution analysis showed that nine varieties had a low tannin content in the kernel, four varieties had mean tannin content in the kernel (1.0–2.0%), one sample had a high tannin content in the kernel (more than 2.0%) (Fig. 3).

There have been identified the low-tannin samples 'ZSK 443/16' (0.18%), 'Zernogradskoe 204/4' (0.17%), and 'Ataman' (0.15%).

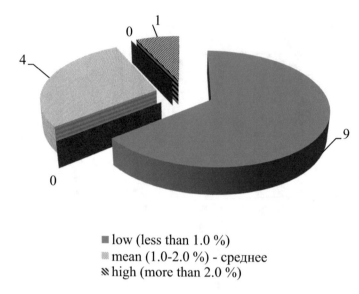

■ low (less than 1.0 %)
▥ mean (1.0-2.0 %) - среднее
▨ high (more than 2.0 %)

Fig. 3 Distribution of the grain sorghum varieties according to tannin content in the kernel, 2018–2020. *Source* Developed and compiled by the authors

The results of the two-factor variance analysis of tannin content in the kernel, presented in Table 3, allow considering the values obtained in the trial as reliable.

Factor A ('genotype') had a greater effect on the tannin content in the kernel (91.0%), the interaction of weather conditions with genotype was 7.5%. Factor B ('year of study') had 0.6% of the effect on the total variation of the trait.

Raw oil is a reserve that is concentrated energy and building reserve for seeds.

Raw oil content in the grain sorghum varieties varied from 3.0% (the variety 'Ataman') to 4.6% (the variety 'Zernogradskoe 88'). The largest raw oil content was identified in the samples 'Zernogradskoe 204/4', 'ZSK 443/16' (4.2% each), and 'Zernogradskoe 88' (4.6%). There has been recommended to utilize these varieties as sources of high raw oil content in the breeding process to grow sorghum varieties and hybrids for feed.

Table 3 Two-factor variance analysis of the grain sorghum varieties according to tannin content in kernel

Source of variance	Sum of squares	Freedom degree	Variance analysis	F_{fact}	F_{tab095}	Effect, %
Factor A	194.5	14	13.9	469.6	1.9	91.0
Factor B	1.2	2	0.6	20.9	3.2	0.6
Interaction AxB	16.1	28	0.6	19.4	1.8	7.5

Source Developed and compiled by the authors

Table 4 Correlation between biochemical indicators of the grain sorghum varieties

	Raw protein,%	Starch,%	Tannin,%	Raw oil,%
Raw protein,%	1	0.04	−0.02	0.42
Starch,%	0.04	1	0.34	0.20
Tannin,%	−0.02	0.34	1	−0.28
Raw oil,%	0.42	0.20	−0.28	1

Source Developed and compiled by the authors

Table 5 Biochemical indicators of the grain sorghum varieties, 2018–2020

Variety	Indicator			
	Raw protein,%	Starch,%	Tannin,%	Raw oil,%
ZSK 443/16	12.5	73.2	0.18	4.2
Zernogradskoe 204/4	12.5	73.5	0.17	4.2

Source Developed and compiled by the authors

There has been conducted a correlation analysis to establish interaction between the studied traits. The analysis showed that raw protein in the kernel of the grain sorghum varieties was in an average positive correlation with raw oil content ($r = 0.42 \pm 0.17$) (Table 4).

There has been identified an average positive correlation between starch and tannin content in the kernel ($r = 0.34 \pm 0.16$). The correlation between starch and raw oil content was weakly positive ($r = 0.20 \pm 0.12$).

There was a weak negative correlation between tannin and raw oil content ($r = -0.28 \pm 0.14$).

According to the study of biochemical indicators, there were identified two-grain sorghum varieties 'ZSK 443/16' and 'Zernogradskoe 204/4', the characteristics of which are presented in Table 5.

These varieties have been recommended for use in the breeding process as the sources of these traits.

4 Conclusions

1. Due to the biochemical analysis of grain sorghum, there have been identified two-grain sorghum varieties 'ZSK 443/16' and 'Zernogradskoe 204/4'.
2. The two-factor variance analysis has determined that Factor B ('year of study') had the main effect on raw protein content in the kernel (47.2%), and Factor A ('genotype') had a slighter effect on it (15.9%). The interaction between factors was 14.3%.

3. The interaction of Factors A and B had the greatest effect on the formation of starch in the kernel (47.3%), while there were 26.2% of genotypic variability (Factor A) and 15.3% belonged to the year of study (Factor B).
4. Factor A ('genotype') had a greater effect on the tannin content in the kernel (91.0%), the interaction of weather conditions with genotype was 7.5%. Factor B ('year of study') had 0.6% of the effect on the total variation of the trait.

References

1. Anisimova IN, Ryabova DN, Malinovskaya EV, Alpatieva NV, Karabitsina YuI, Radchenko EE (2017) Polymorphism according to the traits associated with the CMS-rf genetic system in grain sorghum from the VIR collection. Agric Biol 52(5):952–963. https://doi.org/10.15389/agrobiology.2017.5.952rus
2. Astashov AN (2009) Sorghum as a component of compound feed for broiler chickens. Maize Sorghum 5:13–14
3. Bibi A, Sadaqat HA, Akram HM, Mohammed MI (2010) Physiological markers for screening sorghum (Sorghum bicolor) germplasm under stress conditions. Int J Agric Biol Sci 12(3):451–455
4. Ionova EV, Alabushev AV (2009) Mechanisms of adaptation of grain sorghum plants and biological substantiation of the use of an electromagnetic field of ultrahigh-frequency (EMF UHF). JSC "Rostizdat", Rostov-on-Don, 12
5. Ivanisov MM, Marchenko DM, Nekrasov EI, Rybas IA, Romanyukina IV, Chukhnenko YuYu, Kravchenko NS (2020) Comparative estimation of winter bread wheat varieties in an inter-station trial due to their quality indicators. Grain Econ Rus 4(70):14–18. https://doi.org/10.31367/2079-8725-2020-70-4-14-18
6. Kamthan A, Chaudhuri A, Kamthan M, Datta A (2016) Genetically modified (GM) crops. Milest New Adv Crop Improv Theor Appl Genet 129:39–55
7. Kibalnik OP (2018) Combining ability of CMS-lines of grain sorghum based on A1, A2, A3, A4, 9E, and M-35–1A types of cytoplasmic male sterility. Vavilov J Genet Breed 21(6):651–656. https://doi.org/10.18699/VJ17.282
8. Kovtunov VV, Kovtunova NA, Lushpina OA, Sukhenko NN, Shishova EA, Kravchenko NS (2020) The study of East African grain sorghum samples in the conditions of the Rostov region. Grain Econ Russ 6(72):39–44. https://doi.org/10.31367/2079–8725–2020–72–6–39–44
9. Moreya RS, Hashida Y, Ohsugia R, Yamagishia J, Aoki N (2018) Evaluation of the performance of sorghum varieties grown in Tokyo for sugar accumulation and its correlation with vacuolar invertase genes SbInv1 and SbInv2. Plant Produc Sci 21(4):328–338. https://doi.org/10.1080/1343943X.2018.1510737
10. Nekrasov EI, Marchenko DM, Ivanisov MM, Rybas IA, Grichanikova TA, Romanyukina IV, Kopus MM (2019) Estimation of productivity and grain quality of winter soft wheat varieties in the Rostov Region. Taurida Herald Agrar Sci 4(20):79–85. https://doi.org/10.3352/2542-0720-2019-4-20-79-85
11. Schober TJ, Bean SR (2008) Sorghum and maize. Gluten Free Cere Produc Beverages 1:101–118. https://doi.org/10.1016/B978-012373739-7.50007-1
12. Serna Saldivar SO (2016) Cereals: types and composition. Encycl Food Health 718–723https://doi.org/10.1016/B978-0-12-384947-2.00128-8

13. Shoemaker CE, Bransby DI (2010) The role of sorghum as a bioenergy feedstock. Proceed Sustain Feedstocks Adv Biofuels Workshop 1:149–159
14. Starchak VI, Kibalnik OP, Kameneva OB (2016) The study of the initial material of grain sorghum according to biochemical composition. Maize Sorghum 3:33–36
15. Tan JT, Shang YuF (2011) Biorefinery Engineering. Comprehensive Biotechnology (Second Edition) 2:815–828. https://doi.org/10.1016/B978-0-08-088504-9.00138-0

Technology for Sustainable Agriculture

Study of the Labor Resources of Peasant (Farm) Households by Production Type

Anna V. Ukolova(ID) and **Bayarma Sh. Dashieva**(ID)

Abstract The paper presents the research on highlighting the production types of peasant (farm) households and the study of labor resources in their context. The research subject is a system of statistical indicators. The research object is peasant (farm) households in three regions with different agroclimatic conditions. The information base of the research is the microdata form 1-KFH "Information on productive activities of the heads of peasant (farm) households—individual entrepreneurs." The authors use the grouping method. The scientific novelty of this work consists in the theoretical development and testing of the methodology of allocation of production types of peasant (farm) households and analysis of labor resources by them, which will allow allocating priority directions for the development of small business, labor potential of agriculture, and rural areas. When developing the methodology, the authors studied the experience of the European Union, where the typology developed with the consideration of the specialization of farms was officially approved. The paper characterizes the most common production types at the regional level. Types of peasant (farm) households differ in size, level of intensification, and efficiency of agricultural production. The largest number of permanent workers (including members of peasant (farm) households) per farm and unit of land area and the largest share of hired labor costs in the structure of total costs (from 8 to 11% in the studied regions) are observed in farms with the production direction "dairy cattle breeding." In turn, labor productivity, evaluated by the total income of peasant (farm) households, in each region is the lowest compared to other types. Considering the unresolved tasks of reaching self-sufficiency thresholds for milk and dairy products (in terms of milk), vegetables and melons, fruits and berries, the increased government support and measures to develop small businesses in the labor-intensive sectors of agriculture will help to solve the problems of food security, increase employment, and preserve the rural way of life.

A. V. Ukolova · B. Sh. Dashieva (✉)
Russian State Agrarian University – Moscow Timiryazev Agricultural Academy, Moscow, Russia
e-mail: dashieva.b.sh@rgau-msha.ru

A. V. Ukolova
e-mail: statmsha@rgau-msha.ru

© The Author(s), under exclusive license to Springer Nature Singapore Pte Ltd. 2022   229
E. G. Popkova and B. S. Sergi (eds.), *Sustainable Agriculture*,
Environmental Footprints and Eco-design of Products and Processes,
https://doi.org/10.1007/978-981-16-8731-0_23

Keywords Peasant (farm) household · Grouping method · Production type · Agriculture · Human resources · Typology of peasant (farm) households

JEL codes J43 · Q12 · O13

1 Introduction

Due to the high differentiation of peasant (farm) households (P(F)Hs), the study of labor resources, which are one of the main factors in the development of agricultural production, should be conducted using the method of grouping, including typological grouping. The development of a typology taking into account their specialization is a continuation of the study of the differentiation of P(F)Hs. In previous studies [8], the authors have constructed groupings of P(F)Hs by income using a lognormal distribution for three subjects of the Russian Federation belonging to different groups of Russian regions in terms of agro-climatic conditions determining the specialization of agricultural production, which, in turn, determines the placement of labor resources [2]. The conclusion indicates the possibility of enlarging the analytical groups and establishing uniform boundaries of the income of P(F)Hs for the entire territory of Russia, as is done in the EU in the typology by economic classes.

Another direction of P(F)Hs typification is grouping by production direction. Based on the study of the experience of foreign countries with developed agriculture and the research of Russian scholars, the authors developed a methodology for identifying the production types of collective farms. The testing was conducted on the same subjects of the Russian Federation as in [8] using the annual form 1-KFH "Information on productive activities of the heads of peasant (farm) households—individual entrepreneurs," provided by the recipients of government subsidies to the Ministry of Agriculture of the Russian Federation. The authors understand the production type of collective farms as a group of P(F)Hs with similar production specialization, level of intensity, and production efficiency.

Several Russian scientists have noted the need to distinguish the production types of collective farms. Zubrenkova and Fedotova [9] indicate that "under market conditions, the forms of economic specialization of small farms are variable and unstable, which makes it difficult to typify them. However, it is possible to identify the productive directions and highlight the most typical (frequently occurring) farms from the whole population in certain socio-economic conditions." Itskovich [3] states that "with the help of state regulation, the allocation of production types will contribute to eliminating organizational and production disproportions of collective farms and overcoming their plant-growing one-sidedness." He proposes to distinguish the following production types of P(F)Hs:

- Farms producing cash crops (grains, sunflowers, sugar beets, potatoes, etc.);
- Feed production farms that keep dairy cattle;
- Specialized livestock farms for fattening livestock and poultry;
- Farms focused on the production of vegetables, fruits, and berries;
- Mixed farms.

In contrast to Russia, Germany [7] uses several classifications of agricultural enterprises in conducting and processing the results of the agricultural census, including by production area. Production direction (BWA) is determined based on the industry structure of standardized output (Standardoutputs (SO)). Agricultural enterprises are assigned to one or another production area according to the share of the industry in the total standardized output. The individual production areas (53 Einzel-BWA classes) are combined into 20 classes of the main production area (Haupt-BWA), which, in turn, are aggregated to form nine classes of the general production area (Allgemeine BWA). According to the general direction of production (Allgemeine BWA), the following is distinguished:

- Three classes of crop enterprises specializing in field farming, intensive crop farming, and cultivation of perennial plantations;
- Two classes of livestock enterprises specializing in pasture cattle and fodder production and intensive livestock breeding;
- Three classes of enterprises with a specialization in mixed crop production, mixed livestock production, and crop-livestock enterprises;
- One class of enterprises not covered by the classification system, whose standardized output is 0.

In the UK, the farm typology is based on the Farm Business Survey (FBS) conducted by Rural Business Research, an academic consortium of six university research centers. The study is commissioned by the Department of Environment, Food and Rural Affairs (Defra) and supported by farmers' unions. The results of the study are an authoritative source of information on the economic situation of agricultural enterprises. These results are used in agricultural and environmental policy decision-making and are intended to meet the needs of farmers, government, government partners, public associations, and researchers. On a public website, farmers can identify the type their farm belongs to, anonymously compare the performance of their farm to the average by type or the most efficient farms in their region, and take the findings into account when planning their business.

The study is conducted by sampling the general population formed during the agricultural census. The sample is representative at the national level by type, farms size, and territorial location. More than 2300 farms in England and Wales with revenues of at least €25,000 participate in the FBS survey each year. FBS is a panel survey that retains about 93% of the sample each year.

The classification of farms is based on the EU typology. First, ten aggregated groups are distinguished:

- Cereals farms;
- General cropping farms;
- Horticulture farms;
- Specialized pig farms;
- Specialized poultry farms;
- Dairy farms;
- Farms grazing in "less-favored areas" (LFA);

- Lowland grazing livestock;
- Mixed farms;
- Others (including non-classifiable).

The aggregated groups are further subdivided into 21 basic types. The published reports for nine regions of the country provide unique information detailed by farm type [6].

The Ministry of Agriculture of the Russian Federation has defined the priorities for developing the agro-industrial complex for 2021–2023. One of the priorities in almost all regions of Russia is the development of small-scale farming. Following the order of the Ministry of Agriculture of Russia, priority sub-industries of crop and livestock production are identified for each subject of the Russian Federation.

In the Republic of Buryatia, these sub-industries are as follows:

- Specialty beef cattle breeding;
- Sheep and goat breeding;
- Production of fruit and berry plantations, including planting material, planting, and care of perennial plantations.

In the Lipetsk region, the sub-industries are as follows:

- Production of grain and leguminous crops;
- Production of oilseeds (excluding rape and soybeans);
- Production of fruit and berry plantations, including planting material, planting, and care of perennial plantations;
- Production of milk;
- Development of specialized beef cattle breeding.
- In the Stavropol Territory, the sub-industries are as follows:
- Production of grain and leguminous crops;
- Development of viticulture;
- Production of fruit and berry plantations, including planting material, planting, and care of perennial plantations;
- Sheep and goat breeding.

Thus, it is necessary to introduce the same approach of classifying farms by production type in the Russian Federation. This approach will allow studying each type of farm separately and see the current economic situation in agriculture and the prospects for developing a particular branch.

2   Materials and Methods

Groupings of P(F)Hs were performed in three subjects of the Russian Federation (Republic of Buryatia, Lipetsk Region, and Stavropol Territory) marked with different agroclimatic conditions. These regions were chosen as typical representatives of the three groups of regions, identified by the average sum of temperatures for the period with temperatures above 10 °C: up to 1750, 1750–2600, and above 2600 °C [2]. The Republic of Buryatia is included in the first group of regions with unfavorable natural and climatic conditions. The Lipetsk Region is included in the second group with more favorable conditions for agriculture compared to the first group. The Stavropol Territory is included in the third group with the most favorable agro-climatic conditions but with a predisposition to the formation of medium and strong droughts.

The identification of production types in this paper is carried out according to the data of the departmental form of the Ministry of Agriculture of Russia 1-KFH "Information on productive activities of the heads of peasant (farm) households— individual entrepreneurs." This form shows the structure of revenues from the sale of agricultural products. We excluded the farms that had no land and livestock at the same time and the farms that did not sell agricultural products.

In the classification, a branch was recognized the main if the share of one or another type of product exceeded 50% of the total agricultural production proceeds. A P(F)H was classified as a mixed farm if there was no primary type of production with a share of revenues over 50% and the share of crop or livestock production did not exceed 66%. Otherwise, the farm was recognized as belonging to mixed crop farming or mixed livestock farming. The census of 66% is used in accordance with the recommendations of OKVED-2 for mixed agriculture. "If the gross profit from crop or livestock production is 66% or more of the standard gross profit, the mixed farming activity must be included in crop or livestock production."

As a result of the allocated production types by regions, we conducted a general economic characteristic of the P(F)H and a comprehensive characteristic of labor resources.

3 Results

As a result of the grouping of P(F)Hs by production type, we identified 13 groups in the Republic of Buryatia and the Lipetsk Region and 14 groups in the Stavropol Territory (Table 1). The main types with at least 5% representation of all farms in the region were selected from the obtained groups. Their general economic characteristics were given (Table 2). The characteristic of the labor resources of collective farms is also presented (Table 3).

In the Republic of Buryatia (Tables 2 and 3), the farms specializing in beef cattle breeding (202 farms or 41%) hold the largest share. On average, one farm in this group has 134 hectares of land area (including 17 hectares of crops), 2.1 permanent employees and members of the farm, 154 conditional heads of animals (including 114 heads of cattle), and 3.5 units of agricultural equipment. The level of marketability is low (37.3%). The annual revenue from the sale of livestock products per one head of livestock was 5 thousand rubles. The share of subsidies in the total income of P(F)Hs is 27.6%. Less than half of all P(F)Hs use hired labor. Thus, on average,

Table 1 Production types of P(F)Hs by three constituent entities of the Russian Federation for 2019

Production types of P(F)Hs by constituent entities of Russia		
Republic of buryatia	Lipetsk region	Stavropol territory
Dairy cattle breeding	Dairy cattle breeding	Dairy cattle breeding
Beef cattle breeding	Beef cattle breeding	Beef cattle breeding
Horse breeding	Sheep breeding	Sheep breeding
Sheep breeding	Intensive livestock farming	Intensive livestock farming
Intensive livestock farming	Grain farming	Mixed cattle farming
Mixed cattle farming	Oilseed production	Grain farming
Grain farming	Beet farming	Oilseed production
Potato farming	Potato farming	Vegetable growing and production of gourds
Vegetable farming	Vegetable farming	Beet farming
Fodder production	Fodder production	Potato farming
Mixed crop production	Mixed crop production	Fodder production
Perennial plants	Perennial plants	Mixed crop production
Mixed farming	Mixed farming	Perennial plants
–	–	Mixed farming

Source Compiled by the authors based on [5]

there is one hired worker for each P(F)H member, and labor costs account for 6.3% of total expenses. Labor productivity is 737 thousand rubles per employee. Gross mixed income is only 44 thousand rubles per employee per year.

The second most numerous group of farms in the Republic of Buryatia is the collective farms specializing in dairy cattle farming (128 farms or 26%). The marketability and proceeds from the sale of livestock products per one head of livestock are slightly higher than in the previous group (43.5% and 13.5 thousand rubles, respectively). The share of subsidies in total revenues is lower by 7.8%. Compared with P(F)Hs specializing in beef cattle breeding, labor productivity of this sub-industry is slightly lower and amounts to 646 thousand rubles per employee. This group of farms hire workers and uses hired labor to a greater extent (71.1% of all farms). There are 1.8 hired workers per one member of the P(F)H. On average, there are three workers per farm, which is almost one worker higher than on farms focused on beef cattle breeding.

The third group of farms specializing in sheep breeding includes 60 P(F)Hs (12% of all P(F)Hs of the Republic of Buryatia). This group of farms, along with beef cattle breeding, belongs to one of the sub-industries identified by the Ministry of Agriculture of the Russian Federation as a priority for development. The average number of sheep per farm is 786 heads; income is 1624 thousand rubles; the share of subsidies is 16.8%. The level of marketability is one of the lowest by types—22.3%. Productivity per year does not exceed 1 million rubles per employee. The share of

Table 2 Indicators of size, efficiency, and marketability of collective farms by production type for 2019

Production types of P(F)Hs	Republic of Buryatia			Lipetsk region					Stavropol territory			
	Dairy cattle breeding	Beef cattle breeding	Sheep breeding	Dairy cattle breeding	Beef cattle breeding	Grain farming	Oilseed production	Mixed crop production	Dairy cattle breeding	Beef cattle breeding	Sheep breeding	Grain farming
Number of P(F)Hs	128	202	60	45	48	147	46	20	117	128	221	1202
Per one P(F)H: Income, thousand rubles	1909	1562	1624	6343	7717	15,448	10,016	29,506	2836	3596	2629	9872
Including subsidies	378	431	273	1063	1134	334	14	69	310	540	320	101
Gross mixed income, thousand rubles	95	93	84	1108	1034	2157	1619	6730	442	551	477	1900
Total land area, hectares	111	134	149	83	80	500	351	661	106	167	372	505
Including crops	18	17	11	62	23	441	289	627	32	32	34	358
Average annual number of animals, heads	88	154	190	62	77	15	0	8	56	124	238	11

(continued)

Table 2 (continued)

Production types of P(F)Hs	Republic of Buryatia			Lipetsk region					Stavropol territory			
	Dairy cattle breeding	Beef cattle breeding	Sheep breeding	Dairy cattle breeding	Beef cattle breeding	Grain farming	Oilseed production	Mixed crop production	Dairy cattle breeding	Beef cattle breeding	Sheep breeding	Grain farming
Availability of agricultural machinery, pcs	4.4	3.5	4.8	2.2	1.5	8.6	7.0	13.4	1.6	1.1	0.8	8.2
Proceeds from product sales, thousand rubles												
Crop production per 1 hectare of sown area	1.7	1.2	1.0	7	9	30	32	46	18	24	28	30
Livestock per 1 head of livestock	14	5	4	54	82	17	20	16	45	23	11	18
Material costs per 100 rubles of total income, rubles	51	39	53	32	33	45	51	46	40	39	58	48
Marketability of agricultural products, %	43.5	37.3	22.3	58.2	142.8	81.6	91.6	94.2	62.4	75.8	60.8	81.9

Source Compiled by the authors based on [5]

Table 3 Indicators characterizing labor resources by production type of P(F)Hs for 2019

Production types of P(F)Hs	Number of P(F)Hs	Number of permanent employees of P(F)Hs, people		Number of employees per one member of the P(F)H, people	Share of farms hiring workers, %	Share of labor costs in total expenditures, %	Per 1 permanent employee or member of the P(F)H, thousand rubles		
		Per farm	Per 100 hectares of land area				Income	Including subsidies	Gross mixed income
Republic of Buryatia									
Dairy cattle breeding	128	3.0	2.7	1.8	71.1	9.2	646	128	32
Beef cattle breeding	202	2.1	1.6	1.0	42.6	6.3	737	203	44
Sheep breeding	60	1.8	1.2	0.7	31.7	4.4	902	152	47
Lipetsk region									
Dairy cattle breeding	45	4.6	5.6	3.1	60.0	10.8	1379	231	241
Beef cattle breeding	48	2.8	3.5	1.4	64.6	2.4	2764	406	370
Grain farming	147	3.7	0.7	2.3	64.6	3.1	4213	91	588
Oilseed production	46	2.7	0.8	1.7	65.2	3.1	3686	5	596
Mixed crop farming	20	6.2	0.9	4.9	85.0	5.7	4798	11	1094
Stavropol territory									

(continued)

Table 3 (continued)

Production types of P(F)Hs	Number of P(F)Hs	Number of permanent employees of P(F)Hs, people		Number of employees per one member of the P(F)H, people	Share of farms hiring workers, %	Share of labor costs in total expenditures, %	Per 1 permanent employee or member of the P(F)H, thousand rubles		
		Per farm	Per 100 hectares of land area				Income	Including subsidies	Gross mixed income
Dairy cattle breeding	117	2.9	2.7	1.7	73.5	8.3	970	106	487
Beef cattle breeding	128	2.7	1.6	1.5	70.3	4.7	1323	199	714
Sheep breeding	221	1.6	0.4	0.5	28.1	2.1	1660	202	633
Grain farming	1202	2.9	0.6	1.7	45.8	4.0	3394	35	1599

Source Compiled by the authors based on [5]

farms hiring workers is even lower than in P(F)Hs focused on beef cattle breeding and amounts to 31.7%. The provision of labor in this group is also the lowest—1.2 workers per 100 hectares of land area. There are 0.7 hired workers for each member of the P(F)H. In this regard, the mixed gross income per employee is slightly higher than in the farms specializing in cattle breeding.

Analyzing all production types of P(F)Hs of the Republic of Buryatia, the authors found out that the highest labor productivity is observed in farms with mixed farming and in P(F)Hs growing vegetables, 2121 and 1962 thousand rubles per employee, respectively. The most labor-intensive sub-industries in the Republic of Buryatia in cattle breeding is dairy cattle breeding, in crop production—potato farming. On average, there are 3–3.4 people per farm in these groups of farms. The share of farms hiring workers in dairy cattle farming is 71.1%, in potato farming—85.7%. In the Lipetsk Region (Tables 2 and 3), the largest share is held by the farms growing cereals and leguminous crops (147 P(F)Hs or 41%). The level of marketability of agricultural products in this group of farms is 81.6%.

The group of farms growing sunflowers, which also belongs to the priority sub-industry, is in all respects similar to the group of P(F)Hs growing grain. The highest labor productivity is observed in the P(F)Hs growing sugar beets—10,258 thousand rubles. This group of P(F)Hs has the largest sown area per farm and the largest provision with agricultural machinery per farm. Crop and livestock P(F)Hs have the largest number of employees per farm—12 people. In two priority livestock breeding areas (specialized beef cattle and milk production), the proceeds of livestock production per head exceed 50 thousand rubles. The marketability of agricultural products of dairy cattle P(F)Hs is much lower than that of beef cattle farms. Dairy cattle farming is more labor-intensive compared to beef cattle breeding, and labor productivity is correspondingly lower by 1385 thousand rubles per employee.

In the Stavropol Territory, one of the main priorities for development is the production of grain and leguminous crops and sheep and goat breeding. These P(F)Hs account for the largest number of P(F)Hs—74.5% of all P(F)Hs (Tables 2 and 3). The highest labor productivity is observed in the group of P(F)Hs growing sugar beet—14,806 thousand rubles per employee (this type is not described in the tables provided). Labor productivity in grain farms is 3394 thousand rubles per employee, while there are only 1.7 hired workers per one member of the farm. The provision of the labor force is low—only 0.6 workers per 100 ha of land area.

In sheep farms, labor productivity is less than half the labor productivity of grain farms. The share of farms that hire workers is only 28.1%.

There are only 0.5 hired workers for each member of the farm. The average number of sheep per farm is 1949 heads.

A comparison of three subjects of the Russian Federation with each other with different agroclimatic conditions showed that the Lipetsk Region and the Stavropol Territory exceed the Republic of Buryatia by several times in labor productivity in all major agricultural sub-industries. Comparing the Lipetsk Region and the Republic of Buryatia, we see that subsidies to support beef and dairy cattle breeding are twice as much on average per one P(F)H.

Statistical processing of the primary data allowed us to obtain important data on the efficiency of the activities of P(F)Hs of different production types, particularly on the availability, composition, and efficiency of the use of labor resources of P(F)Hs. The selection of P(F)Hs production types showed differences in specialization by regions depending on natural conditions. Simultaneously, it allowed us to identify the most effective sub-branches in using basic agricultural resources, including labor resources.

The research results show that it is necessary to develop the livestock industry in all studied regions. In P(F)Hs, livestock farming lags far behind the crop sector in terms of income and production efficiency. Particular attention should be paid to the priority areas identified by the Ministry of Agriculture of the Russian Federation. To preserve rural areas and the country's integrity in its eastern part, it is necessary to strengthen measures to support regions with unfavorable agroclimatic conditions.

4 Discussion

Foreign scholars conducted numerous studies related to the typology of farms. For example, scholars from Sweden [1] distinguished farm types by specialization for assessing the impact of changing production activities of farm types on the environment and food production. A farm was classified as a certain type if two-thirds of its standardized output came from a single production. They revealed significant potential for agricultural production under environmentally friendly conditions.

Kong and Castella [4] identified the types of farms in the Rotonak Mondol district of the Battambang province (Cambodia) and characterized the farms according to their resources, productivity, constraints, and ability to innovate. They identified four main types of farms using the method of principal components and cluster analysis: type 1—small farms cultivating highland crops; type 2—large farms cultivating highland crops; type 3—farms with a predominance of off-farm income; type 4—rice farms.

This paper singles out P(F)Hs production types using the structure of proceeds from the sale of agricultural products. It is challenging to characterize the selected types due to the lack of information on the size of agricultural land in the 1-KFH form (including the agricultural land used, production costs, the number of temporary and seasonal workers, the amount of time spent on and off the farm, etc.).

5 Conclusion

The developed methodology can be used to develop agrarian policy measures for the development of small businesses. The identified types of P(F)Hs differ in size, intensification, and efficiency of agricultural production. The most significant number of permanent employees (including members of P(F)Hs) per farm and per unit of the land area is observed in farms focusing on dairy cattle breeding. This production direction is also marked with the largest share of hired labor costs in the structure of total expenditures (8–11% in the studied regions). The productivity of labor, evaluated by the total income of K(F)Hs in each region, is the lowest compared to other types of farms. Considering the unresolved tasks of reaching self-sufficiency thresholds for milk and dairy products (in terms of milk), vegetables, gourds, fruits, and berries, increased state support and measures to develop small businesses in the labor-intensive agricultural sectors will help to address food security issues, increase employment, and preserve rural lifestyles.

References

1. Boke Olén N, Roger F, Brady MV, Larsson C, Andersson GKS, Ekroos J, Clough Y (2021) Effects of farm type on food production, landscape openness, grassland biodiversity, and greenhouse gas emissions in mixed agricultural-forestry regions. Agric Syst 189, 103071. https://doi.org/10.1016/j.agsy.2021.103071
2. Dashieva BSh, Ukolova AV (2021) Analysis of the influence of agricultural climatic conditions on the allocation of labor resources in agriculture. In: Gimbatov ShM, Abdulaeva ZZ, Bashirova AA, Denevizyuk DA (eds). Proceedings of ISPC-CPSLR 2020: The VIII international scientific and practical conference "current problems of social and labor relations." Atlantis Press Publishing, Makhachkala, Russia
3. Itskovich AYu (2013) Peasant (farmer) households: definition, typology, patterns. Probl Mod Econ 3(47):413–417
4. Kong R, Castella J-C (2021) Farmers' resource endowment and risk management affect agricultural practices and innovation capacity in the Northwestern uplands of Cambodia. Agric Syst 190, 103067.https://doi.org/10.1016/j.agsy.2021.103067
5. Ministry of Agriculture of the Russian Federation (2019) Reporting form "Information on productive activities of heads of peasant (farm) households – individual entrepreneurs" (1-KFH). Russia, Moscow
6. The Farm Business Survey (n.d.). Farm Classification in the United Kingdom. https://assets.publishing.service.gov.uk/government/uploads/system/uploads/attachment_data/file/365564/fbs-uk-farmclassification-2014-21oct14.pdf
7. Ukolova AV (2020) On the German agricultural census program. In Borulko VG (ed) Papers of the TSKHA. Russian State Agrarian University–Timiryazev Moscow Agricultural Academy, Moscow, Russia, pp 307–312
8. Ukolova A, Dashieva B (2021) Analysis of the distribution of peasant (farmer) households by income. Account Agric 6https://doi.org/10.33920/sel-11-2106-05
9. Zubrenkova OA, Fedotova OI (2015) Classification of peasant (farmer) farms, factors, and conditions, determine effective functioning. Bull NGII 5(48):38–51

Rural Housing Development Potential

Andrey N. Baidakov, Olga S. Zvyagintseva⬭, Olga N. Babkina⬭, Diana S. Kenina⬭, and Dmitriy V. Zaporozhets

Abstract The main purpose of this work is to study the potential for the development of housing construction in rural areas of Russia. The development potential of housing construction depends on the purchasing power of the rural population, which is formed by the corresponding level of demand for housing. Assessment of demand in the housing market of rural areas takes into account many factors, including the level of income of the population, the price of housing, the level of home improvement, the desire to improve housing conditions for young people, the share of dilapidated, and dilapidated housing. The methodology proposed by the authors for assessing and stimulating demand for housing construction in rural areas is characterized by ease of use and versatility, which is important in the context of the diversity of domestic regions. The results of testing the methodology showed the highest level of demand for housing in the Volga and Siberian Federal Districts, which is due to a larger population than in other regions, and a low level of improvement of residential premises in rural areas. The methodology for assessing the demand for housing is recommended to be used by municipal authorities, in process of making decisions in this area. Based on the data obtained, the administration of the rural area decides on the use of financial and non-financial measures to stimulate demand for housing construction. Also, the technique can be supplemented with predictive studies for a period determined by the accepted permissible terms for improving housing conditions.

Keywords Potential · Rural areas · Housing construction · Demand · Income level

A. N. Baidakov · O. S. Zvyagintseva (✉) · O. N. Babkina · D. S. Kenina · D. V. Zaporozhets
Stavropol State Agrarian University, Stavropol, Russia
e-mail: o-zvyagintseva@yandex.ru

A. N. Baidakov
e-mail: baid21@mail.ru

O. N. Babkina
e-mail: olia-st026@yandex.ru

D. V. Zaporozhets
e-mail: dz44@yandex.ru

JEL Code R21

1 Introduction

The development of housing construction in rural areas is one of the key areas of formation of the socio-economic potential of rural areas of Russia since one of the main conditions for attracting people for permanent residence in them is the formation of a housing stock in the countryside that meets modern quality requirements. This is also superior to urban housing—affordability, environmental friendliness, organic ties with the natural environment, etc. [1].

The rural areas of our country are distinguished by natural, socio-economic, infrastructural, cultural and ethnic diversity, and purchasing opportunities, and the potential of housing construction. In this regard, it's relevant to assess the possibilities of the rural population in terms of purchasing housing or improving housing conditions [9].

Directions for using or building the potential for housing development should be determined, first of all, based on an assessment of the corresponding level of demand from the rural population. The level of demand in the housing market primarily depends on the level of income of the population, which determines the ability of people to purchase housing or participate in mortgage lending programs [2].

Housing demand is chronotopically heterogeneous in rural areas of the country. The main factor determining its value is the level of income of the rural population and its stability. This level correlates with the state of the rural economy and its demographic characteristics.

The second key factor in the formation of the considered demand is the comfort of rural housing—existing and being built. Its increase determines the demand not only for the primary satisfaction of housing needs, but also for the desire to improve the existing housing conditions, even quite comfortable, but no longer meet the current possible level.

The third factor is associated with the desire of many rural residents to proactively solve the housing problem for their children—adults and adolescents. Even if the conditions of their living are quite comfortable.

2 Materials and Methods

In order to differentiate demand for different groups of the population, depending on the level of their income, it's proposed to use the following grouping (Fig. 1).

In Fig. 1, R_{ha} is the housing affordability ratio, which is measured in years.

The first group is a part of the population with a fairly high level of income and generating demand for housing without using borrowed funds. It means that this part of the population can accumulate funds for the purchase of housing in a relatively short period.

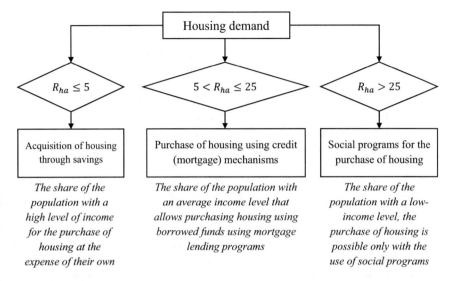

Fig. 1 Grouping of demand for housing depending on the level of income of the population *Source* compiled by the authors

The second group is a part of the population with an average income level. In this case, people don't have the opportunity to accumulate funds for a period that suits them, in connection with which they form the demand for housing by attracting credit (mortgage) resources.

The third group is a part of the population with a low level of income, a socially poorly protected group of the population. The level of income doesn't allow purchasing housing using the people's funds or using borrowed funds. This group forms the demand for social housing using the relevant government programs.

The proposed differentiated approach to the formation and satisfaction of the rural population's demand for improving housing conditions allows carrying out purposeful targeted actions to solve the housing problem in the countryside at all levels—the population, municipal, regional, and federal authorities [4].

The methodology for determining the level of demand in the housing market can be presented in the form of a sequence (algorithm) of the following actions.

1. To assess the level of demand of the population in the housing market, it's necessary to estimate the minimum level of family income with an acceptable value of the period for acquiring housing or improving housing conditions:

$$I_{min} = \frac{C \times A}{R_{ha}^0 \times F_c} + AS \tag{1}$$

where I_{min} is the minimum level of average per capita annual income for assessing current demand, rub;

C is the average market cost of 1 m^2 of housing, rub;

A is the total area of housing necessary to meet the needs of the family, m^2;

F_C—family composition, people;

R_{ha}^0—the maximum permissible value of the period for improving the living conditions of a family in a given rural area, years;

AS—annual subsistence minimum for a given rural area, rub.

Using I_{min} allows estimating the real demand for housing construction in this rural area at an allotted time. In this case, it's possible to set various values of R_{ha}^0, as well as other values, included in the formula (1), based on the predicted estimates of their changes for the coming period of R_{ha}^0 years [12].

In cases where the period for improving the living conditions of a family in a given rural area exceeds 25 years and there are no clear prospects for a decrease in the value of this indicator in the coming years, the issue of classifying this family as poor should be considered [3]. With the subsequent solution of its housing issue through inclusion in one of the categories of socially disadvantaged at the expense of the relevant Programs [6].

The 25-year threshold is determined by the maximum duration of mortgage programs and, in principle, can (must) be reduced.

2. At the next stage, it's necessary to correlate the obtained I_{min} values with the average per capita monetary income of the population of the rural area under consideration, using the distribution of the population by average per capita monetary income (source for the subjects of the Russian Federation - Rosstat, data can also be collected and processed for each rural area).

The generally accepted in statistics gradation according to the level of average per capita money income of the population in rural areas per month [11]:

-   Below 7,000 rub.
-   7,000.1–10,000 rub.
-   10,000.1–14,000 rub.
-   14,000.1–19,000 rub.
-   19,000.1–27,000 rub.
-   27,000.1–45,000 rub.
-   45,000.1–60,000 rub.
-   More than 60,000.1 rub.

Before assignment, it's necessary to divide the obtained I_{min} value by 12 months. By comparing the minimum annual average per capita income required to improve housing conditions with the distribution of the population by levels of average per capita income, the number of people who can afford to improve housing conditions (P_{pay}) is determined:

$$I_{min} \leq I_{low.lim.} \tag{2}$$

where $I_{low.lim.}$ is the lower limit of a certain level of average per capita income for this group of population.

3. After determining the gradation groups and the size of the population that forms them, satisfying the requirements of solvency for purchasing housing or improving housing conditions, it's necessary to subtract the size of the population living in fully comfortable housing available in rural areas from this.

Thus, the current and potential demand for housing, expressed in terms of population, is determined by the formula:

$$D_p = P_{pay} - P_{comf} \tag{3}$$

where D_p is the current demand for housing, expressed in terms of population;

P_{pay}—the current population, whose income allows purchasing or improving housing conditions;

P_{comf}—the current share of the population living in fully comfortable housing, %.

$$P_{comf} = \frac{S_{full} \times S_{full}}{S_{per}} (1 - S_{teen}) \tag{4}$$

S_{total}—the total square of residential premises in rural areas, m^2;

S_{full}—the share of fully equipped living space, %;

S_{teen}—the share of teenagers aged 15 to 34 living in rural areas, %;

S_{per}—living square per 1 inhabitant of the rural area, m^2.

The accounting for the share of teenagers is important because there is a fairly stable tendency to "resettlement" young people, even if they live with their parents in favorable living conditions.

4. To determine the value of demand for housing in square meters, it's necessary to multiply the population that has the financial ability to purchase housing and doesn't have a fully comfortable housing (C_{pp}) by the minimum required number of square meters of living space. And to the obtained value, the area of emergency and dilapidated housing in rural areas, subject to mandatory replacement—formula should be also added (5).

The generally accepted standard for the provision of living space for the population is 18m^2 per person. However, the actual values of the provision of housing space in rural areas have already reached this limit and exceeded it in most constituent entities of the Russian Federation. Therefore, it's advisable to revise the standard for the provision of living space within each municipality, depending on the actual values that have already been achieved.

$$D_{sq.m} = D_p \times N_{per} + S_{tot} \times S_{Emerg/dilap} \tag{5}$$

where $D_{sq.m}$—current demand, expressed in m^2;

$\quad$ N_{per}—norm of provision of living space for 1 person in a given rural area, m^2;

$\quad$ S_{tot}—total square of residential premises in rural areas, m^2;

$\quad$ $S_{emerg/dilap}$—share of emergency and dilapidated housing, %.

To calculate the current level of demand for housing in rural areas, official statistics (Rosstat) were used: the distribution of the population by average per capita income (the entire population excluding federal cities), the average price of 1 m^2 of housing on average in federal districts (primary and secondary housing).

Further, to calculate the current level of demand in rural areas, it's recommended to use statistical data at the local level, taking into account the distribution of income of the rural population in a certain area and the average prices of 1 m^2 of housing in this rural area.

3 Results

While referring the share of the population with an income level that satisfies a particular housing affordability coefficient to the corresponding group, the main attention should be focused on the lower limit of the average per capita income in order to guarantee certain coverage of the demand of the population classified in this category. For example, while calculating the minimum per capita income of the Central Federal District with taking into account the level of housing affordability for 5 years, the value of 34,849.1 rubles was obtained using formula (1). This means that the 5-year housing affordability can be estimated for 22.41% of the population (the share of the population with an average per capita income of over 45,000 rubles).

The share of the population with an income of 27,000 to 45,000 rub falls into the group with the level of housing affordability from 5 to 10 years (Table 1). Thus, the minimum population size that satisfies this category is calculated, while a reserve is created for the development, and stimulation of demand from citizens, who are in the middle of the corresponding range of average per capita incomes of the population of the territory [7, 8].

The share of the solvent population was determined based on the size of the rural population in the context of federal districts.

As can be seen from Table 1, in the South, North Caucasus, Volga, and Siberian Federal Districts, the share of the population, who can purchase housing within 5 years, is higher than in the other four districts. This is due to the lower average price of one m^2 of housing in these territories.

This trend continues for other levels of the housing affordability ratio. However, as a result, the total share of the solvent population at all levels of accessibility doesn't differ much in the context of federal districts (varies from 67 to 75%), which is due to the larger share of the population with low incomes in the above federal districts [13, 14].

Table 1 Determination of the size of the population, the average per capita income of which makes it possible to purchase housing or improve housing conditions in the context of federal districts, 2019 year

Federal district	$R_{ha} < 5$ years			$R_{ha} < 10$ years			$R_{ha} < 15$ years			$R_{ha} < 25$ years		
	I_{min}, thousand roubles	Share of population, %	Population, million people	I_{min}, thousand roubles	Share of population, %	Population, million people	I_{min}, thousand roubles	Share of population, %	Population, million people	I_{min}, thousand roubles	Share of population, %	Population, million people
CFD	34.8	22.4	1.5	23.2	47.7	3.2	19.3	66.8	4.5	16.2	66.8	4.5
NWFD	33.6	20.0	0.4	22.5	46.8	1.0	18.9	67.1	1.4	15.9	67.1	1.4
SFD	27.5	40.7	2.5	19.5	60.1	3.7	16.9	60.1	3.7	14.7	75.4	4.6
NCFD	23.5	30.9	1.5	17.5	50.3	2.5	15.5	50.3	2.5	13.9	67.4	3.3
VFG	26.6	37.8	3.1	19.0	57.3	4.7	16.5	57.3	4.7	14.5	73.1	6.0
UFD	28.9	25.1	0.6	20.2	49.3	1.1	17.3	67.4	1.5	15.0	67.4	1.5
SibFD	26.8	36.4	1.6	19.1	56.5	2.5	16.6	56.5	2.5	14.6	72.9	3.2
FEFD	33.1	26.8	0.6	22.3	52.3	1.2	18.7	70.2	1.6	15.8	70.2	1.6

CFD—Central Federal District.
NWFD—Northwestern Federal District.
SFD—Southern Federal District.
NCFD—North Caucasian Federal District.
VFD—Volga Federal District.
UFO—Ural Federal District.
SibFD—Siberian Federal District.
FEFD—Far Eastern Federal District.
Source calculated by the authors, based on statistical data, and using formulas (1, 2);
I_{min}—the minimum level of average per capita annual income for assessing current demand, rubles;
R_{ha}—the maximum admissible value of the period for improving the living conditions of a family in a given rural area, years.

Table 2 Results of calculations of the current level of demand in the housing market in the context of federal districts, 2019 year

Federal district	S_{full}, %	S_{total}, million m²	S_{per}, m² per person	P_{comf}, million people	P_{pay}, million people	D_p, million people	N_{per}(fact), m²	$D_{sq.m}$, million m²
CFD	43.9	224	32.0	3.1	4.5	1.4	32.0	46.7
NWFD	29.7	75	34.7	0.6	1.4	0.8	34.7	26.8
SFD	45.5	142	23.1	2.8	4.6	1.8	23.1	41.6
NCFD	48.4	97	19.7	2.4	3.3	0.9	19.7	18.6
VFD	37.7	244	29.7	3.1	6.0	2.9	29.7	84.8
UFD	25.7	63	27.7	0.6	1.5	0.9	27.7	26.2
SibFD	15.1	108	24.6	0.7	3.2	2.5	24.6	62.6
FEFD	10.6	51	23.0	0.2	1.6	1.4	23.0	30.3

Source calculated by the authors based on statistical data using formulas (3–5)

S_{full}—the share of fully equipped living space;

S_{tot}—total square of residential premises in rural areas;

S_{per}—living square per 1 inhabitant of the rural area;

P_{comf}—the current share of the population living in fully comfortable housing;

P_{pay}—the current population, whose income allows purchasing or improving housing conditions;

D_p—the current demand for housing, expressed in terms of population;

N_{per}—norm of provision of living space for 1 person in a given rural area;

$D_{sq.m}$—current demand, expressed in m².

Based on the data obtained, an assessment of the level of demand, expressed in population size and m², was carried out (according to points 3 and 4 of the methodology). Table 2 shows the results of the calculation, according to which it can be concluded that the highest level of demand is observed in the Volga and Siberian federal districts (Fig. 2).

This is due to a larger population than in other federal districts and a low level of improvement of residential premises in rural areas. In the calculations, the authors proceeded from the standard of living space provision, which is based on the actual values of this indicator. At the discretion of the administration of the rural area, the value of the standard can be changed.

4 Conclusion

The authors used the values of the variables, which were known at the time of the calculations, although, as has already been mentioned, it's possible to use their predictive estimates for the period under consideration, determined by the accepted permissible terms for improving housing conditions [10]. Then the proposed methodology should include appropriate predictive studies.

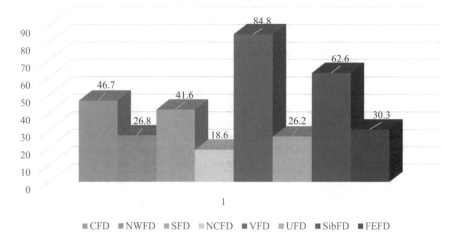

Fig. 2 The results of assessing the level of demand for housing in rural areas in the context of federal districts, 2019 year *Source* compiled by the authors based on statistical data

Based on the data obtained on the existing demand for rural housing, expressed in quantitative terms, the administration of the rural area decides to use financial and non-financial measures (Table 3) to enhance the current and stimulate potential demand for housing [5].

Table 3 Financial and non-financial incentives to stimulate current and potential demand for housing

Decision level	Incentive measures	Content
1	2	3
Municipal	Providing the opportunity to choose the method of acquiring housing (by accumulating or using a mortgage)	Correlation of the shortage of financial resources of a family in the implementation of estimated savings for the purchase of housing with the possibilities of using mortgage lending instruments, according to the method of determining the net present value for the period determined using the method of calculating the indicator "Housing affordability ratio" Mortgage lending instruments will be in demand in the event of a reduction in the shortage of financial resources of the family, while using them, compared to the case of not using them
Federal regional	Differentiation of rates for the rural mortgage program	Differentiation of the rate for the rural mortgage program depending on the energy-saving class of the housing being built. The higher the energy-saving class of the facility, the lower the rate on the mortgage loan

(continued)

Table 3 (continued)

Decision level	Incentive measures	Content
Federal regional	Expanding the content of the rural mortgage program	Dissemination of the program's capabilities for the construction of economic facilities located on the allotted land plot (garage, premises for keeping animals and storing feed, bathhouse, and other capital construction facilities)
Federal regional municipal	Introduction of differentiated electricity charges in rural areas	Development of a program for preferential tariffication of power supply services at night, and for households using electrical appliances, as well as equipment with a three-phase power supply system
Federal regional	Stimulating small businesses in the construction industry in the countryside	Simplification of the procedure for obtaining licenses and permits for small enterprises and micro-enterprises for the production of building materials, elements of building structures, as well as small architectural forms
Regional municipal	Creation of conditions for the development of small businesses in the countryside	Provision of incentives for renting premises, special taxation regimes for small businesses and individual entrepreneurs operating in rural areas, and not related to agricultural production (services to the population, retail trade, and other activities)
Regional municipal	Attracting for permanent residence in rural areas of the population from cities, not necessarily employed in agricultural production	Creation of a favorable image of rural areas and dissemination of information through the media and social networks: ecology; healthy eating; doing business for those types of activities for which the market is saturated in cities; the opportunity for future residents to participate in the design of their own homes and public spaces in rural areas; providing jobs or creating conditions for running a personal subsidiary farm
Regional municipal	Providing benefits for paying utility bills for rural households	Development of a program for subsidizing a part of communal services of rural households, which over a certain period have improved their living conditions in terms of equipment with all life support systems or improvement of the house and the surrounding area. A similar program may apply to migrants who have re-acquired housing in rural areas

Source compiled by the authors

Acknowledgements The research results were partially used in the course of scientific research commissioned by the Ministry of Agriculture of Russia.

References

1. State Program of the Russian Federation (2017) Provision of affordable and comfortable housing and communal services for citizens of the Russian federation, approved by Decree of the Government of the Russian Federation No 1710 dated Dec 30, 2017. http://www.consul tant.ru/document/cons_doc_LAW_286800/. Accessed 24 Aug 20
2. Passport of the National Project (2018) Housing and urban environment (approved by the Presidium of the Council under the President of the Russian Federation for Strategic Development and National Projects, No 16, dated Dec 24, 2018). http://www.consultant.ru/document/ cons_doc_LAW_319211/. Accessed 24 Aug 20
3. Decree of the Government of the Russian Federation of May 31, 2019, No 696 On approval of the state program of the Russian federation. Comprehensive Development of Rural Areas and on Amendments to some Acts of the Government of the Russian Federation (as revised on July 10, 2020). http://www.consultant.ru/document/cons_doc_LAW_326085/. Accessed 24 Aug 20
4. Order of the Government of the Russian Federation of 02.02.2015 No 151-r On approval of the strategy for sustainable development of rural areas of the Russian federation for the period up to 2030. http://www.consultant.ru/document/Cons_doc_LAW_174933/. Accessed 24 Aug 20
5. Order of the Ministry of Agriculture of Russia dated 20.04.2020 No. 214 On approval of the departmental target program "The modern look of rural areas". http://www.consultant.ru/doc ument/cons_doc_LAW_355739/. Accessed 24 Aug 20
6. Frolov VI (2011) Methods of substantiation of programs for sustainable development of rural areas: monograph. Saint Petersburg State University of Architecture and Civil Engineering, p 464
7. Housing in Russia (2016). Stat. Sat. Rosstat, 63.
8. Housing in Russia (2016) Stat Sat Rosstat 78
9. Ilyina IN (2015) The quality of the urban environment as a factor of sustainable development of municipalities. Econ. Manag Nat Econ 5(164):69–82
10. Loginova DA, Strokov AS (2020) Institutional issues of sustainable development of rural areas. https://vgmu.hse.ru/data/2019/06/17/1485071847/Loginova,%20Strokov%202-2019.pdf. Accessed 24 Aug 20
11. Official statistical methodology for the development of indicators for the type of activity "Construction" at the regional and federal levels: approved by Order of Rosstat, dated December 25, 2015, No 654
12. Polyakova NV, Zaleshin VE, Polyakov VV (2020) Diagnostics of the comfort of the living environment in cities: substantiation and formation of the methodology. Bullet Baikal State Univ 30(1):121–129
13. Rural territories of the Russian Federation. Rural territories. https://www.fedstat.ru/. Accessed 24 Aug 20
14. Russian statistical yearbook (2019) Stat Sat Rosstat 708

Expanded Reproduction as the Basis for Agricultural Sustainability: Marketing, Digital Economy, and Smart Technologies

Egor V. Dudukalov⬤, Elena V. Patsyuk, Olga A. Pecherskaya, and Yelena S. Petrenko⬤

Abstract This study puts forward and proves the hypothesis, using the method of regression analysis that at present, expanded reproduction in agriculture is achieved through conveyor-based agro-industrial production, which makes it possible to overcome food shortages, but contradicts the idea of agricultural sustainability. The purpose of this article is to identify the prospects for simultaneously combating hunger and achieving sustainable agriculture in a model of its expanded reproduction. As a result, it is substantiated that to engage in Intensified Productivity, the private sector, which includes farmers, processors, and merchants, requires a sound governmental foundation and management. These are necessary for local agriculture and marketing to compete with imports and for purchasers to approach reasonably priced privately supplied food (Pretty and Bharucha in Ann Bot 114:1571–1596, 2014). Governments must ensure that input procurement, product promotion, and access to routine assets, data, preparation, instruction, and social services are all affordable. This will necessitate sufficient funds for assistance as well as net ventures. Supportable Intensification may also make a significant contribution to mitigating environmental change by increasing carbon retention in well-managed soils and reducing outflows due to increased compost and water system utilisation. There is currently no peaceful agreement or mechanism in place to provide large-scale relief funds to agriculture in non-industrial nations.

E. V. Dudukalov
Russian Presidential Academy of National Economy and Public Administration, Moscow, Russia

E. V. Patsyuk (✉)
Volgograd State Technical University (Sebryakov Branch), Mikhaylovka, Russia
e-mail: elenapatsyuk@yandex.ru

O. A. Pecherskaya
Voronezh State University of Forestry and Technologies Named After G.F. Morozov, Voronezh, Russia
e-mail: olga-pecherskaya@mail.ru

Y. S. Petrenko
Plekhanov Russian University of Economics, Moscow, Russia
e-mail: petrenko_yelena@bk.ru

Keywords Expanded reproduction · Sustainable agriculture · SDGs · Marketing · Digital economy · Smart technologies · Developed countries · Developing countries

JEL Codes Q01 · Q13 · Q18 · O13 · O32

1 Introduction

Global food production must increase to feed a growing global population. Ranchers, in any case, face numerous obstacles. Agriculture should develop the ability to conserve to thrive. However, increased grain production has depleted the traditional asset base of many countries' agribusinesses, jeopardising their future viability [1]. To meet long-term demand, emerging market farmers must quadruple food production over the next 40 years. A task that has been significantly exacerbated by the combined effects of environmental change and increased competition for land, water, and energy [2]. SCPI is a novel paradigm that maximises yields on a smaller amount of land while monitoring assets, mitigating adverse climate effects, increasing average capital, and advancing biological system administrations [3].

Agricultural Production Intensification will be built on expanding frameworks that provide farmers and society as a whole with numerous efficiency, financial, and environmental benefits. Agricultural production that is biologically based strengthens and repairs farmland. SCPI cultivating frameworks will be based on protection horticultural practises, including the use of the high-yielding adapted seed, integrated pest management, plant nutrition based on healthy soils, effective water management, and integration of harvests, trees, and domesticated animals [4]. The concept of controllable Production frameworks is dynamic: they should provide ranchers with a variety of potential practice combinations to investigate and modify in response to their specific local Production circumstances and imperatives. These frameworks contain an abundance of data. SCPI strategies should emphasise growth strategies such as farmer field schools and collaboration with neighbours [5].

Agriculture must return to its roots by recognising the critical nature of healthy soil, relying on natural sources of plant nutrition, and using mineral compost sparingly. Soils rich in biota and natural matter pave the way for increased yield efficiency. The best results are obtained when mineral composts are combined with natural sources, such as trash, nitrogen-fixing harvests, and trees. Mineral composts are cost-effective and ensure that nutrients reach the plant without causing damage to the air, soil, or streams [6]. To promote soil health, protective agriculture, mixed-yield livestock, and agroforestry frameworks that increase soil richness should all be used. They should eliminate incentives that encourage mechanical cultivation and inefficient compost utilisation in favour of farmers' precision methods such as urea profound circumstance and site-specific supplementation.

Farmers will require a diverse group of epigenetically modified enhanced yield varieties that are tolerant of a variety of agro-environments and cultivation methods and resistant to environmental change. Hereditarily enhanced grain assortments accounted for more than half of the increase in yields in recent years [7]. Plant reproducers should strive to achieve comparable results in the future. Optimising the delivery of high-yielding assortments to ranchers, on the other hand, would necessitate significant improvements to the framework connecting Plant Germplasm Collections, Plant Breeding, and Seed Delivery. Nearly 75% of Plant Genetic Resources (PGR) have been depleted over the last century, and by 2050, 33% of currently available varieties may become extinct [7]. Additional funding is required to support the collection, preservation, and use of PGR. Subsidies are also being considered for the reintroduction of public plant breeding programmes. Arrangements should make it easier to connect formal and rancher-preserved seed frameworks and to expand neighbouring seed initiatives.

Sustainable Intensification necessitates more intelligent, precise upgrades to the water system and the adoption of water rationing practices that are environmentally friendly. Urban areas and commercial ventures fiercely compete with agriculture for water resources [8]. Despite its efficiency, the water system is under pressure to mitigate natural consequences such as soil salinisation and nitrate contamination of springs. A critical step towards practical intensification will be the development of an information-based precision water system capable of providing solid and adaptive water application in conjunction with a scarcity water system and wastewater reuse. Measures should be taken to mitigate irrational endowments that incentivise farmers to wastewater. Environmental change is putting a significant number of rainfed small homesteads in jeopardy. Enhancing rainfed efficiency will require the use of more drought-tolerant assortments and water-saving board practices [1].

This study puts forward the hypothesis that at present, expanded reproduction in agriculture is achieved through conveyor-based agro-industrial production, which allows us to overcome food shortages, but contradicts the idea of agricultural sustainability. The purpose of this article is to identify the prospects for simultaneously combating hunger and achieving sustainable agriculture in a model of its expanded reproduction.

2   Materials and Method

When implemented and maintained properly, intensification will result in the "shared benefit" benefits necessary to address the dual challenges of caring for the entire population while conserving the environment. SCPI enables nations to plan, produce, and manage rural production in ways that meet society's needs and desires without jeopardising people's future ability to engage in the full range of ecological labour and products. There has been a decline in the abuse of data sources, such as mineral composts, in tandem with Increased utility is an example of a mutually beneficial relationship that benefits both ranchers and the environment [9–12].

Practical Intensification has a lot to offer small ranchers and their families, who account for more than 33% of the global population, by increasing their efficiency, lowering their costs, increasing their resilience to stress, and strengthening their ability to manage risk. Reduced spending on horticultural information sources will free up funds for ranches' food, health, and education. Increased profits for ranchers will occur at a lower natural cost, resulting in both private and public benefits [13].

Another commitment of preservation horticulture to sustainable production intensification is the avoidance of soil aggravation and the maintenance of harvest buildups on the dirt surface. Traditionalist Agriculture (CA) methods include limited culturing [14], which disturbs only the portion of the soil that will contain the seed column, as well as no-culturing or direct cultivation, in which crops are planted directly into a seedbed that has not been ploughed since the previous crop.

Vulnerability to the cost and availability of energy in the future indicates the importance of efforts to reduce overall ranch force and energy requirements. In comparison to conventional farming, conservation agriculture (CA) through Intensified Production techniques can eliminate up to 60% of those prerequisites [15]. The savings result from the elimination or reduction of the majority of force-concentrated field operations, such as culturing, thereby facilitating work and force bottlenecks, particularly during land readiness. While CA requires an interest in new and proper homestead operations, interest in hardware, most notably the number and size of work vehicles, has vanished completely. Small-scale ranchers who rely on hand labour or animal foothold are also eligible for investment funds.

The empirical basis for this study is the statistics of agricultural sustainability and expanded reproduction in 2020, shown in Fig. 1. Data selected for the countries of the world with the highest values of the Food production index is based on the materials of the World Bank [1].

According to Fig. 1, the highest value of the Food production index in 2020 is observed in Oman (130.1) and Kuwait (126.1), and the highest agricultural stability is in Malawi (54 points) and Burundi (47.3 points). Based on the collected statistics,

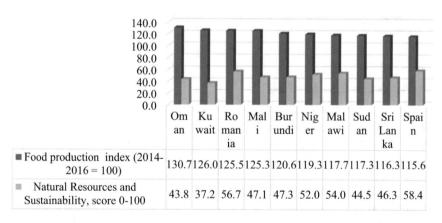

	Om an	Ku wait	Ro man ia	Mal i	Bur undi	Nig er	Mal awi	Sud an	Sri Lan ka	Spai n
■ Food production index (2014-2016 = 100)	130.7	126.0	125.5	125.3	120.6	119.3	117.7	117.3	116.3	115.6
▨ Natural Resources and Sustainability, score 0-100	43.8	37.2	56.7	47.1	47.3	52.0	54.0	44.5	46.3	58.4

Fig. 1 Statistics of agricultural sustainability and expanded reproduction in 2020 *Source* Compiled by the authors based on materials from [1, 16]

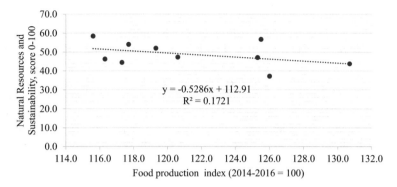

Fig. 2 Regression curve of dependence of agricultural sustainability on expanded reproduction in 2020 *Source* calculated and constructed by the authors

a regression analysis of the data from Fig. 1 and a regression curve was plotted for the dependence of agricultural stability on expanded reproduction in 2020 in Fig. 2.

According to Fig. 2, the reproduction of food products currently contradicts the idea of environmental sustainability of agriculture, as evidenced by negative regression and low correlation. This confirms the hypothesis put forward that it is necessary to adjust the concept and practice of expanded reproduction in agriculture from conveyor-type agro-industrial production to environmentally friendly agriculture.

3   Results

Numerous rural communities make use of a variety of traditional and innovative water collection systems. Ranchers with limited resources employ planting trenches to collect water and remediate contaminated land in preparation for millet and sorghum harvests. The invention increases supplement penetration and accessibility, resulting in critical yield increases, increased soil cover, and reduce downstream flooding. The fourth year of operation brought benefits such as a 400% increase in gross production value, increased soil moisture and richness, and decreased downstream flooding.

Harvest utility is maximised by selecting high-yielding varieties with adequate water availability, soil richness, and harvest security. On the other hand, harvests can thrive despite a scarcity of water. In a shortfall water system, water supply is not always sufficient to meet yield requirements, and moderate pressure is permitted in development plans that are less susceptible to moisture deficiency. The idea is that any loss of production will be minimal and that additional benefits will be realised by diverting stored water to flood future harvests. In any case, utilising a scarcity water system requires a thorough understanding of soil–water and salt management, as well as personal familiarity with crop behaviour, as crop response to water pressure varies significantly [17].

The issue with the public seed system and its capacity to provide ranchers with the high-quality seed of modified assortments is critical to consider when establishing SCPI programmes. The first phase should involve collaboration with all significant partners to develop an appropriate seed method and collection discharge rules [18].

One possible outcome of possible intensification is that local seed producers and economic sectors will play a larger role in assisting ranchers. The critical role of business sectors in preserving variety is becoming clearer. Markets can benefit from initiatives such as neighbouring variety fairs, seed banks, and biodiversity registers, which all promote the preservation and sharing of indigenous resources and encourage quality improvement.

4 Discussion

SCPI will be ineffective unless there is a vibrant and thriving market for data sources and administrations, as well as the final output. The prices ranchers pay for inputs and receive for agricultural harvests may be the most important factor determining the extent, type, and sustainability of harvest intensification. Since SCPI approaches impose a premium on information costs, novel approaches will be required to increase efficacy and influence innovation decisions.

Given the recent volatility of commodity prices, cost adjustment of agricultural yields is an unavoidable precondition for realistic harvest intensification. Price volatility results in significant pay differentials and increased risk for ranchers whose livelihood is based on agriculture [19]. It erodes their ability to invest in viable frameworks and increases the temptation to trade conventional wealth for security. Historically, small-scale plans to address value unpredictability have typically fizzled. Increased visibility and comprehension at the macro-strategic level almost certainly results in more workable arrangements. Existing institutions, such as import financing coordination or limited-risk assurances, could serve as a global safety net.

The absence of market charges for environmental administrations and biodiversity indicates that the benefits of such products are overlooked or undervalued in dynamic markets [20]. Food expenditures in agriculture do not include all costs associated with climate-related food production. There are no offices collecting fees for poor water quality or soil disintegration. If farmgate prices accurately reflected the true cost of production—with ranchers adequately compensated for any environmental damage—food prices would almost certainly increase. Along with charging for insults to plants, schemes may compensate ranchers who practise environmentally responsible ranching [21].

Payments for natural resource management are gaining traction as a component of a more empowered strategy climate for economic agriculture and national development. In any case, attractive arrangements will rely on empowering approaches and frequently absent organisations at the local and global levels.

To maximise their benefits, PES initiatives should span a large number of manufacturers and regions, achieving economies of scale in exchange costs and risk management across the board. It is critical to integrate PES more closely with rural development programmes to minimise exchange costs [22]. Given the open money cutoff levels, novel forms of optional approach execution or increased private subsidisation should be developed, particularly if private PES recipients can be identified.

5 Conclusion

To engage in Intensified Productivity, the private sector, which includes farmers, processors, and merchants, requires a sound governmental foundation and management. These are necessary for local agriculture and marketing to compete with imports and for purchasers to approach reasonably priced privately supplied food [23]. Governments must ensure that input procurement, product promotion, and access to routine assets, data, preparation, instruction, and social services are all affordable. This will necessitate sufficient funds for assistance as well as net ventures.

SCPI requires significant and continuous investment in human, financial, physical, and social capital in developing nations' farming sectors [24]. According to FAO estimates, an annual gross venture of US$209 billion at 2009 prices is required to achieve the required output increases by 2050 in critical horticulture and downstream regions [25]. Additionally, public investment in horticultural innovation, country structure, and social safety nets is necessary. At the moment, interest in agricultural nations' agriculture is undoubtedly insufficient. Since the late 1980s, the absence of domestic subsidies has been exacerbated by a decline in horticulture ODA. These scarcities have combined to produce a historically low level of capital for agricultural events over the last two decades. SCPI's success will require a substantial increase in horticulture speculation [26].

Subsidies for variability and moderation of environmental change are critical for SCPI. Increasing variety in horticulture, for example, through increased plant reproduction and seed frameworks, is a significant component of rational Intensification. SCPI may thus benefit from funding earmarked for reversing environmental degradation. Supportable Intensification may also make a significant contribution to mitigating environmental change by increasing carbon retention in well-managed soils and reducing outflows due to increased compost and water system utilisation. There is currently no peaceful agreement or mechanism in place to provide large-scale relief funds to agriculture in non-industrial nations.

References

1. World Bank (2021) Atlas of sustainable development goals 2017: from world development indicators. The World Bank. https://data.worldbank.org/indicator
2. Tilman D, Cassman KG, Matson PA, Naylor R, Polasky S (2002) Agricultural sustainability and intensive production practices. Nature 418(6898):671–677
3. Petersen B, Snapp S (2015) What is sustainable Intensification? Views from experts. Land Use Policy 46:1–10
4. Tscharntke T, Clough Y, Wanger TC, Jackson L, Motzke I, Perfecto I, Vandermeer J, Whitbread A (2012) Global food security, biodiversity conservation and the future of agricultural intensification. Biol Cons 151(1):53–59
5. Lindenmayer D, Cunningham S, Young A (eds) (2012) Land-use intensification: effects on agriculture, biodiversity and ecological processes. CSIRO publishing
6. Dudley N, Alexander S (2017) Agriculture and biodiversity: a review. Biodiversity 18(2–3):45–49
7. Ray DK, Mueller ND, West PC, Foley JA (2013) Yield trends are insufficient to double global crop production by 2050. PloS one 8(6):e66428
8. Phalan B, Onial M, Balmford A, Green RE (2011) Reconciling food production and biodiversity conservation: land sharing and land sparing compared. Science 333(6047):1289–1291
9. Popkova EG, Sergi BS (2018) Will industry 4.0 and other innovations impact Russia's development? In: Sergi BS (ed) Exploring the future of Russia's economy and markets: towards sustainable economic development. Emerald Publishing Limited, Bingley, UK, pp 51–68
10. Popkova EG, Sergi BS (2020) Social entrepreneurship in Russia and Asia: further development trends and prospects. Horizon 28(1):9–21
11. Sergi BS, Popkova EG, Bogoviz AV, Ragulina YV (2019b) The agro-industrial complex: tendencies, scenarios, and policies. In: Sergi BS (ed) Modelling economic growth in contemporary Russia. Emerald Publishing, Bingley, UK
12. Sergi BS, Popkova EG, Borzenko KV, Przhedetskaya NV (2019) Public-private partnerships as a mechanism of financing sustainable development. In: Ziolo M, Sergi BS (eds) Financing sustainable development: key challenges and prospects. Palgrave Macmillan, pp 313–339
13. Baulcombe D, Crute I, Davies B, Dunwell J, Gale M, Jones J, Pretty J, Sutherland W, Toulmin C (2009) Reaping the benefits: science and the sustainable Intensification of global agriculture. The Royal Society
14. Kuhn NJ, Hu Y, Bloemertz L, He J, Li H, Greenwood P (2016) Conservation tillage and sustainable Intensification of agriculture: regional vs global benefit analysis. Agr Ecosyst Environ 216:155–165
15. Gunton RM, Firbank LG, Inman A, Winter DM (2016) How scalable is sustainable intensification? Nat Plants 2(5):1–4
16. The Economist Intelligence Unit Limited (2021) Sustainable development index 2020. https://foodsecurityindex.eiu.com/Index. Accessed 11 Sept 2021
17. Mueller ND, Gerber JS, Johnston M, Ray DK, Ramankutty N, Foley JA (2012) Closing yield gaps through nutrient and water management. Nature 490(7419):254–257
18. Giller KE, Beare MH, Lavelle P, Izac AM, Swift MJ (1997) Agricultural intensification, soil biodiversity and agroecosystem function. Appl Soil Ecol 6(1):3–16
19. Tittonell P (2014) Ecological intensification of agriculture—sustainable by nature. Curr Opinion Environ Sustain 8:53–61
20. Raven PH, Wagner DL (2021) Agricultural intensification and climate change are rapidly decreasing insect biodiversity. Proc Natl Acad Sci 118(2)
21. Garnett T, Appleby MC, Balmford A, Bateman IJ, Benton TG, Bloomer P, Godfray HCJ (2013) Sustainable intensification in agriculture: premises and policies. Science 341(6141):33–34
22. Tilman D, Balzer C, Hill J, Befort BL (2011) Global food demand and the sustainable Intensification of agriculture. Proc Natl Acad Sci 108(50):20260–20264
23. Pretty J, Bharucha ZP (2014) Sustainable intensification in agricultural systems. Ann Bot 114(8):1571–1596

24. Garnett T, Godfray C (2012) Sustainable Intensification in agriculture. Navigating a course through competing food system priorities. Food climate research network and the Oxford Martin Programme on the future of food, University of Oxford, UK, 51
25. Faostat R (2021) FAOSTAT database. Food Agric Organ UN. http://www.fao.org/faostat/en/#home
26. Knoema E (2021) World data Atlas. https://knoema.com/atlas/ranks/Crop-Production-index

Innovative Model of the Functioning of Consumer Cooperation as an Incentive for Developing the Regional Food Market and Increasing Population Welfare

Natalia I. Morozova⊙, Galina N. Dudukalova⊙, Vladimir V. Dudukalov⊙, Tatiana V. Opeykina⊙, and Larisa V. Obyedkova⊙

Abstract Despite a certain decline in its development, the cooperative sector of the economy still can solve national problems in the field of improving the welfare of the population in rural areas. A cooperative economy, given the stability of its development through the introduction of advanced technology, should stop the outflow of the rural population and make rural areas attractive to the urban population, changing the vector of spontaneous migration and giving impetus to the development of the traditional economic way of life in Russia. Throughout its evolution, consumer cooperation has accumulated considerable experience in the model of production and sale of agricultural products through the association of small and medium-sized producers. This stimulates the creation of a multi-economic agrarian economy and a competitive environment in regional food markets. Competition leads to the need to introduce scientific and technological progress in agriculture, find ways to save resources, and take care of the environment. A "two-story Russia" will bring back the traditional way of life and traditional values, which, unfortunately, are being erased by the "industrial machine" of the big cities. Cooperative formations are not isolated. They are part of a complex spatially heterogeneous system, which complicates the task of choosing the optimal vector of development for these forms of economic management. The problem must be set with a clear identification of the development directions, considering resource constraints. It is necessary to study the real state of

N. I. Morozova (✉) · G. N. Dudukalova · V. V. Dudukalov · T. V. Opeykina
Volgograd Cooperative Institute (Branch), Russian University of Cooperation, Volgograd, Russia
e-mail: miss.natalay2012@yandex.ru

G. N. Dudukalova
e-mail: gdudukalova@ruc.su

V. V. Dudukalov
e-mail: vdudukalov@ruc.su

T. V. Opeykina
e-mail: otv06@bk.ru

L. V. Obyedkova
Volgograd State University, Volgograd, Russia
e-mail: laravik@bk.ru

the object and provide operational and reliable information about future scenarios for its development. Modeling the sustainable development of consumer cooperation is possible only with the use of simulation modeling. Simulation modeling will allow moving to a mechanism of proactive management. Objective possibilities for creating the required model will allow for the creation of computer technology. Integrated computer space can process a network of mathematical matrices, including the collection and analysis of information and the ability to make operational management decisions.

Keywords Agricultural economics · Personnel policy in the system of consumer cooperation · Quality of life of the population · Consumer cooperation · Regional system of consumer cooperation · Innovation · Technology

JEL codes P32 · O33

1 Introduction

Cooperative ideas have centuries of history and unique resources for development. The modern cooperative movement originated in Europe. In the 1860s, it appeared in Russia. There is probably no institution in history that has ever proven so viable and effective over such a long period. Each of the subsequent stages of development made adjustments in cooperation. Nevertheless, the social and economic role of cooperation in the development of the economy remained unchanged.

However, the solution of economic problems and meeting the needs and requirements of shareholders cannot be absolutized since the uniqueness and identity of the idea of cooperation are lost in this case. Consumer cooperatives differ from other organizations by their social purpose and values, among the fundamental ones being equality, justice, personal responsibility, and mutual assistance [1].

The cooperative system is designed to carry out an economic and social mission, including protecting the shareholders' interests and the fight against social inequality, poverty, and other adverse phenomena of the market economy. According to the famous English scholar A. Marshall, "Some forms of activity serve purposes of a social nature; others have a business basis, and only cooperation combines both." According to the Russian researcher of the cooperative sector of the economy, L. E. Fine, cooperation has shown and shows itself as a kind of "civilizer" of market relations. Thus, consumer cooperation contributes to the movement of the market economy toward socialization around the world.

2 Materials and Methods

At the early stage of economic development, cooperative relations were formed under the influence of self-regulating economic mechanisms and were spontaneous, sometimes chaotic formations. In turn, nowadays, there is a need to regulate this process; priority task becomes the translation of the cooperative sector of the economy on the course of technological renewal, which requires the study, generalization, and proposal of rational, effective, and advanced scientific technologies and mutually beneficial partnerships and forms of cooperation in the system of development of cooperation.

Cooperative formations are not isolated. They are part of a complex spatially heterogeneous system, which complicates the task of choosing the optimal vector of development for these forms of economic management. The problem must be set with a clear identification of the development directions, considering resource constraints. It is necessary to study the real state of the object and provide operational and reliable information about future scenarios for its development in unstable conditions [2–8]. In this situation, it is necessary to turn to simulation tools that transfer information about the functioning of the studied object from the category of inert material in the evaluation process, thereby increasing the efficiency and validity of management decisions.

As a result, there are objective possibilities to create a model of functioning and development of consumer cooperation as a system that considers the specifics of a particular territory, helping to automate the collection and analysis of information on the current state of the object and carry out pre-active planning with the ability to make strategic and operational management decisions by management bodies and all interested stakeholders.

3 Results

The world-historical experience is convincing evidence of the weighty contribution of the consumer cooperation system to the socio-economic development of the national economy and the improvement of the population's welfare. According to the World Cooperative Monitor, in 2020, the turnover of cooperative organizations in the world was $2.146 billion. The cooperative sector dominates the agricultural sector (104 enterprises) and the insurance sector (101 enterprises). The second place is occupied by the wholesale and retail trade sectors, including retail cooperatives (33 enterprises). Consumer cooperatives (21 companies) represent the third largest sector of the economy.

Most of the large cooperatives in the top 300 rankings are located in the most industrialized countries, such as the USA (74 companies), France (44 companies), Germany (30 companies), Japan (24 companies), the Netherlands (17 companies), and Italy (12 companies) (Fig. 1).

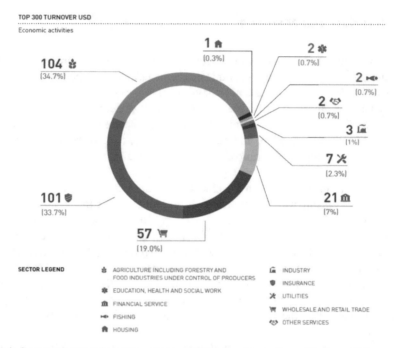

Fig. 1 Sectors of economic activity of cooperative enterprises in the world. *Source* [9]

What is the reason for the success of cooperative organizations in the global economy? First of all, the advantages of cooperation are manifested in the deepening of specialization and the social division of labor, which increases the production efficiency. This fact was mentioned by A. Smith, who wrote, "The division of labor in any craft, however much it is introduced, causes a corresponding increase in labor productivity" [10, pp. 80–81]. In Marx's "Capital," the doctrine of cooperation was placed "on that high pedestal which it lacked in the works of other economists." It was K. Marx who gave the doctrine of cooperation "a new, extremely important meaning that other economists had passed over" [11, pp. 404–405].

The Russian economist N. I. Zieber also proves the decisive importance of cooperation. He indicates that cooperation allows explaining all other manifestations of social life. The scientist believed that the key knowledge is the concept of social cooperation, which is a macroeconomic model of the organization of all social production. This concept should not be equated with the theory of the social division of labor. N. I. Zieber proposed a model of economic development in general and believed that "civilization is cooperation" since "union and freedom are its factors" [11].

Going deeper into Russian history, one cannot help but remember the incredible growth of cooperative farms at the turn of the nineteenth–twentieth centuries, when Russia was the world's leader in the number of farms and the number of shareholders. It was a fortunate direction of finding a format of market relations acceptable to Russia, considering national features and mentality. A favorable social environment allowed for the development of the traditional Russian artel, which had its roots in the thirteenth–fourteenth centuries, bringing to it the principles of the European cooperative movement.

Despite the modest share of the cooperative economy in the modern development of Russia, it is necessary to note the positive dynamics of its growth rate, which inspires confidence in the possible renaissance of this institution. Considering the sectoral structure of the modern economy burdened by the sanctions, the issues of food security and the development of the domestic agro-industrial complex become critical.

It is still possible to restore the system of consumer cooperation in rural areas. In cities, the competitive advantage is in the hands of large retail chains, whose financial capabilities are immeasurably higher.

4 Discussion

Throughout its evolution, consumer cooperation has accumulated a wealth of experience in the model of production and sale of agricultural products by uniting small and medium-sized producers, stimulating the creation of a multi-economic agrarian economy, and developing a competitive environment in regional food markets.

Competition leads to the need to introduce the achievements of scientific and technological progress in agriculture, find ways to save resources, and take care of the environment, which will lead to an increase in indicators such as the population's welfare, the quality of goods and services produced, and production efficiency. We believe that the high quality of products will be the main competitive advantage of cooperatives [12].

The association of cooperative organizations will form a value chain, including the purchase, storage, and processing of agricultural products and bringing them to the end consumer. The correlation of interests (rather than antagonistic confrontation) of agricultural producers, processing enterprises, and the market of products will allow forming an effective mechanism of consumer cooperation system functioning [13].

The cooperative can also protect the interests of its members from monopolistic pressure from supply chains, banks, other entities, and outside interference.

Additionally, most of the shareholders of the system of consumer cooperation are villagers. Thus, the creation of the cooperative stimulates the expansion of jobs and the growth of each shareholder's standard of living. We can make a figurative comparison. As the child of need and the mother of wealth, cooperation has always intensified under challenging periods of economic and social life [14].

Thus, the cooperative associations form a unified economic mechanism, providing a closed production and technological cycle and reducing production risks. Processing enterprises provide themselves with a reliable raw material base and a guarantee of sales of finished products.

An innovative model of formation and functioning of consumer cooperation should have the open nature of a logistics pilot, focused on the development and consolidation of cooperative formation according to the following algorithm:

1. Formation of the structure:

    -   Formation of the resource base;
    -   Formation of a base of logistical capabilities;
    -   Formation of a database of potential investors.

2. Formation of technological links;
3. Functional structure with identified patterns;
4. Development of a clear perspective on the development of commodity policy, allowing the engagement of priority and the most profitable activities in the perspective of the development program of cooperation at the micro-, meso- and macro levels;
5. Mechanism of generalization and implementation in consumer cooperation of the experience of advanced enterprises and international experience in the development of the sphere.
6. Development of the mechanism of prospective state regulation of the development of cooperation in Russia.

The starting point for implementing the proposed model can be the formation of a basis for designing a vector of economic development (in a particular case, certain production), namely—the creation of maps-bases of the underlying regional resources or the creation of a unified regional base for the development of consumer societies. The determination of the structure of this resource will give the main impetus for the rational formation of a particular production cycle. In turn, this will allow solving several acute social problems of the region.

5 Conclusion

Nowadays, the cooperative system has not exhausted its potential for development. It is the most massive socio-economic movement that confidently demonstrates a steady development trend in the world economy. Russian cooperation is in dissonance with the global trend. Unfortunately, Russia has lost its leading positions and many reference points on which the international cooperative movement is based.

The system of consumer cooperation can play a specific role in the revival of rural areas and improve rural residents' welfare [15]. For objective reasons, in this economic niche, it is impossible to make a super-profit. However, it is vital for certain segments of the population and the sustainable development of the country.

Additionally, the cooperative economy, given the sustainability of its development, should stop the outflow of the rural population, become attractive to the urban population, change the vector of spontaneous migration, and give impetus to the development of a traditional economic way of life in Russia.

It is critical to revive and strengthen the cooperative movement while the idea of collectivism is still preserved in the public consciousness. It is necessary to use the cooperative movement to the full for establishing a prosperous socially-oriented Russian state. A "two-story Russia" will bring back the traditional way of life and traditional values, which, unfortunately, are being erased by the "industrial machine" of the big cities.

References

1. Tinyakova VI, Morozova NI, Gunin VK, Kireeva OI (2019) Revival of the system of consumer cooperation in Russia; sustainable development of the territory and growth of quality life. Amazon Investiga 8(18):351–358
2. Bellman R (1957) On the computational solution of programming problems involving almost block diagonal matrices. Manage Sci 3(4):403–406
3. Brown RG (1971) Economic order quantities for materials subject to engineering changes. Prod Invent Manag 2:89–91
4. Czarnecki MT (1999) Managing by measuring: how to improve your organization's performance through effective benchmarking. American Management Association, New York, NY
5. Dorfman R, Samuelson PA, Solow RM (1958) Linear programming and economic analysis. McGraw-Hill, New York, NY
6. Fleming QW, Hoppelman JM (1996) Earned value project management. PMI, New York, NY
7. Morse P (1958) Queues, inventories and maintenance. Wiley, New York, NY
8. Tijms HC (1994) Stochastic models—an algorithmic approach. Wiley, New York, NY
9. International Cooperative Alliance (2020) Word cooperation monitoring, 2020. https://monitor.coop/sites/default/files/publication-files/wcm2020web-final-1083463041.pdf
10. Smith A (2000) An inquiry into the nature and causes of the wealth of nations (P. N. Klukin Transl. from English; A. Merkuryeva Ed.). EKSMO-Press, Moscow, Russia (Original work published 1776)
11. Zieber NI (1959) David Ricardo and Karl Marx in their social and economic studies. In: Selected economic works, vol. 1. USSR, Moscow, pp 404–405
12. Porter ME (2001) On competition (O. L. Pelyavsky, E. L. Usenko, & I. A. Shishkina Transl. from English). Publishing House "Williams", Moscow, Russia (Original work published 1998)
13. Tinyakova VI, Morozova NI, Konovalova OV, IY, Proskurina, Falkovich EB (2020) The cluster form of organization and the prospects for its application to provide the sustainable development of cooperative entrepreneurship. Revista Gênero Direito 9(4):1092–1103
14. Vakhitov KI (2017) Cooperation: theory, history, and practice. Selected sayings, facts, materials, comments. Dashkov and K, Moscow, Russia
15. Tinyakova VI, Morozova NI, Ziroyan MA, Falkovich EB (2018) Monitoring of human resources and a new educational structure for training specialists as key factors to reactivate the system of consumer cooperation in Russia. Amazon Investiga 7(17):353–359

Application of Micro Preparations as an Element of Agrobiotechnology for Soybean Cultivation in the Conditions of the Central Federal District

Kristina Yu. Zubareva⬥, Pavel Vas. Yatchuk⬥, Irina L. Tychinskaya⬥, and Evgeny Yu. Korolev⬥

Abstract The paper presents the results of studying the effectiveness of cultivation elements of promising soybean varieties Osmon and Zusha, selected by the Federal Scientific Center of Legumes and Groat Crops, which differ in the type of growth. The analysis of the growth of sown areas, gross yields, and yields of soybeans in the world, Russia, and in the Central Federal District of the Russian Federation allows us to speak about the possibility of producing up to 8–10 million tons of soybean in Russia and up to 2–2.5 million tons in the Central Federal District. To achieve such indicators, it is necessary to increase the area of sowing and the productivity of soybeans by implementing modern breeding methods and improving technology for different natural and climatic zones of the country. Plant vegetation and productivity management by implementing micro regulators and growth stimulators from Shchelkovo Agrohim are of great importance for consistently high yields of soybeans. Thus, the field experiments showed that the application of these products increased the number of beans per plant compared to the control without seed and plant treatment by 27.3% in the variety Zusha and by 44.5% in the variety Osmon. Grain weight per plant increased by 34.9% and 18.6%, respectively. Yields increased by 1.9 c/ha (6.1%) and 1.4 c/ha (4.2%). Additionally, the content and yield of protein per unit area increased by 85.02 and 77.02 kg or 7.6% and 6.7%, respectively.

Keywords Soybean · Micro fertilizers · Biostimulants · Yield structure · Yield · Yield quality · Raw protein

K. Yu. Zubareva (✉) · P. Vas. Yatchuk · I. L. Tychinskaya · E. Yu. Korolev
Federal Scientific Center of Legumes and Groat Crops, Streletsky village, Orel Region, Russia
e-mail: kristi_orel@bk.ru

P. Vas. Yatchuk
e-mail: pridatko1990@mail.ru

I. L. Tychinskaya
e-mail: pridatko1990@mail.ru

E. Yu. Korolev
e-mail: korolev.ew.91@mail.ru

JEL code Q1

1 Introduction

Optimization of agricultural production is aimed at more economically, energetically, and environmentally efficient crop cultivation [1]. The successful solution to this problem is facilitated by the production of less energy-intensive crops in crop production with science-based elements in the agricultural technologies of their cultivation, including the use of micro fertilizers, microbiological preparations, and bio-preparations. In this aspect, it is of great scientific importance to increase the cultivated areas of such a valuable crop as soybean, which has food, fodder, technical, and medicinal value [2] in the implementation of the UN World Food Program, in general, and the Doctrine of Production Security of the Russian Federation [3], in particular.

Soybean has a unique chemical composition, the characteristics of which allow creating protein-balanced foods and products with functional, therapeutic, and prophylactic properties for many diseases [4]. Compared with the proteins of grain crops, soybean protein provides a higher collection of easily digestible protein and essential amino acids per unit of the cultivated area by its amino acid composition and the total amount of protein content.

For soybeans, an essential mechanism of interaction with rhizobacteria is their ability to stimulate the formation of nitrogen-fixing symbiosis with nodule bacteria [5, 6].

These circumstances have led to an increase in the volume of this crop, both in Russia and worldwide (Table 1). Soybean acreage in Russia has been rapidly growing in recent years, thus being one of the targets of developing the country's agricultural sector. According to the preliminary data of the Federal State Statistics Service (Rosstat), the sown area of soybean amounted to 2,859.5 thousand hectares in 2020. In 2010, this figure equaled 1,209.3 thousand hectares. That is, in 10 years, there was an increase in crop area by 2.4 times. The soybean yield per hectare also increased: from 11.8 c/ha in 2010 to 15.9 c/ha in 2020. The gross production of soybean was 4,283 thousand tons in 2020, which is 1.7 times higher than the annual average for the period from 2010 to 2019 [7–9].

It should be noted that soybean acreage in Russia has been growing steadily. Nevertheless, it decreased by 7.7% in 2020 (219.5 thousand hectares less than in 2019). This happened for two reasons:

- The purchase price of soybeans from major processors has fallen to a record low level;
- A ban was imposed on soybean exports to China from the Far East.

Increasing the yield indicator has a complete functional–linear direct relationship to the increase in gross yield of crops (the statistical relationship of the two values is close to unity) [10]. According to Rosstat, the soybean yield in the Orel Region

Table 1 Actual sown areas of soybeans, gross yields, and yields of soybeans (at weight after processing) in the Russian Federation on farms of all categories

	2010–2019 (annual average)	2020 in % to 2010–2019	2019	2020*	2020 in % to 2019
Sown area, thousand hectares	2,051.1	171.7	3,079	2,859.5	92.9
Gross yields, thousand tons	2,534.6	159.2	4,360	4,283	92.8
Yield, centner per hectare of harvested area	13.4	118.7	15.7	15.9	101.3
Including in the Orel region					
Sown area, thousand hectares	184.6	56.0	119.2	103.3	86.7
Gross yields, thousand tons	205.5	91.8	195.5	188.7	96.5
Yield, centner per hectare of harvested area	14.1	131.9	16.7	18.6	111.4

Note *Preliminary data. *Source* Rosstat [7–9]

in 2020 was 18.6 c/ha, which is 111.4% higher than in 2019 and 131.9% higher than in 2010–2019. Simultaneously, the soybean yield in the advanced farms of the region reached up to 30.4 c/ha (Dubovitskoe LLC, the Maloarkhangelsk district), 27.6 c/ha (Exima-Agro LLC, the Pokrovsky district), and 21.7 c/ha (Sosnovka JSC, the Livensk district) [11]. This fact shows the real possibilities of increasing yields by improving technology and implementing new breeding achievements.

One of the directions to improve some aspects of the technology of soybean cultivation is the treatment of seeds before sowing with inoculants and bio stimulators, as well as leaf feeding of vegetative plants with micro fertilizers, which provides highly productive agrocenosis and stable yields of high quality [12]. Therefore, the developed science-based agricultural technologies of cultivating released soybean varieties in specific soil and climatic conditions are critical in crop production. In this regard, our research aims to study the response of new promising soybean varieties Osmon and Zusha to comprehensive pre-sowing seed treatment and implementation of foliar micro fertilizer during cultivation in the Orel Region. The research task is to study the features of the formation of elements of the yield structure, its quantity, and quality with the wide-row method of sowing.

2 Materials and Methods

Experimental studies were conducted on the experimental field of the Federal Scientific Center of Legumes and Groat Crops in the 2019–2020 field rotation on a fallow forecrop.

The soil in the experimental field is dark gray forest, medium-loamy, medium-humus (4.2%), slightly acidic (pH 5.1), with an average content of exchangeable potassium (12.2%, according to Maslov's method), and shallow content of mobile phosphorus (18.0%, according to Chirikov's method).

The research objects were two contrasting varieties:

- Osmon, an early maturing variety of indeterminant types of growth and development (in the State Register of the Russian Federation since 2018) [13];
- Zusha, a mid-early variety of semi-determinant type of growth and development (in the State Register of the Russian Federation since 2015) [14].

We studied the effectiveness of micro fertilizers and a fungicide seed dressing from Shchelkovo Agrokhim company [15], marked with low application rates and relatively low cost.

Rizoform is a liquid soybean seed inoculant based on a strain of the specialized soybean bacteria *Bradyrhizobium japonicum* (2–3 × 109 CFU/ml). It is used before sowing, provides biological nitrogen to plants during the most critical phases of crop growth and development, increases soil fertility, and activates soil microbiota, positively affecting the quantity and quality of yield and subsequent crops in the rotation. It was used at a dose of 3.0 L/t of seed.

Biostim Start is an amino acid biostimulant for germination, development of root system at the initial stage of ontogenesis, and relieving "transplanting" stress. It was used at a dose of 1.0 l/t of seed.

Biostim Oily is a micro fertilizer and bio stimulator with microelements for the foliar dressing of vegetative soybean plants to eliminate the deficiency of NPK and microelements during growth and development. This fertilizer increases drought tolerance, frost resistance, and resistance to diseases and stress. It was used at a dose of 1.0 l/ha of crops.

Intermag Pro Legumes and Pods is a multicomponent micro fertilizer for the foliar dressing of vegetative soybean plants. It effectively maintains the balance of macro- and microelements during the critical periods of growth and development of the crop. This micro fertilizer contains a growth activator to increase the quality of absorption of vital components of the soil solution. Moreover, it increases plant resistance to biotic and abiotic stresses. It was used at a dose of 1.0 l/ha of crops.

Scarlet, ME is a fungicide seed dressing containing active components at a dose of 100 g/l of imazalil and 60 g/l of tebuconazole. It provides long-lasting, fast, and high levels of fungicidal activity against a wide range of diseases. Moreover, it stimulates the formation of a strong root system and increases frost and drought resistance. It was used at a dose of 0.4 l/t of seed.

Fig. 1 Sowings of soybean variety Zusha before the first leaf dressing in the 1–2 triple leaves phase: on the right—control, untreated seeds; on the left—experimental version where seeds were pre-treated before sowing. *Source* Photographed by the authors

Seeds were pre-treated before sowing with a tank mixture of *Scarlet, ME* and *Biostim Start*. On the day of sowing, the seeds were inoculated with *Rizoform* and a stabilizer adhesive. The seeding rate of germinated seeds per one hectare is generally accepted for zoning; in quantitative terms, it equals 0.6 million units. The method of sowing is wide-row with a row spacing of 45 cm (Fig. 1). The counting area of the plots is 10 m^2, the repetition is four times, and the placement is randomized. Leaf dressing was carried out by spraying crops with a tank mixture *Biostim Oily* and *Intermag Profi* in the phase of the third triple leaf and the phase of budding; the rate of working fluid—300 l/ha. Before harvesting, plant samples were taken from the plots to analyze the structural elements of the potential biological yield. Harvesting was carried out by SAMPO-130 direct harvester in the ripening phase when the optimum moisture content of the grain was 16–18%. Yield data are given at standard moisture content and 100% purity.

The studies were accompanied by phenological observations, analysis of crude protein content in the grain [16] on an Infratec 1241 device, and statistical processing of the obtained data using Microsoft Office Excel and Dispersion 3.01.

In 2019–2020, meteorological conditions were favorable for growing and developing soybean plants. The accumulation of active temperature sums greater above 10 °C was 2,597.2 °C for the soybean harvest of 2019. In 2020, this figure was 5.1% lower and amounted to 2,463.8 °C.

Soybeans are very demanding to heat. The optimal temperature for soybean germination is 15–20 °C. In 2020, the sowing and sprouting period was longer since the average daily temperature was 11.2 °C. In 2020, only some single soybean sprouts appeared on day 19, compared to day 6 in 2019.

The flowering period occurred in optimal temperature conditions (21.4 °C), which eliminated the abortion of flowers and led to the formation of full-grown beans.

Soybeans are especially demanding of moisture in the soil during seed formation and ripening. The hydrothermal coefficient (HTC) was characterized as dry (HTC = 0.31) in 2020 and as torrid (HTC = 0.71) in 2019.

The yield decreases if soybean plants experience moisture deficits during the critical phases of growth and development that require water consumption. This situation was observed in August 2020, when the period of seed ripening was extended.

However, soybeans poorly tolerate overwatering of soil, especially before flowering. In 2020, the hydrothermal coefficient was interpreted as containing excessive moisture (HTC = 2.03). In overwatered soil, the aeration of the root system is disturbed. In this case, the conditions for the formation of nodule bacteria worsen, the growth and development of plants are slowed, which results in the reduced number of flowers and full beans. The vital activity of nodules is inhibited due to the development of more anaerobic processes [5].

3   Results

The productivity of soybean yield was largely influenced by the following:

- Treatment of soybean seeds before sowing with a tank mixture of fungicidal dressing *Scarlet, ME* and bio stimulator amino acid *Biostim Start*;
- Inoculation with microbial fertilizer *Rizoform* and stabilizer-adhesive (Static, 0.85 l/t) (at the day of sowing);
- Subsequent double treatment of crops with a tank mixture of a multicomponent micro fertilizer *Intermag Pro Legumes and Pods* and bio stimulator *Biostim Oily* at the three triple leaves phase and budding phase (Table 2).

Thus, these micro preparations increased the number of beans per plant compared to the control without seed and plant treatment by 27.3% in the variety Zusha and 44.5% in the variety Osmon. In the variety Zusha, grain weight per plant increased by 34.9% and amounted to 8.9 g. In the variety Osmon, grain weight per plant increased by 18.6% and amounted to 8.3 g. There is a variability in the index of 1000 seeds weight in the direction of increasing from the control variant to the experimental variant.

A feature that determines the efficiency of mechanized harvesting of leguminous crops and soybeans is the height of the bottom bean attachment, which can be increased by using growth stimulants and organic and mineral fertilizers [17, 18]. Micro preparations used to treat seeds and vegetative plants influenced the average stem length of soybean plants by 2.9 cm in the variety Zusha compared to the control. The plants reached 112.6 cm, and the height of the bottom bean attachment was 14.2 cm, which is 3.0 cm more than the control variant.

Quantification of productivity components affects the yield value. Table 3 shows the grain yield of promising soybean varieties Zusha and Osmon depending on research options.

Table 2 Effect of agricultural technologies on the structure of soybean yield, 2019–2020

Variant	Stem length, cm	Height of bottom bean attachment, cm	Number of beans per plant, pcs	Graininess of the bean, pcs	Weight of 1,000 seeds, g
Zusha variety					
Control	109.7	11.2	21.6	1.8	180.0
Field experiments	112.6	14.2	27.5	1.9	185.2
Osmon variety					
Control	135.3	22.2	20.0	2.3	143.3
Field experiments	136.0	22.4	28.9	2.3	149.7
$HCP_{0,5}$	$HCP_{05A} = 11.9$; $F_{05B} < F_t$; $F_{05AB} < F_t$	$HCP_{05A} = 4.2$; $F_{05B} < F_t$; $F_{05AB} < F_t$	$HCP_{05B} = 4.8$; $F_{05B} < F_t$; $F_{05AB} < F_t$	$HCP_{05A} = 0.1$; $F_{05B} < F_t$; $F_{05AB} < F_t$	$HCP_{05A} = 4.0$; $F_{05B} < F_t$; $F_{05AB} < F_t$

Note Factor A—variety; Factor B—variant. *Source* Compiled by the authors

Table 3 Effect of micro fertilizers on soybean yield (2019–2020)

Variants	Yield, c/ha			
	2019	2020	Average for two years	Yield increase, c/ha (averaged for two years)
Zusha variety				
Control	31.0	27.3	29.2	-
Field experiment	31.7	30.5	31.1	1.9
Osmon variety				
Control	35.8	27.6	31.7	-
Field experiment	36.2	30.1	33.1	1.4
2019—$HCP_{05A} = 0.5$; $F_{05B} < F_t$; $F_{05AB} < F_t$, Factor A—variant, 2020—$HCP_{05B} = 0.7$; $F_{05A} < F_t$; $F_{05AB} < F_t$, Factor B—variety				

Source Compiled by the authors

Thus, over two years of research, the yield in the variant with the application of micro preparations was higher than in control by 1.9 c/ha (6.1%) for the variety Zusha and by 1.4 c/ha (4.2%) for the variety Osmon. The yield amounted to 31.1 c/ha and 33.1 c/ha, respectively.

Micro preparations used in the cultivation of soybeans in the treatment of seeds and crops positively affected the qualitative composition of soybean grain, namely, on the formation of the protein complex (Table 4).

Table 4 Effect of preparations on the quality of soybean grain (2019–2020)

Variants	Raw protein		± to the control
	content, %	harvest, kg/ha	kg/ha
Zusha variety			
Control	38.2	1115.44	-
Experiment	38.6	1200.46	85.02
Osmon variety			
Control	36.1	1144.37	-
Experiment	36.9	1221.39	77.02

Source Compiled by the authors

Seed and crop treatment increased the collection of raw protein from a unit of the cultivated area of soybean variety Zusha by 85.02 kg/ha (7.6%) compared to the control. For the variety Osmon, the increase is 77.02 kg/ha (6.7%).

4 Conclusion

The research results indicate the effectiveness of the complex micro preparations of Shchelkovo Agrohimkak on certain elements of yield structure and the overall productivity of soybeans of Zusha and Osmon varieties in the Orel Region. The soybean productivity increased by 1.9 c/ha (6%) for the Zusha variety and by 1.4 c/ha (4%) for the Osmon variety due to the following actions:

- Pre-sowing treatment of seeds with *Scarlet, ME* (a dose of 0.4 l/t) and *Biostim Start* (a dose of 1.0 l/t) (in advance);
- Treatment with *Rizoform* with Static adhesive (a dose of 3.0 l/t) (on the day of sowing);
- Spraying of the vegetative plants at the phase of the third triple leaf and the phase of budding with a tank mixture *Biostim Oily* (a dose of 1.0 l/ha) and *Intermag Profi Legumes and Pods* (a dose of 1.0 l/ha).

Additionally, protein content and yield per unit area increased by 85.02 and 77.02 kg/ha, relative to the control.

References

1. Kochurko VI, Abarova EE (2014) Role of microelements in the formation of soybean yield. Zemledelie 8:30–32
2. Zhuchenko AA (2004) Ecological genetics of cultivated plants and agrosphere problems: theory and practice. Agrorus, Moscow, Russia

3. Presidential Executive Office (2020) Executive order on approval of the food security doctrine of the Russian Federation (January 21, 2020 No. 20). Moscow, Russia. https://www.garant.ru/products/ipo/prime/doc/73338425/. Accessed 18 May 2021
4. Petibskaya VS (2012) Soybean: chemical composition and uses. Polygraph-UG, Maykop, Russia
5. Beregovaya YuV, Tychinskaya IL, Petrova SN, Parahin NV, Puhalsky JV, Makarova NM, Shaposhnikov AI, Belimov AA (2018) Cultivar specificity of the rhizobacterial effects on nitrogen-fixing symbiosis and mineral nutrition of soybean under agrocenosis conditions. Agric Biol 53(5):977–993. https://doi.org/10.15389/agrobiology.2018.5.977rus
6. Provorov NA, Onishchuk OP (2019) Ecological and genetic bases for construction of highly effective nitrogen-fixing microbe-plant symbioses. Ecol Genetics 17(1):11–18. https://doi.org/10.17816/ecogon17111-18
7. Federal State Statistics Service [Rosstat] (2020) Planted areas, gross yields, and yields of crops in the Russian Federation in 2020 (preliminary data). https://rosstat.gov.ru/compendium/document/13277. Accessed 4 Feb 2021
8. Rosstat regional office of Orel Region (n.d.) Agriculture, hunting, and forestry. https://orel.gks.ru/sh_ohota_lh?print=1. Accessed 18 May 2021
9. Laikam KE (2019) Agriculture in Russia: statistical digest. Rosstat, Moscow, Russia
10. Bayanova OV (2020) Econometric study of grain production indicators in the Russian Federation. Modern Econ Success 1:120–125
11. Zotikov VI, Sidorenko VS, Matveichuk PV (2020) Productivity and grain quality of winter wheat and soybean varieties at LLC "Dubovitskoe." Legumes Groat Crops 1(33):92–98. https://doi.org/10.24411/2309-348X-2020-11162
12. Piskov VB, Chernyshev VP, Karakotov SD (2016) M-dinitroaromaticmoiety as a fragment of biologically active compounds. Pharm Chem J 49(11):724–734
13. Zadorin AM, Zelenov AA, Mordvina MV (2019) Achievements of selection of federal scientific center of legumes and groat crops in the aspect of the growth of soybean production in Russia. Legumes Groat Crops 2(30):53–56. https://doi.org/10.24411/2309-348X-2019-11088
14. Schelkovo Agrohim (n.d.) Catalog. https://eng.betaren.ru/catalog/. Accessed 18 Май 2021
15. Schelkovo Agrohim (n.d.) Seeds of grain legumes, oilseeds, and cereals. https://eng.betaren.ru/catalog/seeds/. Accessed 18 May 2021
16. Feeds, mixed feeds, and raw material. Determination of mass fraction of nitrogen and calculation of the mass fraction of crude protein. Part 1. Kjeldahl method (2014) (July 1, 2014 GOST R 32044.1–2012 [ISO 5983–1:2005]). Standardinform, Moscow, Russia
17. Akulov AS, Vasilchikov AG (2019) Study of the effectiveness of the use of growth promoter. Legumes Groat Crops 2(30):72–77. https://doi.org/10.24411/2309-348X-2019-11092
18. Marakaeva TV (2020) The suitability of selection lenil specimens for mechanized harvesting. Bull KSAU 9:41–45. https://doi.org/10.36718/1819-4036-2020-9-41-45

Marker-Assisted Selection of Pea Interspecific Hybrids with Introgressive Alleles of Convicilin

Sergey V. Bobkov and **Tatyana N. Selikhova**

Abstract The paper studies the peculiarities of selection of interspecific pea hybrids with introgressive alleles of convicilin by marker bands (convicilin isoforms) of electrophoretic spectra of seed proteins. Accessions of wild pea species *Pisum fulvum* from the Federal Research Center N. I. Vavilov All-Russian Institute of Plant Genetic Resources (VIR) were used as donors of valuable convicilin alleles. Hybridization of wild accessions k-6070 and k-2523 with cultivated pea allowed us to obtain 78 seeds of interspecific hybrids F_2, F_3, BC_1F_1, BC_1F_2, BC_2F_1, BC_2F_4. Seeds of interspecific hybrids were split into two cotyledons. One cotyledon with a bud was used to produce plants, and the other was subjected to electrophoretic analysis of storage proteins. The cotyledon with a bud was either germinated in advance or placed in the soil in a dry state, where it was grown to whole plants. In 2020, 78 dry isolated cotyledons were planted in the soil, and 21 plants were obtained (26.9%). Seventy isolated cotyledons of hybrid seeds were subjected to electrophoretic analysis. Sixteen interspecific hybrids were isolated as carriers of "wild" convicilin isoforms alleles. Of these, we labeled five hybrids containing only "wild" convicilin isoforms and, accordingly, alleles encoding them, but without "cultivated" isoforms of this protein. These include two hybrids F_3, one hybrid BC_2F_1 from the crossing of cultivated pea with a wild accession k-6070, and two hybrids F_2 obtained in crosses with the accession k-2523. We obtained the seeds from two labeled hybrid plants BC_1F_2 carrying the alleles of the "wild" convicilin isoforms of accession k-6070.

Keywords Pea · *Pisum sativum* · *P. fulvum* · Wild relative · Introgression · Protein · Convicilin · Isoform

JEL codes Q1

S. V. Bobkov (✉) · T. N. Selikhova
Federal Scientific Center of Legumes and Groat Crops, Orel Region, Streletsky village, Russia

T. N. Selikhova
e-mail: tat.selihowa@yandex.ru

283

1   Introduction

Pea (*Pisum sativum* L.) is a valuable leguminous crop. Pea grains contain high-quality protein. Compared to soy protein, pea protein is better absorbed by animals and humans due to the lower content of proteinase inhibitors and lectins [1, 2]. In 2019, in the world, pea was grown on 7.1 million hectares and 14.2 million tons of grain were produced [3]. In Russia, the area and production of grain amounted to 1.2 million hectares and 2.4 million tons, respectively. In 2019, Russia's share in global pea production was 16.9%.

The protein content of pea varieties is aproximately 26%, while in wild pea it exceeds 30% [4, 5]. Studies using 2-D electrophoresis revealed 88 storage and 68 non-storage proteins in the seeds of pea cultivar Cameor [6]. Storage proteins in pea are mainly salt-soluble globulins, while water-soluble albumin represents the second important class of reserve proteins [7]. Globulins include legumin, vicilin, and convicilin [8, 9]. The content of albumin is 10%–20% of the total protein content. The content of legumin and vicilin is 65%–80% [10]. Convicilin is contained in smaller quantities. A study of 59 lines and samples of pea showed that the variation in the content of legumin, vicilin, and convicilin in the protein extract was 5.9%–24.5%, 26.3% and 52.0%, and 3.9%–8.3%, respectively [11]. In another study, electrophoretic analysis under reducing conditions showed that the protein extracts had 4.2–4.8 higher vicilin content than convicilin [12].

In general, pea protein is rich in lysine but contains little methionine and tryptophan. There are differences in the amino acid composition between the individual groups of proteins. Albumins contain more tryptophan, lysine, and methionine compared to globulins [7, 13]. Vicilin contains no sulfur-containing amino acids (methionine and cysteine). Nevertheless, it is labeled with high levels of aspartic acid and lysine [7, 14]. Legumin contains more valine, glycine, and alanine. Convicilin differs from both vicilin and legumin in its amino acid composition. Unlike vicilin, convicilin contains sulfur-containing amino acids and differs from legumin in its higher lysine content [14].

In 1980, R. R. Croy, J. A. Gatehouse, M. Tyler, and D. Boulter isolated convicilin [14]. After that, it began to be considered as the third separate globulin protein of pea [15]. Unlike vicilin, it has its own coding genes [6, 15]. Convicilin has extensive homology to the amino acid sequence of vicilin. It additionally contains an extended N-terminus with more residues of acidic amino acids [15, 16]. Different concentrations of vicilin and legumin in the mixture, and separately, can form good gels from protein isolates, while convicilin hinder gelation [9]. Therefore, pea varieties enriched with vicilin and legumin but not containing convicilin may serve as a desirable material for the food industry [11, 17]. However, no genotypes have been found in cultivated pea that do not contain convicilin in their seeds [1]. At the same time, isoforms of convicilin with a lower molecular weight were detected in accessions k-6070 and k-2523 of wild pea species *Pisum fulvum* from the VIR collection [18].

Currently, there is an increasing interest in using wild relatives of pea as a source of new alleles of economically valuable traits [19, 20]. The introgression of "light" convicilin isoform alleles from accessions k-6070 and k-2523 into elite varieties and breeding lines is a good prospect for pea breeding to obtain high-quality isolated proteins. Studying the effect of wild pea convicilin isoforms on the physicochemical properties of protein isolates is a prerequisite for the use of introgressive pea lines in the breeding process.

The paper aims to investigate the peculiarities of selection of pea interspecific hybrids with introgressive alleles of convicilin by marker bands (convicilin isoforms) of seed protein electrophoretic spectra.

2   Materials and Methods

Plants of *P. fulvum* accessions k-2523 and k-6070 with valuable convicilin alleles from the collection of the Federal Research Center N. I. Vavilov All-Russian Institute of Plant Genetic Resources (VIR), interspecific hybrids of the first and second generations, and backcrossed lines were grown in the greenhouse box in 2019 and 2020. During the growing season, interspecific hybrids were crossed with cultivated pea plants.

Seeds of interspecific pea hybrids were separated into two cotyledons. The cotyledon with a bud was planted in the soil culture, and the other was subjected to electrophoretic analysis. Pea seed proteins were separated in polyacrylamide gel using SDS-PAGE electrophoresis [21]. Extraction of storage proteins was performed for 20 h with electrode buffer (TRIS, glycine, and sodium dodecyl sulfate) in the refrigerator at 4 °C, pH = 8.3. After centrifugation, 10 μl of the extract was mixed with an equal volume of application buffer (TRIS–HCl, glycerol, sodium dodecyl sulfate, β-mercaptoethanol, and bromophenol blue). Ten microliters of the resulting mixture were placed in cells of 5% stacking gel filled with electrode buffer in a vertical electrophoresis chamber VE-4 (Helicon, Russia). Protein separation occurred in a 12.5% gel.

Soybean variety Lancetnaya was used as an external group to interspecific hybrids. The positions of the pea proteins were compared with the positions of the reference protein bands of the soybean spectrum. We used a point-based assessment to estimate the intensity of band staining: one point corresponded to the weak intensity of staining, two points corresponded to intensive staining, and three points corresponded to highly intensive staining. Localization of storage proteins on electrophoretic spectra was determined using a set of markers with a molecular mass of 6.5–200 kDa (Sigma-Aldrich, USA). Bands with a molecular weight of ~70 kDa indicated the localization of convicilin isoforms [11, 15].

3 Results

To obtain interspecific pea hybrids, wild pea species accessions k-6070 and k-2523, donors of valuable convicilin alleles, were crossed with varieties Stabil, Sofia, Temp, and Rodnik and breeding lines PAP 485/4 and L-375. Backcrosses were carried out in the combination of *P. sativum* × *P. fulvum* (k-6070). Totally, seventy-eight seeds of interspecific hybrids F_2, F_3, BC_1F_1, BC_1F_2, BC_2F_1, and BC_2F_4 were used for marker-assisted selection of pea plants with introgressive convicilin alleles (Table 1). Nine cotyledons of interspecific hybrids F_2 *P. sativum* × *P. fulvum* (k-2523) and 69 hybrid cotyledons F_2, F_3, BC_1F_1, BC_1F_2, BC_2F_1, and BC_2F_4 *P. sativum* × *P. fulvum* (k-6070) were planted in soil.

After coat removing, the seeds of interspecific hybrids were separated into two cotyledons. One cotyledon (without bud) was used for extraction and separation of storage protein by SDS-PAGE electrophoresis. Two approaches were used to acclimatize isolated cotyledons with buds into a soil culture. In the first approach, seedlings grown from cotyledon buds on moist filter paper in Petri dishes were planted in the soil. In the second approach, dry isolated cotyledons with a bud were placed in the soil.

In the present experiment, dry isolated cotyledons of interspecific pea hybrids were planted in polypropylene vessels with soil. The planted cotyledons were numbered. Out of 78 cotyledons planted, 21 seedlings were obtained (Fig. 1). The efficiency of obtaining plants from isolated cotyledons was 26.9%.

Seventy interspecific pea hybrids F_2, F_3, BC_1F_1, BC_1F_2, and BC_2F_1 obtained from the crossing of cultivated pea with the accessions of *P. fulvum* k-6070 and k-2523 were subjected to electrophoretic analysis (Table 2, Fig. 2).

Electrophoretic spectra of seed proteins were used to analyze segregation of interspecific hybrids by convicilin isoforms. Plants with introgressive isoforms of convicilin grown from a cotyledon located in the same seed as the analyzed one were selected for further hybridization with cultivated pea and subsequent use in pea breeding on high protein quality.

Table 1 Pea seeds used for marker-assisted selection of interspecific hybrids with introgressive alleles of convicilin

Interspecific hybrid	Generation	Sown seeds
Pisum sativum × *P. fulvum* (k-2523)	F_2	9
P. sativum × *P. fulvum* (k-6070)	F_2	2
	F_3	9
	BC_1F_1	7
	BC_1F_2	27
	BC_2F_1	21
	BC_2F_4	3
Total		**78**

Source: Compiled by the authors

Fig. 1 Sprouting of interspecific pea hybrids from cotyledons placed dry in a soil culture: **a** semi-leafless interspecific hybrids BC_1F_2 (PAP 485/4 × k-6070) × Stabil (A10, A11); **b** leafy interspecific hybrids BC_2F_1 Temp × ((PAP 485/4 × k-6070) × Stabil) (D1), BC_2F_1 (Stabil × k-2523) × Stabil (C9), BC_1F_2 (PAP 485/4 × k-6070) × Rodnik (C10), BC_1F_2 (PAP 485/4 × k-6070) × L-375 (C7). *Source* Photographed by the authors

Of 70 cotyledons of pea interspecific hybrids subjected to electrophoretic analysis, 16 were labeled as carriers of alleles of introgressive convicilin isoforms (Table 2).

It should be noted that interspecific pea hybrids may have introgressive and traditional convicilin alleles, as well as two types of alleles together. For example, in crosses of cultivated pea with the *P. fulvum* accession k-6070 revealed two hybrids F_3 and one BC_2F_1, which had wild but no traditional convicilin alleles. Crossing of cultivated pea with the accession k-2523 provided two hybrids F_2 containing introgressive but not traditional convicilin alleles. In hybrid combination with accession k-6070 from the two labeled cotyledons BC_1F_2, we obtained pea plants that bloomed and formed seeds. The indicated plants contained both introgressive and traditional convicilin alleles. Thirty-three hybrid seeds BC_1F_3 were collected from these plants (Table 3).

Table 2 Isoforms of convicilin in cotyledons of interspecific pea hybrids revealed with SDS-PAGE electrophoresis

Name of labeled interspecific hybrid	Wild parent	Cotyledon generation	Convicilin isoforms	
			Introgressive	Traditional
A1	k-6070	BC_1F_2		Yes
A2	k-6070	BC_1F_2		Yes
A3	k-6070	BC_1F_2		Yes
A4	k-6070	BC_1F_2		Yes
A5	k-6070	BC_1F_2		
A6	k-6070	BC_1F_2		Yes
A10	k-6070	BC_1F_2		Yes
A11	k-6070	BC_1F_2		Yes
A12	k-6070	BC_1F_2		Yes
B1	k-6070	F_3		Yes
B2	k-6070	F_3		Yes
B3	k-6070	F_3		Yes
B4	k-6070	F_3	Yes	
B5	k-6070	F_3		Yes
B6	k-6070	F_3		Yes
B7	k-6070	F_3	Yes	
B10	k-2523	F_2	Yes	Yes
B11	k-2523	F_2	Yes	Yes
B12	k-2523	F_2		Yes
C2	k-6070	BC_1F_2	Yes	Yes
C3	k-6070	BC_1F_1	Yes	Yes
C4	k-6070	BC_1F_1	Yes	Yes
C5	k-6070	BC_1F_1	Yes	Yes
C6	k-6070	BC_1F_2		Yes
C7	k-6070	BC_1F_2	Yes	Yes
C8	k-2523	BC_1F_2	Yes	Yes
C9	k-2523	BC_2F_1	Yes	Yes
C10	k-6070	BC_1F_2	Yes	Yes
C12	k-6070	BC_1F_2		Yes
D1	k-6070	BC_2F_1		Yes
D2	k-6070	BC_2F_1		Yes
D3	k-6070	BC_2F_1		Yes
D4	k-6070	BC_2F_1		Yes
D7	k-6070	BC_2F_1		Yes

(continued)

Table 2 (continued)

Name of labeled interspecific hybrid	Wild parent	Cotyledon generation	Convicilin isoforms	
			Introgressive	Traditional
D8	k-6070	BC_2F_1		Yes
D9	k-6070	BC_2F_1		Yes
D10	k-6070	BC_2F_1		Yes
D11	k-6070	BC_2F_1		Yes
D12	k-6070	BC_2F_1		Yes
E1	k-6070	BC_2F_1		Yes
E2	k-6070	BC_2F_1	Yes	
E3	k-6070	BC_2F_1		Yes
E4	k-6070	BC_2F_1		Yes
E5	k-6070	BC_2F_1		Yes
E6	k-6070	BC_2F_1		Yes
E7	k-6070	BC_2F_1		Yes
E8	k-6070	BC_2F_1		Yes
E9	k-6070	BC_1F_1		Yes
E10	k-6070	BC_1F_1		Yes
E11	k-6070	BC_1F_1		Yes
E12	k-6070	BC_1F_1		Yes
F1	k-6070	BC_1F_1		Yes
F2	k-6070	BC_1F_1		Yes
F3	k-6070	BC_1F_2		Yes
F4	k-6070	BC_1F_2		Yes
F9	k-6070	BC_1F_2		Yes
F10	k-6070	BC_1F_2		Yes
F11	k-6070	BC_1F_2		Yes
F12	k-6070	BC_1F_2		Yes
G1	k-6070	BC_1F_2		Yes
G2	k-6070	BC_1F_2		Yes
G3	k-2523	F_2		Yes
G4	k-2523	F_2		Yes
G5	k-2523	F_2	Yes	
G6	k-2523	F_2		Yes
G7	k-2523	F_2		Yes
G8	k-2523	F_2		Yes

(continued)

Table 2 (continued)

Name of labeled interspecific hybrid	Wild parent	Cotyledon generation	Convicilin isoforms	
			Introgressive	Traditional
G9	k-2523	F_2	Yes	
G10	k-2523	F_2		Yes
G12	k-2523	F_2	Yes	Yes

Source Compiled by the authors

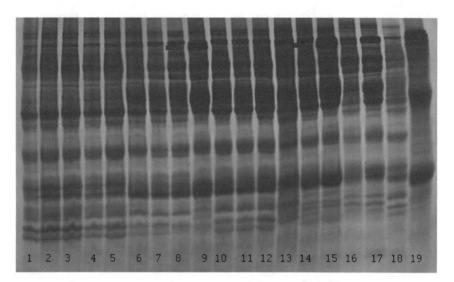

Fig. 2 Seed protein spectra of *P. fulvum* accessions k-6070 and k-2523 from the VIR collection, variety Stabil, and interspecific hybrids: 1–15—interspecific hybrids, 16—*P. fulvum* accession k-2523, 17—*P. fulvum* accession k-6070, 18—variety Stabil, 19—spectrum of the soybean (variety Lancetnaya). Red arrows show introgressive convicilin isoforms in interspecific hybrids BC_1F_2. *Source* Obtained by the authors

In total, only 33 (19.6%) out of 168 seeds of labeled pea interspecific hybrids were obtained from plants that inherited alleles of introgressive convicilin isoforms.

4 Conclusion

Crosses of cultivated pea with wild *P. fulvum* accessions k-6070 and k-2523 allowed us to obtain 78 seeds of interspecific hybrids F_2, F_3, BC_1F_1, BC_1F_2, BC_2F_1, BC_2F_4. Seeds of interspecific hybrids were split into two cotyledons. One was used to produce plants, and the other was used for electrophoretic analysis of storage proteins. Seventy-eight dry isolated cotyledons with a bud were planted in the soil, including

Table 3 Seeds harvested from labeled plants of pea interspecific hybrids

Name of labeled interspecific hybrid	Seed generation	Number of seeds	Alleles of wild convicilin isoforms
Seeds of labeled plants			
A1	BC_1F_3	11	
A2	BC_1F_3	10	
A10	BC_1F_3	32	
C7	BC_1F_2	17	Yes
C10	BC_1F_2	16	Yes
D1	BC_2F_2	3	
D2	BC_2F_2	6	
D3	BC_2F_2	4	
D10	BC_2F_2	5	
D11	BC_2F_2	10	
E9	BC_1F_2	2	
F1	BC_1F_3	7	
F3	BC_1F_3	19	
F4	BC_1F_3	17	
G3	F_3	2	
G10	F_3	7	
Total		168	
Seeds with alleles of "wild" convicilin isoforms		**33 (19.6%)**	

Source Compiled by the authors

nine cotyledons of interspecific hybrids F_2 *P. sativum* × *P. fulvum* (k-2523) and 69 hybrid cotyledons of F_2, F_3, BC_1F_1, BC_1F_2, BC_2F_1, and BC_2F_4 *P. sativum* × *P. fulvum* (k-6070). Cotyledons planted in the soil produced 21 plants (26.9%). Seventy cotyledons of hybrid seeds were subjected to electrophoretic analysis. Sixteen interspecific hybrids were isolated as carriers of the alleles of introgressive convicilin isoforms. Of these, five hybrids contained no traditional but only introgressive convicilin isoforms and, accordingly, the alleles encoding them. These include two hybrids F_3, one hybrid BC_2F_1 from the crossing of cultivated pea with a wild accession k-6070, and two hybrids F_2 obtained in crosses with the accession k-2523. We obtained the seeds of the next generation from two labeled hybrid BC_1F_2 plants carrying the alleles of the "wild" convicilin isoforms of accession k-6070.

Acknowledgements The authors would like to express their sincere gratitude to E. V. Semenova (Federal Research Center N. I. Vavilov All-Russian Institute of Plant Genetic Resources) for providing accessions of wild pea k-6070 and k-2523 from the VIR collection.

References

1. Bobkov SV, Selikhova TN (2019) Obtaining pea interspecific hybrids for use in marker assistant selection on high protein quality. Legumes Groat Crops 3(31):15–22. https://doi.org/10.24411/2309-348X-2019-11108
2. Domoney C, Duc G, Ellis THN, Ferrandiz C, Firnhaber C, Gallardo K, Hofer J, Kopka J, Kuster H, Madueno F, Salon C (2006) Genetic and genomic analysis of legume flowers and seeds. Curr Opin Plant Biol 9(2):133–141. https://doi.org/10.1016/j.pbi.2006.01.014
3. FAOSTAT. (n.d.). Crops. http://www.fao.org/faostat/en/#data/QC. Accessed 21 March 2021
4. Bobkov SV, Uvarova OV (2012) Peas use perspectives for production of storage proteins isolates. Zemledelie 8:47–48
5. Novikova NE, Kosikov AO, Bobkov SV, Zelenov AA (2017) The effects of late foliar fertilization and application of growth regulators on the yield and protein productivity of pea (Pisumsativum L.). Agrohimia 1:32–40
6. Bourgeois M, Jacquin F, Savois V, Sommerer N, Labas V, Henry C, Burstin J (2009) Dissecting the proteome of pea mature seeds reveals the phenotypic plasticity of seed protein composition. Proteomics 9(2):254–271. https://doi.org/10.1002/pmic.200700903
7. Lam A, Can Karaca A, Tyler R, Nickerson M (2016) Pea protein isolates: Structure, extraction, and functionality. Food Rev Intl 34(2):126–147. https://doi.org/10.1080/87559129.2016.1242135
8. Duranti M, Gius C (1997) Legume seed protein content and nutritional value. Field Crop Res 53:31–45. https://doi.org/10.1016/S0378-4290(97)00021-X
9. O'Kane F, Happe R, Vereijken J, Gruppen H, van Boekel M (2004) Characterization of pea vicilin. 2. Consequences of compositional heterogeneity on heat-induced gelation behavior. J Agric Food Chem 52(10):3149–3154. https://doi.org/10.1021/jf035105a
10. Stone A, Avarmenko N, Warkentin T, Nickerson M (2015) Functional properties of protein isolates from different pea cultivars. Food Sci Biotechnol 24(3):827–833. https://doi.org/10.1007/s10068-015-0107-y
11. Tzitzikas EN, Vincken JP, Groot J, Gruppen H, Visser RG (2006) Genetic variation in pea seed composition. J Agric Food Chem 54(2):425–433. https://doi.org/10.1021/jf0519008
12. Barac M, Cabrilo S, Pesic M, Stanojevic S, Zilic S, Macej O, Ristic N (2010) Profile and functional properties of seed proteins from six pea (Pisum sativum) genotypes. Int J Mol Sci 11(12):4973–4990. https://doi.org/10.3390/ijms11124973
13. Rubio L, Pérez A, Ruiz R, Guzmán M, Aranda-Olmedo I, Clemente A (2013) Characterization of pea (Pisum sativum) seed protein fractions. J Sci Food Agric 94(2):280–287. https://doi.org/10.1002/jsfa.6250
14. Croy RR, Gatehouse JA, Tyler M, Boulter D (1980) The purification and characterization of a third storage protein (convicilin) from the seeds of pea (Pisum sativum L.). Biochem J 191(2):509–516. https://doi.org/10.1042/bj1910509
15. O'Kane F, Happe R, Vereijken J, Gruppen H, van Boekel M (2004) Characterization of Pea Vicilin. 1. Denoting Convicilin as the α-Subunit of the Pisum vicilin family. J Agric Food Chem 52(10):3141–3148. https://doi.org/10.1021/jf035104i
16. Bown D, Ellis T, Gatehouse J (1988) The sequence of a gene encoding convicilin from pea (Pisum sativum L.) shows that convicilin differs from vicilin by an insertion near the N-terminus. Biochem J 251(3):717–726. https://doi.org/10.1042/bj2510717
17. Casey R, Domoney C (1999) Pea globulins. In: Shewry RP, Casey R (eds) Seed proteins. Kluwer Academic Publishers, Amsterdam, The Netherlands, pp 171–208
18. Selikhova TN, Bobkov SV (2013) Polymorphism of storage proteins in accessions of pea wild taxa. Rep Russ Acad Agric Sci 5:20–22
19. Bobkov SV, Lazareva TN (2012) Band composition of electrophoretic spectra of storage proteins in interspecific pea hybrids. Russ J Genet 48(1):47–52

20. Bobkov SV, Selikhova TN (2017) Obtaining interspecific hybrids for introgressive pea breeding. Russ J Genetics Appl Res 7(2):145–152. https://doi.org/10.1134/S20790597170 20046

21. Konarev VG (2000) Identification of varieties and registration of the gene pool of cultivated plants by seed proteins. N. I. Vavilov All-Russian Institute of Plant Genetic Resources (VIR), St. Petersburg, Russia

Changes in the Agrochemical Indicators of Light Gray Forest Soil and the Yield of Grain Crops During the Rotation of Crop Rotation Under the Influence of Various Systems of Its Processing in the Conditions of the South-East of the Volga-Vyatka Region

Alexey V. Ivenin◉**, Yulia A. Bogomolova**◉**, and Alexander P. Sakov**◉

Abstract The main purpose of the research is to study the effect of fertilizers in the rotation of the seven-pole grain crop rotation and straw destructors in various systems of processing light gray forest soil on the productivity of grain crops and changes in soil fertility. The research was carried out in the Nizhny Novgorod Region in the 2014–2020 years. The studied tillage systems, the applied fertilizers, and straw destructors contributed to a decrease in the humus content by 0.03–0.24% compared to the initial indicator of its content during the field experiment, to a deficit-free balance of mobile phosphorus, with an increase in its content according to options by 35.0–89.8 mg/kg and exchangeable potassium with an increase in its content according to options by 24.2–85.6 mg/kg in comparison with the initial data. No-till technology allows reducing the loss of humus in comparison with the steel studied tillage systems. It was revealed that deep soil cultivation systems, carried out by a plough with dumps and without them, provide the highest productivity of crop rotation -15.93 and 15.97 t/ha f.u., i.e., 5.47–0.18 t/ha higher than for other studied field experiments. The use of mineral doses ($N_{60}P_{60}K_{60}$) increases the productivity of crop rotation by 7.16–16.51 t/ha f.u., compared to the natural fertility of the soil. The use of Stimix®Niva, like a straw destructor, contributes to an increase in the productivity of crop rotation on natural soil fertility (by 1.01 t/ha f.u.) and against the background of the use of mineral fertilizers in a dose of $N_{60}P_{60}K_{60}$ (by 0.04 t/ha f.u.).

A. V. Ivenin (✉) · Y. A. Bogomolova · A. P. Sakov
Nizhny Novgorod Research Institute of Agriculture - Branch of the Federal Agrarian Scientific Center of the North-East Named After N. V. Rudnitsky, Nizhny Novgorod, Russia

Y. A. Bogomolova
e-mail: nnovniish@rambler.ru

A. P. Sakov
e-mail: nnovniish@rambler.ru

Keywords No-till technology · Biological product · Straw destructor · Humus ·
Exchangeable potassium · Mobile phosphorus

JEL Codes R01 · F22 · F63 · J15 · J61 · Q01 · Q15

1 Introduction

Today, the introduction of resource-saving technologies for grain production helps to
reduce the cost of agricultural production [8, 9, 18]. It becomes possible with taking
into account all the positive and negative results of their impact on the crops and the
soil. While growing crops, they can be influenced by the means of their root system
through a soil-absorbing complex (SAC). In modern conditions (the use of modern
high-performance and high-precision agricultural equipment, modern plant protection systems, the use of straw as organic fertilizer, and calculated doses of mineral
fertilizers), it is also possible to qualitatively, and most importantly, to accurately
influence the SAC to create optimal conditions for the growth and development of
cultivated plants. These techniques should be studied in the complex of using various
tillage systems in the dynamics of alternation of crop rotations [1, 6, 17].

It is necessary to use shredded straw as an organic fertilizer in large volumes, which
is a source of carbon replenishment for the formation of humus, with a mandatory
admission—the use of destructors [3, 4, 7, 16, 20].

In the process of agricultural production, it is necessary to influence the agrochemical properties of the soil to ensure a deficit-free balance of its nutrients. Obtaining
high yields cannot be achieved without taking into account the reproduction of soil
fertility. And this, in turn, will help to ensure more stable production of agricultural
products over the years [2, 15, 19].

2 Materials and Methods

Research has been carried out since 2014 on an experimental field; the accounting
area of the field is 132 m^2. The soil is light gray forest, slightly acidic (pH$_{KCl}$ 5,6),
the content of humus -1.5%, exchangeable potassium -140 mg/kg, and mobile
phosphorus -253 mg/kg. The repetition is fourfold with a systematic placement of
options [10].

The research was carried out in grain crop rotation:

1. Mustard for seeds;
2. Winter wheat;
3. Soy;
4. Spring wheat;
5. Peas;
6. Oats [11, 12].

In 2014, white mustard was used as equalizing sowing; and in 2020, quantitative indicators for this crop were recorded. The varieties of the studied crops: white mustard—*Raduga*, winter wheat—*Moskovskaya-39*, soybeans—*Light*, spring wheat—*Ester*, peas—*Krasivy*, oats—*Yakov* [11].

The field experiment scheme included 5 tillage systems (factor A), differing in many points:

I.   Traditional moldboard (control)—autumn ploughing with a PN-3–35 plough by 20–22 cm;
II.   Moldboard "deep"—ploughing with a plough without dumps by 20–22 cm;
III.   Moldboard "shallow"—autumn tillage with the Pottinger Synkro 5030 K chisel cultivator to a depth of 14–16 cm;
IV.   Minimal—autumn disking with the Discover XM 44,660 not had harrow to a depth of 10–12 cm;
V.   Zero tillage (No-till)—sowing with the Sunflower 9421–20 seeder [10].

The system of pre-sowing tillage of crop rotation was generally accepted for the Nizhny Novgorod Region and has similar results in all the studied variants of the field experiment (except for the No-till variant).

For each tillage system, the use of mineral fertilizers and straw destructors (factor B) was studied according to the following scheme:

1.   Straw without fertilizers (control);
2.   Straw + N_{10} (introduced into the field experiment scheme in 2015);
3.   Straw + $N_{60}P_{60}K_{60}$;
4.   Straw + $N_{60}P_{60}K_{60}$ + N_{10};
5.   Straw + $N_{60}P_{60}K_{60}$ + Stimix®Niva;
6.   Straw + Stimix®Niva [13, 14].

Mineral fertilizers were applied according to the scheme of the field experiment in options 3, 4, 5 (according to factor B) for spring cultivation in dosage of $N_{60}P_{60}K_{60}$ kg d./ha for each investigated crop rotation, in the form of a mixture of diammofoska and ammonium nitrate.

3   Results

Weather conditions were not the same in all study years. Data on weather conditions in 2014–2021 studies are presented in Fig. 1.

In 2014, 2017, and 2018, the weather conditions for the growth and development of cultivated plants were generally unfavourable, the hydrothermal coefficients (HTC), respectively, amounted to values for the growing season -1.01, -1.18, and -1.11, which is less than the average long-term value of this indicator (1.24).

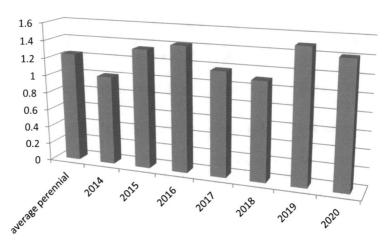

Fig. 1 Hydrothermal coefficients for research of 2014–2020 years. *Source* [17]

From the data in Fig. 1, it can be seen that the weather conditions for the periods from 2015 to 2020 years completely met the requirements of the growth and development of the studied crops, the HTC, respectively, which amounted to 1.35; 1.42; 1.50; 1.41 [4].

The indicator of the humus content in the soil by the end of the alternation of the crop rotation varies within the range of 1.29–1.37%, which is 0.13–0.21% lower in comparison with the initial content during the establishment of the experiment (1.50%) (Table 1).

Thus, the incorporation of plant residues (straw + stubble-root residues) of cultivated grain and leguminous crops is not enough to maintain a deficit-free humus balance on light gray forest soil within the framework of grain crop rotation in the South-East of the Volga-Vyatka Region [3, 4, 13].

On average, according to factor A, direct sowing technology allows, during one rotation of the studied crop rotation, reducing the loss of humus in comparison with the steel studied soil cultivation systems—the decrease in humus content is the lowest in the field experiment—by 0.13% of the initial one. The traditional tillage system contributes to the greater mineralization of humus during the study period (the average content by factor A is 1.29%, which is 0.21% lower than the initial value). At the same time, a decrease in the humus content was noted in the variant of cultivation of grain crops by autumn ploughing with a plough with dumps by 0.06–0.08% (HCP_{05} according to factor A-0.06) in comparison with shallow resource-saving variants (variants 3.4) and No-till technology. Use of ammonium nitrate at a dose of 10 kg per 1 ton of straw, as its destructor, according to the background of the use of mineral fertilizers in a dose of $N_{60}P_{60}K_{60}$, allows reducing losses in the humus content by 0.09% of its initial value and provides its highest content in the field experiment (1.41%), which by 0.07–0.11% is higher (HCP_{05} according to factor B-0.07) in comparison with other studied options. So, the use of a biological product

Table 1 Change in agrochemical indicators of the topsoil depending on various systems of its processing, fertilizers, and biological products (end of crop rotation—autumn 2020)

Processing system (factor A)	Fertilizers (factor B)						Average for factor A
	1.Straw (control)	2.Straw + N_{10}	3.Straw + $N_{60}P_{60}K_{60}$	4.Straw + $N_{60}P_{60}K_{60} + N_{10}$	5.Straw + $N_{60}P_{60}K_{60}$ + Stimix®Niva	6.Straw + Stimix®Niva	
Humus %							
I. Traditional (control)	1.31	1.28	1.30	1.35	1.24	1.26	1.29
II. Moldboard "deep"	1.33	1.26	1.34	1.43	1.27	1.29	1.32
III. Moldboard "shallow"	1.37	1.33	1.33	1.41	1.37	1.32	1.36
IV. Minimal	1.32	1.35	1.37	1.41	1.32	1.31	1.35
V. Zero/No-till	1.27	1.29	1.36	1.47	1.44	1.37	1.37
Average for factor B	1.32	1.30	1.34	1.41	1.33	1.31	–
HCP_{05} (factor A) 0.06							
HCP_{05} (factor B) 0.07							
HCP_{05} (factor AB) 0.15							
P_2O_5, mg/kg							
I. Traditional (control)	291.4	315.0	311.2	317.9	322.8	303.9	310.4
II. Moldboard "deep"	265.5	312.2	313.5	308.6	364.3	320.8	314.2
III. Moldboard "shallow"	295.6	318.6	325.4	319.0	322.7	310.9	315.4
IV. Minimal	304.4	341.8	280.2	303.1	343.7	320.9	315.7
V. Zero/No-till	283.1	318.8	349.8	365.5	360.5	343.6	336.9
Average for factor B	288.0	321.3	316.0	322.8	342.8	320.0	-

(continued)

Table 1 (continued)

HCP$_{05}$ (factor A) 35.4

HCP$_{05}$ (factor B) 38.7

HCP$_{05}$ (factor AB) 86.6

K_2O, mg/kg

I. Traditional (control)	158.5	212.0	270.0	257.5	197.5	154.0	*208.3*
II. Moldboard "deep"	128.0	136.0	182.5	176.5	220.5	361.0	*200.8*
III. Moldboard "shallow"	239.5	186.5	195.5	179.0	212.5	136.0	*191.5*
IV. Minimal	166.0	223.0	254.0	218.5	179.0	188.5	*204.8*
V. Zero/No-till	129.0	218.0	226.0	246.5	212.5	172.0	*165.7*
Average for factor B	*164.2*	*195.1*	*225.6*	*215.6*	*204.4*	*202.3*	–

HCP$_{05}$ (factor A) 30.7

HCP$_{05}$ (factor B)/ LSD$_{05}$ 33.6

HCP$_{05}$ (factor AB)/LSD$_{05}$ 75.1

Source compiled by the authors

in terms of the background of mineral nutrition (humus content is 1.33%) and in terms of the zero background of fertilization (1.31%) does not affect the humus content in comparison with the control variant of the experiment with natural soil fertility (1. 32%) (HCP_{05} to factor B-0.07). The use of ammonium nitrate as a straw destructor for a zero-mineral background (1.30%), and separately the background $N_{60}P_{60}K_{60}$ (1.34%) also does not affect the changes in the humus content in comparison with the content in the variant with natural soil fertility (1.32%) (HCP_{05} to factor B-0.07).

The content of mobile phosphorus by the end of the first alternation of the grain crop rotation is in the range of 310.4–336.9 mg/kg (HCP_{05} according to factor A- 35.4), which is 57.4–83.9 mg/kg higher than the initial content (253 mg/kg) (Table 1). This indicates that the incorporation of straw, plant residues of previous crops, the introduction of mineral fertilizers and straw destructors contribute not only to maintaining a deficit-free balance of phosphorus but also to its accumulation in the topsoil [4, 5].

The use of Stimix®Niva on the background of $N_{60}P_{60}K_{60}$, on average to factor B, increases the content of mobile phosphorus in comparison with the option with natural soil fertility by 54.8 mg/kg by the end of the first alternation of the crop rotation (HCP_{05} to factor B—38.7).

This is due to the low removal of this element by the yield of agricultural crops of the crop rotation (the yield from this option is one of the highest in the field experiment, second only to the option of using ammonium nitrate on the background $N_{60}P_{60}K_{60}$ (Table 2), and also to the ability of this biological product to translate inaccessible forms of phosphorus available with the activation of biological processes in the soil.

The provision of the arable soil layer with exchangeable potassium by the fall of 2020, on average for the studied tillage systems, is estimated as increased and is in the range of 165.7–208.3 mg/kg, which is 25.7–68.3 mg/kg higher than the content of exchangeable potassium in the original soil in 2014 (140 mg/kg) (Table 1). Direct sowing technology provides a positive balance of this nutrient for the rotation of grain crop rotation, but the accumulation rate is the lowest among all studied tillage systems (except for the option of non-moldboard "shallow" tillage with a chisel cultivator-191.5 (NSR_{05} to factor A-30.7) −165.7 mg/kg, which is 25.7 mg/kg higher than the initial value [4, 11, 12].

The use of background doses in combination with the studied straw destructors (ammonium nitrate and the biological product Stimix®Niva), and also the use of the preparation Stimix®Niva, according to the zero-mineral background, increase the content of exchangeable potassium in comparison with the variant with natural soil fertility by 38.1–61, 4 mg/kg (HCP_{05} by factor B-33.6) by the end of the first rotation of grain crop rotation (Table 1) [3].

One of the important indicators based on which it can be concluded about the effectiveness of a particular technology for the production of crop products is the yield of agricultural crops.

The results of research on the yield of crops cultivated in grain-new crop rotation are presented in Table 2.

Table 2 Yield of agricultural crops depending on various systems of soil cultivation, fertilizers, and biological products per crop rotation (2015–2020), t/ha

Processing system (factor A)	Fertilizers (factor B)	Winter wheat	Soy	Spring wheat	Peas	Oats	Mustard white	Total yield of agricultural crops, t/ha f.u
I. Traditional (control)	1.Straw (control)	1.65	1.41	1.87	2.31	3.34	0.24	12.17
	2.Straw + N10	– *	1.60	2.45	2.70	3.85	0.37	– [1]
	3.Straw + N60P60K60	2.97	1.86	3.41	2.49	4.65	0.52	18.40
	4.Straw + N60P60K60 + N10	4.60	1.84	3.13	2.76	4.82	0.41	19.70
	5.Straw + N60P60K60 + Stimix®Niva	2.91	1.75	3.02	2.48	4.42	0.44	16.83
	6.Солома + Stimix®Niva	1.91	1.63	2.06	2.58	3.64	0.31	13.37
	Average for I (factor A)	*2.81*	*1.68*	*2.66*	*2.55*	*4.12*	*0.38*	*15.93*
II. Moldboard "deep"	1	1.75	1.39	1.78	1.96	3.18	0.28	11.61
	2	–	1.83	2.27	2.55	3.88	0.34	–
	3	3.32	1.87	3.41	2.54	4.58	0.47	18.16
	4	4.53	2.04	3.32	2.48	4.89	0.52	19.94
	5	3.51	1.82	3.28	2.38	4.36	0.39	17.67
	6	1.82	1.63	1.93	1.92	3.75	0.28	12.70
	Average for II (factor A)	*2.99*	*1.76*	*2.67*	*2.31*	*4.11*	*0.38*	*15.97*
III. Moldboard "shallow"	1	2.02	1.12	1.87	2.04	2.81	0.27	11.38
	2	–	1.37	2.26	2.18	3.54	0.41	–
	3	3.47	1.81	3.73	2.40	4.14	0.52	18.06
	4	3.48	1.96	3.99	2.58	4.56	0.47	19.15
	5	3.22	1.83	3.37	2.55	3.93	0.46	17.30
	6	1.93	1.71	1.84	1.82	3.23	0.27	12.18
	Average for III (factor A)	*2.82*	*1.63*	*2.84*	*2.26*	*3.70*	*0.40*	*15.35*
IV. Minimal	1	1.80	1.15	1.69	1.88	2.93	0.28	10.91
	2	–	1.47	2.32	1.86	3.33	0.35	–

(continued)

[1] Excluding winter wheat data in 2015.

Table 2 (continued)

Processing system (factor A)	Fertilizers (factor B)	Winter wheat	Soy	Spring wheat	Peas	Oats	Mustard white	Total yield of agricultural crops, t/ha f.u
	3	3.79	1.60	3.68	2.43	4.07	0.40	17.95
	4	4.27	2.17	3.44	2.68	4.74	0.58	20.10
	5	3.73	2.20	3.52	2.42	4.11	0.42	18.52
	6	2.03	1.77	1.89	1.95	3.76	0.31	13.14
	Average for IV (factor A)	*3.13*	*1.73*	*2.76*	*2.20*	*3.82*	*0.39*	*15.79*
V. Zero No-till	1	1.49	0.40	1.23	0.86	1.67	0.17	6.46
	2	–	0.49	1.48	0.48	1.53	0.32	–
	3	3.42	0.48	3.28	1.38	2.87	0.46	13.22
	4	4.18	0.85	2.99	1.21	3.12	0.71	14.54
	5	3.57	0.70	2.69	2.24	3.78	0.52	14.99
	6	1.30	0.74	1.17	0.74	1.15	0.08	5.88
	average by V (factor A)	*2.79*	*0.61*	*2.14*	*1.15*	*2.35*	*0.38*	*10.50*
Average for factor B	1.Straw (control)	*1.74*	*1.09*	*1.69*	*1.81*	*2.79*	*0.25*	*10.51*
	2.Straw + N10	–	*1.35*	*2.16*	*1.95*	*3.23*	*0.36*	–
	3.Straw + N60P60K60	*3.39*	*1.52*	*3.50*	*2.25*	*4.06*	*0.47*	*17.02*
	4.Straw + N60P60K60 + N10	*4.21*	*1.77*	*3.37*	*2.34*	*4.43*	*0.54*	*18.68*
	5.Straw + N60P60K60 + Stimix®Niva	*3.39*	*1.66*	*3.18*	*2.41*	*4.12*	*0.45*	*17.06*
	6.Straw + Stimix®Niva	*1.80*	*1.50*	*1.78*	*1.80*	*3.11*	*0.25*	*11.52*
HCP_{05}/ LSD_{05}	*Factor A*	0.21	0.07	0.24	0.15	0.29	0.04	–
	Factor B	0.21	0.07	0.22	0.17	0.32	0.06	–
	Factors AB	1.39	0.82	*0.56*	0.51	0.67	0.12	–

Source compiled by the authors

The use of autumn disking as the main tillage, on average for factor A, contributes to an increase in the yield of winter wheat, in comparison with other studied soil cultivation systems, up to 3.13 t/ha, which is 0.31–0.34 t/ha (HCP$_{05}$ by factor A −0.21), except for the option with non-moldboard "deep" tillage with a plough without dumps (2.99 t/ha). Direct sowing technology provided the same level of winter wheat yield as traditional tillage −2.79–2.81 (HCP$_{05}$ to factor A −0.21). Natural soil fertility (control) provided the yield of the studied crop in the range of 1.49–2.02 t/ha (HCP$_{05}$ to the AB factor −1.39). The application of N$_{60}$P$_{60}$K$_{60}$ increased the yield of winter wheat by 1.32–1.99 t/ha. At the same time, the use of a biological product (Stimix®Niva) in terms of the mineral background did not increase the yield of winter crops in the weather conditions of 2014–2015. The maximum average crop yield by factor B was observed in the variant with the combined application of N$_{60}$P$_{60}$K$_{60}$ and ammonium nitrate at a dose of 10 kg of a.i. per 1 ton of straw −4.21 t/ha, which is 0.82–2.47 t/ha higher than other options (HCP$_{05}$ to factor B −0.21) (Table 2) [3, 11, 13].

The highest average yield of soybean grain by factor B was obtained according to the background of mineral nutrition (N$_{60}$P$_{60}$K$_{60}$) with the use of ammonium nitrate as a straw destructor −1.77 t/ha, which is 0.11–0.68 t/ha higher in other studied variants of field experience (HCP$_{05}$ for factor B −0.07). The studied systems of mechanical tillage (option 1–4) ensure the yield of soybean grain in 1.64–1.76 t/ha, and the No-till technology allowed obtaining it at the level of 0.61 t/ha, which is 1.03- 1.15 t/ha lower (HCP$_{05}$ by factor A- 0.07) (Table 2).

The use of autumn ploughing as the main tillage (with dumps and without them) contributed to the formation of the yield of spring wheat against the background of natural soil fertility at the level of 1.78–1.87 t/ha. The introduction of mineral doses (N$_{60}$P$_{60}$K$_{60}$) increased the yield by 1.3–1.9 times.

The average yield of white mustard according to factor A was the same for all studied soil cultivation systems and was in the range of 0.38–0.40 t/ha (HCP$_{05}$ according to factor A −0.04).

The maximum yield of mustard grain was obtained by a zero-tillage system with the combined use of mineral fertilizers at a dose of N$_{60}$P$_{60}$K$_{60}$ and N$_{10}$ (ammonium nitrate -10 kg per ton of straw) −0.71 t/ha, which is 0.13–0.63 t/ha higher than the rest of the field experiment options (HCP$_{05}$ according to the AB factor −0.12).

The mustard cultivation according to the natural fertility of light gray forest soil reduces its yield to minimum values for all studied technologies with the use of the biological product Stimix®Niva. The average yield for factor B was 0.25 t/ha, which is 0.11–0.29 t/ha lower than the rest of the options for the use of fertilizers and straw destructors (HCP$_{05}$ by factor B −0.06) (Table 2).

To analyze the influence of the studied tillage systems and the use of mineral fertilizers and straw destructors on the yield in general of all grain crops during the rotation of the crop rotation, it is necessary to bring the yield of various crops to a single denominator—the yield expressed in-feed units. This indicator is equivalent to measuring different types of crop production. One kg of oat grain is taken as the main feed unit. The conversion factors are winter and spring wheat—1.14; soy—1.31; peas—1.18; oats—1.0; white mustard—0.98.

The total yield of crop rotation crops in feed units is presented in Table 2.

The use of ammonium nitrate at a dose of 10 kg a.i. per 1 ton of straw in terms of natural soil fertility (the second variant of the field experiment according to factor B) was introduced into the scheme of the field experiment in 2015, so this option will not be taken into account by the authors in the process of analyzing the total yield in fodder units of agricultural crops for the first rotation of grain crop rotation.

From the data in Table 2, it can be seen that deep soil cultivation systems, carried out by a plough, with dumps and without them, provide the highest total yield of grain crops for a rotation of crop rotation—15.93 and 15.97 t/ha f.u., which is 5.47–0.18 t/ha f.u. higher than for other studied tillage systems (Table 2).

The use of $N_{60}P_{60}K_{60}$ (factor B) allows increasing the productivity of crop rotation crops up to 17.02–18.68 t/ha a.u., in comparison with the natural soil fertility (0.51–11.52 t/ha f.u.).

The use of the biological product Stimix®Niva, like a straw destructor, contributes to an increase in the productivity of grain crops of crop rotation on natural soil fertility (by 1.01 t/ha f.u.) and against the background of the use of mineral fertilizers at a dose of $N_{60}P_{60}K_{60}$ (by 0.04 t/ha f.u.), according to Table 2.

4 Conclusion

It was found that for the rotation of the studied six-field grain rotation of the tillage system, the applied fertilizers and straw destructors do not provide a deficit-free balance of humus: its amount is in the range of 1.26–1.47%, which is 0.03%. It is 0.24% lower than the original content at the start of the rotation (1.50). The No-till technology allows, during one rotation of the studied crop rotation, reducing the loss of humus in comparison with the steel studied tillage systems. The use of ammonium nitrate, as a destructor of chopped straw, against the background $N_{60}P_{60}K_{60}$, allows reducing losses in the humus content by 0.07–0.11% in comparison with other studied options (HCP$_{05}$ to factor B-0.07).

It was also found that for the rotation of a six-field grain rotation, the studied tillage systems and the applied fertilizers and straw destructors contribute to a deficit-free balance of mobile phosphorus (the content is 35.0–89.8 mg/kg higher in comparison with the initial content at the time of laying the field experiment (253 mg/kg)) and exchangeable potassium (the content is 24.2–85.6 mg/kg higher than the content of exchangeable potassium in the original soil (140 mg/kg)).

The use of the biological product Stimix®Niva, according to the background $N_{60}P_{60}K_{60}$, and on average according to factor B, contributes to an increase in the content of mobile phosphorus in comparison with the variant with natural soil fertility by 54.8 mg/kg by the end of the first alternation of the crop rotation (HCP$_{05}$ to factor B-38.7).

The application of doses of mineral fertilizers $N_{60}P_{60}K_{60}$ in combination with the studied straw destructors (ammonium nitrate and the biological product Stimix®Niva), and also the use of Stimix®Niva separately, according to the zero-mineral background, increase the content of exchangeable potassium in comparison with the variant with natural soil fertility by 38.1- 61.4 mg/kg (HCP_{05} by factor B-33.6) by the end of the first grain rotation.

It was revealed that deep tillage systems carried out by a plough with dumps and without them provide the highest total yield of grain crops for an alternation of crop rotation −15.93 and 15.97 t/ha f.u., which is 5.47–0.18 t/ha f.u. higher than for other studied soil cultivation systems. The use of mineral fertilizers in a dose of $N_{60}P_{60}K_{60}$ increases the productivity of crop rotation by 7.16–16.51 t/ha f.u. compared to the natural fertility of the soil. The use of the biological product Stimix®Niva, like a straw destructor, contributes to an increase in the productivity of crop rotation on natural soil fertility (by 1.01 t/ha f.u.) and on the background of the use of mineral fertilizers at a dose of $N_{60}P_{60}K_{60}$ (by 0.04 t/ha f.u.).

References

1. Antonov VG, Ermolaev AP (2018) The effectiveness of long-term use of minimal methods of tillage in crop rotations. Agric Sci Euro-North-East 4(65):87–92. https://doi.org/10.30766/2072-9081.2018.65.4.87-92
2. Antonov VG, Ermolaev AP (2018) The effectiveness of long-term use of minimal methods of tillage in crop rotations. Agric Sci Euro-North-East 4(65):87–92
3. Bogomolova YuA, Sakov AP, Ivenin AV (2018) The influence of the systems of the main processing of light gray forest soil and fertilizers on its agrochemical indicators in the link of the grain crop rotation of the Nizhny Novgorod Region. Agroch Bull 5:32–39
4. Bogomolova YuA, Sakov AP, Ivenin AV (2018) Energy and economic efficiency of growing agricultural crops in the rotation of grain crop rotation using various systems of processing light gray forest soil and the use of fertilizers and biological products in the conditions of the Volga-Vyatka Region. Agrarian Sci 4:49–50
5. Bogomolova YuA, Sakov AP, Ivanin AV (2018) The influence of tillage and fertilizers on changes in its agro physical properties and the yield of soybeans in the link of grain crop rotation. Agrarian Sci Euro-North-East 3:62–69
6. Dzyuin AG (2018) The influence of straw in crop rotation on the number of microorganisms and biological activity of the soil. Agric Sci Euro-North-East 1(62):58–64
7. Hallam MJ, Bartholomen WV (1953) Influence of rate of plant residue addition in accelerating the decomposition of soil organic matter. Soil Sci Soc Amer J 17:365–368
8. Ivenin VV, Mikhalev EV, Krivenkov VA (2017) The efficiency of spring wheat cultivation against the background of complete mineral fertilization with the introduction of resource-saving No-till technology in grain-grass crop rotation on light gray forest soils of the Nizhny Novgorod Region. Agric Sci 11–12:22–25
9. Ivenin VV, Ivenin AV, Shubina KV, Mineeva NA (2018) Comparative efficiency of technologies for cultivation of grain crops in the link of crop rotation on light gray forest soils of the Volga-Vyatka Region. Bull Chuvash State Agric Acad 3(6):27–31
10. Ivenin AV, Sakov AP (2019) Influence of systems for processing light gray forest soil and the use of fertilizers and biological products on the weediness and yield of peas in the Nizhny Novgorod Region. Agrarian Sci 2:77–80
11. Ivenin AV, Sakov AP (2020) The influence of the systems of processing light gray forest soil on the yield and energy efficiency of growing grain crops for the rotation of grain crop rotation in the Volga-Vyatka Region. Bulletin of the Kazan State Agrarian University 15, 2(58):14–19

12. Ivenin AV, Sakov AP, Bogomolova YuA (2019) The influence of the systems of processing light gray forest soil and fertilizers on the productivity of agricultural crops in the link of grain crop rotation in the Volga-Vyatka Region. Agrarian Russia 1:9–14
13. Ivenin AV, Sakov AP (2020) Influence of systems for processing light gray forest soil on the yield and quality of oat grain in the Nizhny Novgorod Region. Agrarian Sci Euro-North-East 5:580–588
14. Ivenin AV, Sakov AP (2020) The influence of the systems of processing light gray forest soil and the use of fertilizers and biological products on the root supply and productivity of agricultural crops in the link of grain crop rotation in the Volga-Vyatka Region. Agrarian Sci 3:81–86
15. Karabutov AP, Solovichenko VD, Nikitin VV, Navolneva EV (2019) Reproduction of soil fertility, productivity and energy efficiency of crop rotations. Agriculture 2:3–7. https://doi.org/10.24411/0044-3913-201910201
16. Kozlova LM, Popov FA, Noskova EN, Ivanov VL (2017) Improved resource-saving technology of tillage and the use of biological products for spring crops in the central zone of the North-East of the European part of Russia. Agric Sci Euro-North-East 3(58):43–48
17. Kozlova LM, Popov FA, Noskova EN, Denisova AV (2018) Application of the main elements of resource-saving environmentally friendly technologies in the cultivation of spring grain fodder crops in the central zone of the North-East of the European part of Russia. Problems of intensification of animal husbandry taking into account environmental protection and the production of alternative energy sources, including biogas: a collection of articles. Warsaw: Institute of Technological and Natural Sciences in Falenty, 67–74
18. Pilipenko NG, Dneprovskaya VN (2012) The effectiveness of resource-saving technologies for pre-sowing tillage in field crop rotation. Agriculture 4:29–30
19. Shapovalova NN, Menkina EA (2018) Agrochemical state and biological activity of the soil in the aftereffect of prolonged use of mineral fertilizers. Bull Orenburg State Agrarian Univ 5(73):43–46
20. Vejan P, Abdullah R, Khadiran T, Ismail S, Nasrulhaq A (2016) Boyce role of plant growth promoting Rhizobacteria in agricultural sustainability—a review. Molecules 21:1–17

The Influence of Agrotechnology on Grain Quality and Yield of Winter Wheat of the Yuka Variety in the Conditions of the Western Ciscaucasia

Irina V. Shabanova ⓘ**, Nikolay N. Neshchadim** ⓘ**, and Aleksandr P. Boyko**

Abstract The experiment was conducted under the conditions of a typical grain-row crop rotation on the leached chernozem of the Kuban. The use of fertilizers at a dose of N120P600K40 and 400 t/ha of manure against the background of chemical protection agents for various types of tillage allows you to get a harvest of winter wheat Yuca 70–73 t/ha with a protein content of 14.3–14.9% and gluten 24–25%. A further increase in the doses of applied fertilizers does not correspond to a mathematically reliable increase in yield. The grain of winter wheat contained manganese, zinc, copper, cobalt, cadmium, and lead were below the limit values for food of the adult population.

Keywords Winter wheat · Fertilizers · Trace elements · Leached chernozem · Yield · Quality

JEL Codes Q10 · Q14 · Q15 · Q24 · Q51

1 Introduction

Winter wheat is one of the main crops grown on the territory of the Krasnodar Territory. The gross harvest of wheat reaches more than 60% of the total number of cereals. Despite the wide distribution in the territory of Kuban, the cultivation of high-grade winter wheat grains is difficult, primarily due to unfavorable weather conditions associated with low precipitation in the initial growth phases. The result of a study

I. V. Shabanova (✉) · N. N. Neshchadim
Kuban State Agrarian University Named After I. T. Trubilin, Krasnodar, Russia

N. N. Neshchadim
e-mail: neshhadim.n@kubsau.ru

A. P. Boyko
Federal Research Center All-Russian Institute of Plant Genetic Resources Named After N. I. Vavilov, Adler, Sochi, Russia
e-mail: aos.vir@mail.ru

© The Author(s), under exclusive license to Springer Nature Singapore Pte Ltd. 2022   309
E. G. Popkova and B. S. Sergi (eds.), *Sustainable Agriculture*,
Environmental Footprints and Eco-design of Products and Processes,
https://doi.org/10.1007/978-981-16-8731-0_30

by the Lithuanian Institute of Agriculture showed that in dry years, the yield of the wheat grain is significantly lower, but the quality of grain meets the requirements for baking [1]. To increase productivity while maintaining product quality, it is necessary to use modern agricultural technologies, including a combination of maintaining soil fertility through applied fertilizers, plant protection, and tillage, and to improve the processes of aeration and nutrition of plants, especially in hot and arid climates [13, 14]. The use of various mineral fertilizers and manure can significantly vary the level of fertility, which affects the yield and quality of the grain of the winter wheat [3, 4, 6, 9]. The maximum yield of winter wheat grain can be achieved by applying nitrogen fertilizers in doses of N150-N210, using loosening to a depth of 22–24 cm, and varying the precursors in the crop rotation [2, 5].

It should be noted that the desire of manufacturers to increase the dose of fertilizers could lead to the accumulation of harmful substances in the products, for example, heavy metals. Studies by Lubyte et al. [7] showed that long-term use of fertilizers caused accumulation in winter wheat grain (mg/kg): cadmium 0.02–0.06, lead 0.02–0.05, nickel 0.08–0.12, chromium 0.11–0.16, copper 2.24–4.10, zinc 16.1–24.9, manganese 14.6–20.2, and iron 37.3–60.2. The highest correlation coefficients of 0.87–0.97 between the content of heavy metals and the fertilizer application doses were observed for Cd, Cr, Zn, Ni, Fe, and Mn; the absence of dependence is typical for Cu and Pb [7].

2 Methodology

Within the framework of many years of experience, laid down in 1991 at the experimental station of the Kuban State Agrarian University in grain-grass-row crop rotation, the influence of agricultural technology on the quality and yield of the obtained products was studied [8, 10, 15]. We investigated the effect of four factors on the yield and quality of winter wheat variety Yuka: method of soil treatment, the regulation of fertility by making of deposit of the manure and doses of mineral fertilizers, and means of plant protection.

In the soil of the experimental station, the leached chernozem soil of the Western Ciscaucasia, the pH is close to neutral, which increases its buffer properties in relation to food elements, including trace elements [11, 12].

The method of tillage included: *the soil protection method* with the flat-cutting machines working bodies at a depth of no more than 10 cm; *the recommended method for the Krasnodar Territory*, including loosening by 22–24 cm; *the intensive-implying method* of deep loosening by 70 cm twice in an 11-year crop rotation for corn and sugar beet.

Control variants (0) for different methods of tillage of the soil without the use of fertilizers and plant protection products.

Option 1—the manure once per rotation 200 t/ha, N60P30K20, a biological product of fungicidal action "Baksis, Zh" 0.4 l/ha and entomological mixture 3.0 l/ha;

Option 2—the manure once per rotation 400 t/ha, N120P60K40, chemicals "Falkon, KE" 0.6 l/ha and "BI-58 New" 1.0 l/ha;

Option 3—the manure once per rotation of 600 t/ha, N240P120K80, with chemical preparations, the herbicide "Sekator VDG" 0.2 l/ha was used.

The quality of winter wheat was studied by IR spectrometry, the content of heavy metals was studied by atomic absorption method on the Kvant 2 AT device, and the content of residues of herbicides and pesticides was determined by the chromatographic method.

3 Results

The applied agricultural technologies made it possible to obtain a high yield of winter wheat of the Yuka variety, reaching 79 c/ha with all types of tillage and with a high dose of fertilizer (Table 1). On the control and with the low doses of fertilizers (N120P60K40), the obtained winter wheat corresponded to class 3 in terms of protein content (13–14%) and gluten content (22–23%) for all tillage methods. Increasing the doses of mineral fertilizers to N120P60K40-N240P120K80 allowed increasing the content of raw gluten to 25–26%, and protein to 14.9%, which contributed to the increase of the grain class to the second.

Table 1 Quality and yield of winter wheat grain of the Yuka variety under uses of various agricultural technologies (2017–2019)

Option	Yields		Protein		Gluten	
	t/ha	Δ	%	Δ	%	Δ
The soil protection method						
0	5.2	–	13.6	–	22.3	–
1	6.1	0.9	13.9	0.3	23.4	1.1
2	7.0	1.8	14.9	1.6	25.3	3.0
3	7.9	2.7	14.8	1.8	26.6	4.3
The recommended method for the Krasnodar Territory						
0	5.4	–	13.5	–	22.3	–
1	6.2	0.8	14.1	0.6	23.6	1.3
2	7.3	1.9	14.5	1.0	24.1	1.8
3	7.9	2.5	14.9	1.4	25.5	3.2
The intensive-implying method						
0	5.2	–	13.6	–	22.3	–
1	5.9	0.7	14.3	0.7	23.8	1.5
2	7.0	0.8	14.8	1.,2	25.5	2.2
3	7.9	2.7	14.9	1.3	25.4	3.1
LSD$_{05}$	0.5	–	0.3	–	0.15	–

Source Developed and compiled by the authors

Regression analysis of data on grain quality and yield showed a coefficient of determination R2 of 0.92–0.99; therefore, changes in the dependent variables of gluten by 92%, protein-by 96%, and yields by 99% are explained by an increase in the applied doses of fertilizers (Table 3). The regression coefficients of protein and gluten for the methods of tillage and plant protection significantly differ from 0 (P < < 0.001).

The content of heavy metals in winter wheat grain is manganese and zinc 0.2 MPC, copper, cobalt, and zinc not higher than 0.5 MPC, and cadmium on variants with high doses of fertilizers 1.0 MPC for baby food (Table 2). The accumulation of cadmium in the grain is associated with an insufficient content of zinc and copper in the soil; therefore, the vacancies are occupied by cadmium, which is similar in chemical and physical properties.

The results of statistical processing allowed us to identify the dominant factors affecting the environmental safety of the grown grain (Table 3). Regression analysis of data on the content of heavy metals in grain from four factors showed a deter-mination coefficient R2 of 0.55–0.65; therefore, changes in the dependent variables are not sufficiently reliably explained by an increase in the doses of fertilizers and manure. The regression coefficients for the methods of tillage and plant protection

Table 2 The content of heavy metals in the grain of winter wheat of the Yuka variety under uses of various agricultural technologies, mg/kg

Option	Cu	Mn	Zn	Pb	Cd	Co
The soil protection method						
0	4.9	23.4	28.1	0.17	0.05	0.04
1	3.3	18.3	24.6	0.04	0.05	0.02
2	2.9	18.8	24.1	0.05	0.05	0.02
3	2.7	21.9	24.4	0.06	0.06	0.03
The recommended method for the Krasnodar Territory						
0	4.9	20.0	28.1	0.14	0.04	0.03
1	3.1	20.2	26.1	0.06	0.04	0.04
2	2.9	21.9	25.7	0.07	0.05	0.03
3	3.1	22.1	27.1	0.08	0.08	0.06
The intensive-implying method						
0	5.0	26.1	27.9	0.06	0.05	0.02
1	3.5	23.6	21.7	0.07	0.03	0.04
2	3.8	27.1	26.7	0.09	0.04	0.06
3	3.2	24.2	27.0	0.07	0.04	0.02
HCP$_{05}$	0.2	0.7	0.3	0.01	0.01	0.01
LOC	10	110	50	0.5/0.3*	0.1/0.06*	1.0
*for feeding children						

Source Developed and compiled by the authors

Table 3 Regression dependence of the yield and quality of winter wheat grain of the Yuka variety on the factors of the applied technology

Indicator	Free term of the equation	Share effect and the regression coefficients for the factors				R^2
		Treatment of the soil	Manure	Mineral fertilizers	Plant protection	
Yield, t/ha	−14.1	14.3 −0.100	1.4 0.010	45.1 −0.333	39.2 0.292	0.99
Protein, %	22.3	9.3 0.075	1.3 0.010	68.6 −0.549	20.8 −0.163	0.96
Gluten, %	−61.3	13.6 0.188	1.0 0.010	39.8 −0.550	45.6 −0.635	0.92
Cu, mg/kg	−92.9	11.9 −0.150	1.0 0.010	0.1 0.001	87.0 1.101	0.58
Mn, mg/kg	−350.7	8.9 0.325	0.1 −0.028	0.1 0.014	90.9 3.282	0.62
Zn, mg/kg	−52.4	23.5 0.201	4.6 −0.039	7.8 −0.073	64.1 0.545	0.56
Pb, mg/kg	2.10	41.6 −0.010	4.2 0.001	4.2 −0.001	50.0 −0.012	0.27
Cd, mg/kg	5.2	1.0 0.001	1.0 −0.001	8.0 0.010	90.0 0.091	0.65
Co, mg/kg	10.4	1.0 0.001	1.0 −0.001	8.0 0.010	90.0 −0.090	0.55

Source Developed and compiled by the authors

significantly differ from 0 ($P \ll 0.001$) in terms of the accumulation of heavy metals in the grain. The use of fertilizers and manure does not have a significant effect on the content of heavy metals in grain; the regression analysis showed low beta values (Table 3).

The data of multiple regression analysis showed that the increase in yield, protein, and gluten content in grain is significantly affected by the application of mineral fertilizers and plant protection (Table 3). As expected, the use of manure twice per crop rotation does not significantly affect the quality of products.

The influence of agricultural technology components on the content of heavy metals in grain is ambiguous. Thus, the content of cadmium, manganese, and cobalt in the grain of winter wheat of the Yuka variety is influenced by the use of plant protection products. Zinc accumulates in the grain with an increase in the depth of loosening of the soil, which is associated with a high background content of it in the soil and migration from the subsurface layer. The increase of lead and copper in the grain is affected to the same extent and method of tillage and crop protection. The negative values beta of the regression analysis is explained by the decrease of acquisitions of metals by the healthy plants in soil.

The application of fertilizers and changing the method of tillage contribute to an increase in the content of protein and gluten in the grain; the points on the normal probability graph are compactly located along the theoretically expected straight line, so the application of the linear regression model is correct (Fig. 1). For heavy metals, there is no clear pattern of the effect of the doses of applied fertilizers and methods of tillage on their content in winter wheat grain. For cobalt and manganese, the graphs are concave, and for Pb, Cd, Zn, and Cu, they are convex, which is due to the similarity of the chemical properties of these elements (Fig. 1).

The use of plant protection products, including pesticides, herbicides, and fungicides, did not contribute to the accumulation of toxic substances in the grain. The content of Mycotoxins (aflatoxin B1, deoxynivalenol, zearalenone, ochratoxin A) was below the MPC, and the accumulation of T-2 toxin at the level of 0.1 mg/kg indicated the presence of fungal infection in the soil. Pesticides (spiroxamine, propiconazole, tebuconazole, thiabendazole, triadimenol) and Herbicides (amidosulfuron, sodium iodosulfuron-methyl, mefenpyr-diethy) in winter wheat grain were found in the form of "traces," except for organomercury pesticides and derivatives of 2.4-D, which were not used, and their presence is associated with previous soil contamination.

The use of plant protection products, including pesticides, herbicides, and fungicides, did not contribute to the accumulation of toxic substances in the grain. The content of Mycotoxins (aflatoxin B1, deoxynivalenol, zearalenone, ochratoxin A) was below the MPC, the accumulation of T-2 toxin at the level of 0.1 mg/kg indicated the presence of fungal infection in the soil. Pesticides (spiroxamine, propiconazole, tebuconazole, thiabendazole, triadimenol) and Herbicides (amidosulfuron, sodium iodosulfuron-methyl, mefenpyr-diethy) in winter wheat grain were found in the form of "traces." The discovery of traces of organomercury pesticides and derivatives of 2.4-D, which were not used, is associated with previous soil contamination.

4   Conclusion

The use of mineral fertilizers in high doses of N120P60K40-N240P120K80 allows increasing the yield of winter wheat grain to 70–79 t/ha with high indicators of the quality of grain. However, an increase in mineral nutrition above N120P60K40 did not affect the quality of the grain in terms of gluten and protein content. The change in the method of tillage practically does not affect the quality of grain products, which is due to both the relatively small depth of fixation of the root system of winter wheat, and the deep loosening by 70 cm twice in a 11-year crop rotation for corn and sugar beet.

The accumulation of trace elements and heavy metals in winter wheat grain under different agricultural technologies had a different character. For Mn, Cu, Zn, and Pb, a decrease in their content in grain was observed with an increase in fertilizer doses compared to the control, which was explained by an increase in yield and increased soil buffering concerning these metals. For cobalt, a clear pattern of the effect of

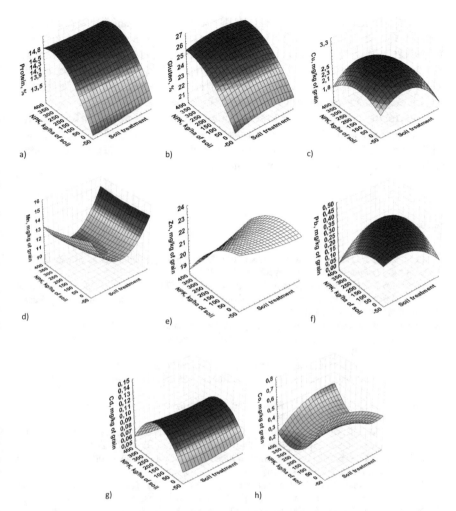

Fig. 1 Dependence of the content of heavy metals, yield, and quality of winter wheat grain of the Yuka variety on the method of tillage and the dose of fertilizers (x—tillage (loosening depth, cm), y—NPK, kg/ a of soil): a) Protein, $\% = 268 - 5 \cdot x + 0.02 \cdot y + 0.1 \cdot x^2 - 0.0002 \cdot x \cdot y - E^{-5} \cdot y^2$; b) Gluten, $\% = -3,577 + 69 \cdot x + 0.02 \cdot y - 0.3 \cdot x^2 - 2E^{-5} \cdot x \cdot y - 2 E^{-5} \cdot y^2$; c) Cu, content, mg/kg of grain $= -5,016 + 97 \cdot x - 0.03 \cdot y - 0.5 \cdot x^2 + 0.0003 \cdot x \cdot y - 2E^{-6} \cdot y^2$; d) Mn, content, mg/kg of grain $= 24,406 - 473 \cdot x - 0.2 \cdot y + 2.3 \cdot x^2 + 0.002 \cdot x \cdot y + 2E^{-6} \cdot y^2$; e) Zn, content, mg/kg of grain $= 1,637 - 31 \cdot x - 0.3 \cdot y + 0.15 \cdot x^2 + 0.003 \cdot x \cdot y - 2E^{-5} \cdot y^2$; f) Pb, content, mg/kg of grain $= -709 + 14 \cdot x - 0.1 \cdot y - 0.07 \cdot x^2 + 0.001 \cdot x \cdot y - 2E^{-6} \cdot y^2$; g) Cd, content, mg/kg of grain $= 26 - 0.513 \cdot x + 0.1 \cdot y + 0.003 \cdot x^2 - 1.3E^{-5} \cdot x \cdot y - 9E^{-7} \cdot y^2$; h) Co, content, mg/kg of grain $= -575 - 11.3 \cdot x + 0.05 \cdot y + 006 \cdot x^2 - 0.001 \cdot x \cdot y - 4.5E^{-7} \cdot y^2$. *Source* Developed and compiled by the authors

fertilizers was not revealed; this may be due to the low demand for wheat for this element. Concerns are caused by the accumulation of cadmium on the variant with N240P120K80 when loosening the soil to a depth of 22–24 cm and soil treatment to a depth of no more than 10 cm. The buffering capacity of leached chernozems concerning cadmium is low, and the lack of zinc and copper synergists in the soil contributes to the accumulation of this element in the grain.

Thus, for the cultivation of environmentally safe winter wheat grain of high quality, it is recommended to limit the use of mineral fertilizers at the level of N120P60K40 with any method of tillage, and also it is permissible to use chemical protective agents in the doses recommended by manufacturers.

References

1. Ceseviciene J, Leistrumaite A, Paplauskienė V (2009) Grain yield and quality of winter wheat varieties in organic agriculture. Agron Res 7(I):217–223
2. Gimbatov AS, Muslimov MG, Ismailov AB, Alimirzaeva GA, Omarova EK (2016) The role of mineral fertilizer in increasing the productivity and quality of winter wheat grain. Res J Pharm Biol Chem Sci 7(5):1304–1310
3. Kvashin AA, Neshchadim NN, Gontcharov SV, Gorpinchenko KN (2019) Economic efficiency and bioenergetic assessment of predecessors and fertilizer systems in the sunflower cultivation. Helia 42(70):101–109
4. Kvashin AA, Neshchadim NN, Yablonskay EK, Gorpinchenko KN (2018) Crop yield and the quality of sunflower seeds in the use of fertilizers and growth-regulating substances. Helia 41(69):227–239
5. Litke L, Gaile Z, Ruža A (2018) Effect of nitrogen fertilization on winter wheat yield and yield quality. Agron Res 16(2):500–509
6. Logvinov AV, Mishchenko VN, Moiseev VV, Neshchadim NN, Tsatsenko LV (2019) Problems of creating a three-way cross hybrid of sugar beet. Eurasian J BioSci 13(2):1291–1293
7. Lubyte J et al (2008) The concentration of heavy metals in the grains of differently fertilized winter wheat crops. Zemdirbyste 95(2):61–71
8. Nenko NI, Neshchadim NN, Yablonskay EK, Sonin KE (2016) Prospects for sunflower cultivation in the Krasnodar region with the use of plant growth regulators. Helia 39(65):197–211
9. Neshchadim NN, Slusarev VN, Kravtsov AM, Hurum HD (2018) The influence of prolonged cultivation of crops with various technologies on the properties of leached chernozem of Western Ciscaucasia. J Pharm Sci Res 10(9):2328–2331
10. Neshchadim NN, Tsatsenko LV, Koshkin SS, Kazartseva AT, Fedulov YP (2017) Criteria for assessing the reproductive potential of traditional varieties of winter soft wheat and the possibility of their use in the selection process. J Pharm Sci Res 9(12):2590–2595
11. Neshchadim NN, Shabanova IV, Kvashin AA, Fedulov YP, Tsatsenko LV (2020) The effect of agricultural technologies on the dynamics of the content of Mn, Zn, Cd Co, Pb, and Cu in leached back soil of western Ciscaucasia and maize grains. Int J Emerg Technol 11(2):978–984
12. Shabanova I, Neshchadim N, Gorpinchenko K, Boyko A (2020) Mycotoxins, pesticides, and heavy metals content in the winter wheat grain at different cultivation technologies on leached Kuban chernozem. E3S Web of Conferences, 203, 02012. https://doi.org/10.1051/e3sconf/202 020302012

13. Šíp V, Vavera R, Chrpová J, Kusá H, Růžek P (2013) Winter wheat yield and quality related to tillage practice, input level, and environmental conditions. Soil Tillage Res 132:77–85
14. Song X, Wang J, Huang W, Yan G, Chang H (2009) Monitoring spatial variance of winter wheat growth and grain quality under variable-rate fertilization conditions by remote sensing data. Nongye Gongcheng Xuebao 25(9):155–162
15. Vasil'Ko VP, Garkusha SV, Naidenov AS, Neshchadim NN, Kravtsov AM (2017) Status and optimization of arable land fertility in low-lying kettle cultivated lands of the central zone of the Krasnodar territory. Asian J Microbiol Biotechnol Environ Exp Sci 19(1):66–72

Changes in the Fertility of Agrogenic Soil During Chemical Reclamation

Olga V. Gladysheva⬤, Elena V. Gureeva⬤, and Vera A. Svirina

Abstract The paper aims to evaluate the effect of dolomite flour on the fertility of gray forest soils and the yield of crops in the grain-grass-tilled crop rotation. The authors present the results of the experiment on the impact of chemical ameliorants on the fertility of the intensively used gray forest soils and the yield of crops. The experiment was conducted in 2012–2017 in the Ryazan Region. The authors show that the ameliorant had a positive effect on the physical and agrochemical properties of the soil over six years. The greatest effect of ameliorant, expressed in the reduction of acidity, was established on the second crop of the crop rotation. This effect continues until the end of crop rotation. After that, the processes of acidity increase, and the content of calcium in the soil and absorption capacity decrease. During the rotation, there was a significant increase in humus by 2.2% in the absence of mineral fertilizers and by 2.6% in the variant with the application of mineral fertilizers. Dolomite flour enriched the soil with magnesium by 23%–36%, reduced soil density by 3%–6%, and improved soil structural conditions by increasing water-tight aggregates by 1.4%–3.1% and porosity by 1.8%–2.7%. Ameliorant increased the productivity of the crop rotation in the variant without fertilizers by 6.4 quintals of fodder units, in the variant with the use of fertilizers—by 10.2 quintals of fodder units, and in the variant with the complex application of fertilizers and ameliorant— by 19.9 quintals of fodder units. Further periodic liming of arable soils is a promising method to increase crop productivity and maintain soil fertility.

Keywords Chemical amelioration · Soil acidity · Fertility · Dolomite powder · Crop rotation

O. V. Gladysheva (✉) · E. V. Gureeva · V. A. Svirina
Institute of Seed Production and Agrotechnologies, Federal State Budgetary Scientific Institution "Federal Scientific Agroengineering Center VIM", Ryazan, Russia
e-mail: GladyshevaOV@yandex.ru

E. V. Gureeva
e-mail: gureeva@bk.ru

V. A. Svirina
e-mail: podvyaze@bk.ru

© The Author(s), under exclusive license to Springer Nature Singapore Pte Ltd. 2022   319
E. G. Popkova and B. S. Sergi (eds.), *Sustainable Agriculture*,
Environmental Footprints and Eco-design of Products and Processes,
https://doi.org/10.1007/978-981-16-8731-0_31

JEL Code Q16

1 Introduction

The primary field production in the Ryazan region is concentrated on gray forest (37%) and chernozem soils (43%). During the use of arable land to grow crops, there occurs soil acidification even without applying physiologically acidic mineral fertilizers. According to agrochemical service stations, 69% of the arable land in the region had an acidic environment in 2005 and 74.3% in 2019, including 31.6% of strongly and moderately acidic soils.

Soil acidity is one of the main factors preventing high yields of most crops. The solution to the problem of increased soil acidity in the region is largely promoted by increasing the financial interest of agricultural producers in agrochemical land reclamation and government support measures with a significant share of cost compensation. Thus, chemical reclamation activities were carried out on an area of 479 hectares in 2016, on 1,438 hectares in 2017, on 4,126 hectares in 2018, and 14,332 hectares in 2019.

The Institute of Seed Production and Agrotechnologies—a branch of the Federal Scientific Agroengineering Center VIM—has been researching chemical reclamation of dark gray forest soils and its impact on improving the sustainability of agriculture since 2011.

2 Materials and Methods

In the twentieth century, there were symptoms of a global crisis of intensive farming methods due to systemic soil degradation [1].

Several works [2, 3, 5, 7] note that unreasonable farming, reduction of application of mineral and organic fertilizers, violation of principles of crop rotation, and the cessation of application of lime-containing materials led to an increase in the areas of soils with excessive acidity, increased degradation processes, deterioration of the agrophysical state of arable lands, and reduction of agrochemical indicators of fertility.

Soil acidity can be considered an indicator determining the direction of the use of arable land and its productivity.

The works [8–14, 16–19] show that the strategic goal of the development of the Russian AIC is the conservation, restoration, and improvement of the efficiency of soil use. Liming or chemical reclamation is the basic and most effective measure to achieve this goal in some Russian regions. These measures help eliminate excessive acidity with a lasting optimizing effect on the basic agrochemical features of the soil. This improves the availability of nutrients to the growing plants, increases their resistance to stress, increases the return on fertilizers used and their payback, and increases crop productivity.

The research aims to evaluate the effect of dolomite flour on the fertility of gray forest soils and the yield of crops in the grain-grass-tilled crop rotation.

The field experiment was conducted in the Ryazan Region using the following crop rotation: barley with oversowing of clover—clover (first year of use) –oatmeal mixture—winter wheat—corn for silage–spring wheat. Two factors were taken into account during the experiment: A—mineral fertilizer background $N_0P_0K_0$ and $N_{90}P_{90}K_{90}$ and B—chemical ameliorant background. When laying experiments and conducting measurements, the analyses were based on the approved methods [4, 14, 15]. The experiment was laid in a fourfold repetition on the plot with the following indicators of fertility:

- Salinity (pH) −4.98 (without fertilizers) and 4.87 (with fertilizers);
- Hydrolytic acidity (A_h) −4.69 mg-eq/100 g of soil (without fertilizers) and 5.86 mg-eq/100 g of soil (with fertilizers);
- Humus content (Tyurin's method) −3.05% (without fertilizers) and 3.15% (with fertilizers);
- Content of mobile phosphorus P_2O_5 (Kirsanov's method) − 10.6 mg/100 g of soil (without fertilizers) and 19.0 mg/100 g of soil (with fertilizers);
- Content of exchanged potassium K_2O (Kirsanov's method) −9.2 mg/100 g of soil (without fertilizers) and 12.3 mg/100 g of soil (with fertilizers);
- Total absorbed bases (S) −20.5 mg-eq/100 g of soil (without fertilizers) and 18.5 mg-eq/100 g of soil (with fertilizers);
- Absorption capacity (V) −81.3%–75,9%;
- Content of exchanged Ca − 16.9–17.5 mg-eq/100 g of soil;
- Content of exchanged Mg −2.4 mg-eq/100 g of soil.

When laying the experiment in 2011, the soil was treated with a chemical ameliorant at the rate of 1.5 A_h with 55% of Ca and 33% of Mg, corresponding to GOST 14,059–93 Limestone meal (dolomite) grade B, grade 1 [6].

3   Results

Dolomite flour contributed to the reduction of saline and hydrolytic acidity. As a result, moderately acidic soils moved into the group of slightly acidic (Table 1). In the first year in the variant without fertilizers, the ameliorant contributed to changes in acidity pH from 4.98 to 5.43 units, and $CaCO_3$ consumption for an acidity shift of 0.1 units was 2.21 t/ha. In the first year in the variant with fertilizers, the acidity changed from 4.87 from 5.32 units to 2.7 t/ha of $CaCO_3$ consumption.

The most significant effect of liming was achieved two years after applying ameliorant with a decrease in the level of its consumption. The value of pH_{saline} changed by 0.68 and 0.6 and equaled 5.66 and 5.47. In the next four years, the processes of acidification of the soil environment began. For the year, the increase in acidity in the

Table 1 Change in soil fertility under the influence of dolomite flour, average for a rotation, 0–30 cm

Fertilizer and liming system	Acidity		Content in soil							Amount of absorbed bases, mg-eq/100 g soil	Absorption capacity, mg-eq/100 g soil
	Saline, pH_{saline}	Hydrolytic, A_h	Total humus, %	Nitrate nitrogen, mg/kg of soil	P_2O_5, mg/100 g of soil	K_2O, mg/100 g of soil	magnesium (Mg), mg-eq/100 g soil	calcium (Ca), mg-eq/100 g soil		
$N_0P_0K_0$	5.00	3.72	3.03	3.47	10.9	10.0	2.64	15.80	19.7	25.9
$N_0P_0K_0$ + $CaCO_3$	5.47	2.70	3.098	4.47	10.9	10.7	3.29	15.26	22.9	28.7
Change of indicators	+0.47	−1.02	+0.068	+1.0	0	+0.7	+0.65	−0.54	+3.2	+2.8
$N_{90}P_{90}K_{90}$	4.82	4.55	3.127	3.81	20.2	12.3	2.49	15.40	19.0	25.6
$N_{90}P_{90}K_{90}$ + $CaCO_3$	5.34	3.25	3.208	5.26	21.7	13.2	3.40	16.00	23.0	28.3
Change of indicators	+0.52	−1.3	+0.081	+1.45	+1.5	+0.9	+0.91	+0.6	+4.0	+2.7
HCP_{05} factor A	0.45	0.82	0.14	0.72	1.53	1.11	0.34	0.46	2.49	2.49
HCP_{05} factor B	0.20	0.49	0.056	0.83	1.66	1.54	0.28	0.50	1.50	1.50

Source Compiled by the authors

variant without fertilizers averaged 0.07 units and 0.06 units in the variant with fertilizers. In the variant with mineral fertilizers without ameliorants, the acidification of the arable soil horizon by 0.05 units per rotation was noted.

The conducted studies indicate the high efficiency of liming against the background of mineral fertilizers and its prolonged effect on maintaining soil fertility. The highest amount of humus was recorded in the fifth year after the application of ameliorant. In the sixth year, there was a slight downward trend of 0.002%–0.003%. On average, the humus content increased by 0.068% in the variant without mineral fertilizers and by 0.081% in the variant with fertilizers during the crop rotation.

There is a positive effect of dolomite flour on the amount of nitrate-nitrogen in the soil—it is significantly higher than variants without ameliorant. In the experiment, there was a slight increase in the soil layer (30 cm) of mobile forms of phosphorus and potassium under the influence of dolomite flour, especially in the second and third years after its introduction. The application of lime is initially accompanied by a sharp increase in the concentration of calcium in the arable soil solution. Subsequently, crop cultivation and downward movement of the soil profile with water infiltration influence the decrease of Ca, especially in the layer 0–20 cm. Liming the soil with the systematic use of mineral fertilizers slows the decrease of calcium in the arable layer. After the first rotation in the experiment, the calcium content in this option is significantly higher—by 0.60 mg/100 g of soil.

Most soils in the Ryazan Region have a shallow content of magnesium due to the increased acidity of the soil solution. Dolomite flour has 33% of magnesium in its content. The application of dolomite flour contributed to the enrichment of the soil with magnesium by 23%–36%. During the rotation, its highest amount was noted in the second year after reclamation—4.03–4.06 mg/100 g of soil.

The use of mineral fertilizers accelerates the loss of absorbed bases in the soil. This process can be slowed down by adding dolomite flour.

The greatest effect of dolomite flour on the saturation of the bases was in the period of 2–4 years of the action of dolomite flour—25.0–27.5 mg-eq/100 g of soil.

This figure corresponded to 23.0 mg-eq/100 g soil in the variant with mineral fertilizers and with ameliorant on average for the rotation. In the variant without fertilizers and without ameliorant, the saturation with bases was 19.7 mg-eq/100 g soil.

The absorption capacity in the initial year of application of dolomite flour was 22.4–22.7 mg-eq/100 g of soil. Subsequently, under the third and fourth crops in the crop rotation, there was an increase in the value of the indicator by 9.4–10 mg-eq/100 g of soil. By the end of crop rotation (in the sixth year of ameliorant action), indicators of absorption capacity returned to the level of the first year of dolomite flour action. On average, this indicator in variants with $CaCO_3$ was 10% higher than without the use of calcium-containing material. In dynamics, the soil absorption capacity increased by 41%–45% in variants with the ameliorant and by 58%–61% without it.

The biological activity of soil responds positively to lime treatment (Table 2). This is confirmed by the rate of flax decay, which is 13%–57% faster than the variant without fertilizer and ameliorant.

Table 2 Changes in physical indicators of soil fertility under the influence of dolomite flour, average for a rotation, 0–30 cm

Fertilizer and liming system	CO_2 release from the soil, mg CO_2 per 1 m^2/h	Soil density, g/cm^3	Content of water-tight aggregates, %	Total porosity, %
$N_0P_0K_0$	125.6	1.497	43.1	47.7
$N_0P_0K_0$ + $CaCO_3$	196.2	1.405	44.5	49.5
Change of indicators	+70.6	−0.092	+1.4	+1.8
$N_{90}P_{90}K_{90}$	166.7	1.42	43.2	48.4
$N_{90}P_{90}K_{90}$ + $CaCO_3$	243.6	1.370	46.3	51.1
Change of indicators	+76.9	−0.05	+3.1	+2.7
HCP_{05} factor A	56.65	0.04	3.30	2.12
HCP_{05} factor B	43.9	0.023	1.28	1.60

Source Compiled by the authors

Additionally, the variants with $CaCO_3$ increased carbon dioxide emissions from the soil by 46%–56%. This indicates an increase in the rate of decomposition of plant residues in the soil and, as a rule, a greater number and activity of microorganisms involved in the decomposition of organic matter.

Dolomite flour affects the agrophysical indicators of soil fertility—its density and structure. It was found that the ameliorant shows a decompaction effect on the gray forest soils, reducing its density by 3%–6% compared to the value without ameliorant. There was a reliable increase in the number of agronomically valuable aggregates by 1.4%–3.1% and soil porosity by 1.8%–2.7%. Improved water resistance of aggregates increases the ability of this type of soil to resist the damaging effects of precipitation and water flow, optimizes the air exchange and water permeability of the soil profile, and increased water-holding potential. Without fertilizer, aeration porosity (non-capillary) equaled 13.3%. The application of dolomite flour increased it to 14.6%. The most optimal non-capillary porosity is noted in the variant with ameliorant and mineral fertilizer—16.3%, which is 1.4% higher than without the application $CaCO_3$ (14.9%).

Mineral fertilizers, applied with dolomite flour, intensified their effect on the crop rotation and ultimately contributed to increased yields (Table 3).

In the rotation, the average increase in crop productivity from the application of ameliorant was 6.4 quintals of fodder units (without fertilizer) and 10.2 quintals of fodder units (with fertilizer). The increase in the productivity from the application of mineral fertilizers was 9.7 quintals of fodder units and 19.9 quintals of fodder units from the complex application of fertilizers and ameliorants (Fig. 1).

Calculation of economic efficiency showed that the highest additional net income was obtained when combining chemical reclamation with mineral fertilizers. The

Table 3 Crop yields in crop rotation under the influence of dolomite flour, t/ha

Fertilizer and liming system	Barley with oversowing of clover	Clover (first year of use)	Oatmeal mixture	Winter wheat	Corn for silage	Spring wheat
$N_0P_0K_0$	2.19	39.6	16.6	3.64	26.4	2.70
$N_0P_0K_0$ + $CaCO_3$	2.38	44.5	18.4	3.96	30.7	3.13
Change of indicators	+0.19	+4.9	+1.8	+0.32	+4.3	+0.43
$N_{90}P_{90}K_{90}$	3.17	41.3	20.2	4.91	28.2	3.74
$N_{90}P_{90}K_{90}$ + $CaCO_3$	3.60	47.2	23.3	5.44	33.3	4.40
Change of indicators	+0.43	+5.9	+3.1	+0.53	+5.1	+0.66
HCP_{05} factor A	0.164	0.66	2.30	0.054	2.53	0.58
HCP_{05} factor B	0.097	0.465	1.36	0.284	2.53	0.404

Source Compiled by the authors

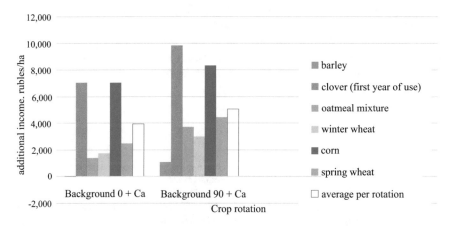

Fig. 1 Effectiveness of using dolomite flour on dark gray forest soils. *Source* Compiled by the authors

average income for the rotation equaled 5,049 rubles/ha, depending on the crop—from 1,062 rubles/ha to 9,818 rubles/ha. The use of dolomite flour without fertilizer provided an average income per rotation of 3,924 rubles/ha (from −59 rubles/ha to 7,049 rubles/ha).

4   Conclusion

Our experiment revealed a direct effect of dolomite flour on reducing soil acidity. Dolomite flour serves as a tool in creating favorable conditions to increase the productivity of alternating crops in a crop rotation. Simultaneously, a significant factor positively affecting the fertility of long-used soil is a rapid optimization of its basic physical and chemical properties and an increase in the intensity of biological activity with prolonged action of ameliorant. The most significant economic effect is observed in the application of chemical reclamation in combination with mineral fertilizers.

The authors found that the maximum effect of the ameliorant was manifested in two years after the application. This effect lasted until the end of the crop rotation. After that, there began the processes of increasing acidity, reducing the calcium content in the soil, and decreasing absorption capacity. This indicates the need for further supportive liming of arable soils.

The solution to the problem of increased soil acidity is a significant increase in the volume of liming of land in the Ryazan Region and bringing these volumes up to 200 thousand hectares annually at a dose of lime material 7.3–10.5 t/ha. To maintain an optimal balance of calcium in the future, it is necessary to establish the frequency of liming—once every six to eight years, depending on the intensity of using arable land and receipt of organic matter in the soil.

The advantage of the practical application of the acquired knowledge lies in overcoming excessive acidity, increasing the efficiency of the crop industry, obtaining additional products, and preserving soil fertility for future generations.

References

1. Aparin BF, Ye, Sukhacheva Yu (2019) Husbandry: its past, present, and future. Biosfera 11(3):109–119. https://doi.org/10.24855/biosfera.v11i3.507119
2. Barber HS (1998) Greps for cave-inhabiting. Sci Sos. 46:259–266
3. Degtyareva IA, Lomako EI, Yapparov AH (2003) Effect of different doses of lime on the biological activity of leached chernozem. Agrochem Herald 4:24–26
4. Derzhavina LM, Bulgakova DS (Ed) (2003) Methodological guidelines for comprehensive monitoring of agricultural land fertility. Moscow, Russia: All-Russian Research Institute of Automation
5. Gladysheva OV, Pestryakov AM, Svirina VA (2015) Effect of liming on the physicochemical properties of clay loamy dark gray forest soil and the productivity of crops. Plodorodie 6(87):17–19
6. Gladysheva OV, Pestryakov AM, Svirina VA, Krasnikov NG (2013) Determination of optimum $CaCO_3$ rate to neutralize the acidity of heavy loam dark grey forest soil. Plodorodie 6(75):46–47
7. Gladysheva OV, Svirina VA, Suhryakova OA (2018) Current state of crop production in Russia and Kazakhstan. Agrarian Sci 7–8:62–65
8. Gvozdev VA, Gureeva EV, Ovsyannikova MV, Markova VE (2020) Influence of liming on soybean productivity in the conditions of the Central Region of the Non-Chernozem Zone. In: Vinogradov DV, Yevsenina MV (eds) Ecological state of the natural environment and scientific and practical aspects of modern agricultural technologies. Ryazan State Agrotechnical University, Ryazan, Russia, pp 81–84

9. Kalichkin VI, Minina IN (2002) Influence of lime treatment on agrochemical properties of sod-podzolic soils and productivity of crop rotations. In: Soil liming issues (pp 85–91). Moscow, Russia: Agroconsult

10. Kirpichnikov NA, Shilnikov IA, Akanova NI, Chernyshkova LB (2014) Phosphate status of soddy-podzolic soil depending on the application of lime and phosphate fertilizers. Plodorodie 4(79):21–23

11. Kononova MM (1984) Organic matter and soil fertility. Eurasian Soil Sci 8:6–20

12. Korchenkina NA, Makhalov RM (2015) Effect of mineral fertilizers and aftereffect of lime material on the dynamics of mobile potassium in a light gray forest soil. Plodorodie 3(84):8–10

13. Kulakovskaya TN (1990) Optimization of the agrochemical system of soil nutrition of plants. Moscow, USSR: Agropromizdat

14. Lykov AM (1982) Reproduction of soil fertility in the Non-Chernozem Soil Zone. Moscow, USSR: Rosselkhozizdat

15. Methodology of field experiments to develop rates of lime fertilizers to shift the reaction of the soil environment to the optimal pH level on different types of soils (1987) Moscow, USSR: Pryanishnikov Institute of Agrochemistry

16. Pestryakov AM (2011) Dynamics of agrochemical properties of dark gray forest soils with the use of fertilizers in different crop rotations. In Sychev VG (Ed), Results of long-term research in the system of the geographic network of experiments with fertilizers of the Russian Federation (for the 70th anniversary of Geoset) (pp 72–94). Moscow, Russia: All-Russian Research Institute of Automation

17. Shilnikov IA, Akanova NI (2011) Issues of liming in contemporary conditions. Plodorodie 3(60):22–24

18. Shilnikov IA, Lebedeva LA (1987) Soil liming. Moscow, USSR: Agropromizdat

19. Shilnikov IA, Sychev VG, Akanova NI (2013) The state and efficiency of the chemical reclamation of soils in the agriculture of the Russian Federation. Plodorodie 1(70):9–13

Adaptive Capacity of Strawberries in Autumn

Zoya E. Ozherelieva⬡, Pavel S. Prudnikov⬡, Marina I. Zubkova, and Sergey D. Knyazev⬡

Abstract Strawberry cultivars are studied for physiological and biochemical changes that characterized the condition of the plants in autumn. The studies found that in autumn, on the background of lowering the level of hydration, for plants of strawberry, there was a characteristic increase in bound water and reduction of free water in the leaves. At the end of autumn, the most ratio of colloidal to free water, lower level lipid peroxidation, and reactive oxygen species were determined in Solovushka, Tzaritza, Sara, and Korona. The latter indicates that these strawberry varieties have a high adaptive ability in the climatic conditions of Central Russia.

Keywords Strawberry · Varieties · Adaptation · Colloidal water · Free water · Malondialdehyde (MDA) · Hydrogen peroxide (H_2O_2) · Superoxide dismutase (SOD) · Catalase

JEL Codes Q01 · Q1 · F64

1 Introduction

Strawberry is a valuable and favorite berry crop culture, which is appreciated for its early ripening, excellent taste, and high productivity [1–3].

However, strawberry has low winter hardness, which can significantly decrease productivity. Resistance to unfavorable abiotic factors of the winter period is one

Z. E. Ozherelieva (✉) · P. S. Prudnikov · M. I. Zubkova · S. D. Knyazev
Russian Research Institute for Fruit Crop Breeding (VNIISPK), Zhilina, Russia
e-mail: ozherelieva@vniispk.ru

P. S. Prudnikov
e-mail: prudnikov@vniispk.ru

M. I. Zubkova
e-mail: zubkova@vniispk.ru

S. D. Knyazev
e-mail: ksd_61@mail.ru

© The Author(s), under exclusive license to Springer Nature Singapore Pte Ltd. 2022
E. G. Popkova and B. S. Sergi (eds.), *Sustainable Agriculture*,
Environmental Footprints and Eco-design of Products and Processes,
https://doi.org/10.1007/978-981-16-8731-0_32

of the most important characteristics, which can determine the economic value in the areas of horticultural crop cultivation [4]. Resistance to low temperatures is achieved when plants transit into a state of dormancy and passes through the hardening phases [7]. A significant role of hardening and snow cover was found for the successful wintering of strawberries [15]. The hardening is a difficult physiological and biochemical process, which links with an increase of bound water and changes in membrane lipids. At the same time, the permeability of the plasmalemma to water is increasing and it prevents the formation of intracellular ice in the protoplast. At the same time, plants significantly increase their frost resistance [17]. It was shown that at the end of the autumn period, the amount of colloidal water in plants increases sharply, and the content of free water decreases [8]. At the same time, it is noted that the ratio of easy-to-hard-to-recover water can characterize the resistance of plants to low temperatures in the autumn–winter period because the winter-hardy varieties have lower ratio of free/bound water in most cases than in non-winter-hardy varieties [8, 11, 13].

2 Methodology

Based on the Russian Research Institute for Fruit Crop Breeding (VNIISPK), some physiological and biochemical processes of strawberry adaptability in the autumn period were studied. The research involved Russian varieties—Kokinskaya early, Solovushka, Tsaritsa, Urozhainaya TzGL—Russia; Swedish—Sara; Italian—Alba, and Dutch—Korona, Sonata. Strawberry plants were planted randomly in a threefold number of replications in the area in 2016. There were 30 bushes in each replication. The scheme of planting was 90 × 20 cm.

The values of the absolute maximum and minimum air temperatures of the autumn period according to the data of the VNIISPK weather station for the years of research are shown in Table 1.

The fractional composition of water in strawberries leaves was determined by the Okuntsov-Marinchik method [9]. The method is based on changing the concentration of sucrose solution when the leaf tissue is immersed in it. We poured 2 ml of 30% sucrose solution into the buckets weighed on the scales and they were weighed. Strawberries leaves were crushed in a laboratory mill and 0.4 g was immersed in 30% sucrose solution. Based on the initial volume of the solution, its initial and final

Table 1 Air temperature in autumn 2017–2018

T, °C	2017			2018		
	September	October	November	September	October	November
Max	28.0	15.5	9.0	29.8	21.5	10.6
Min	−1.5	−4.8	−9.2	−1.0	−2.8	−18.5

Source Compiled by the authors

concentration, and the amount of water taken from the leaf tissue by the solution was determined. The concentration of sucrose in the solution was determined using a digital PAL-1 Refractometer [13].

To determine the intensity of peroxidation of membrane lipids (POL), malonic dialdehyde (MDA) [14] and hydrogen peroxide [6] were found.

A modified technique was used to determine the activity of superoxide dismutase (SOD) [5]. Catalase activity was determined by the method [16].

Significant differences between the cultivars (LSD_{05}) were determined with a reliable probability of 95%.

3 Results

In September 2017, precipitation fell only in the first decade of the month—13.7 mm. At the end of September, the low level of the water content of strawberries leaves was noted—from 32.0–42.5%. At the same time, in September, free water prevailed, which varied from 21.0–30. %. The amount of bound water was in the range of 3.0–19.3%. In the first week of October, the amount of precipitation increased (27.3 mm), which affected the water content of the leaves, which increased during this period by 12.5–13.0%. Despite the increase in the level of common water content, an increase of bound water by 17.6–20.7% was observed in the leaves of the varieties in October. In November, free water decreased by 6.7–11.6% in strawberry leaves. In the first decade of the month, a deviation below the norm of the average daily air temperature was registered—1.5 °C, little precipitation fell—3.9 mm. The prevailing weather conditions affected the water content of strawberry leaves, which decreased by 5.6–16.2%. The amount of bound water in the leaves was in the range of 21.5 to 33.2%. The varieties Solovushka, Tsaritsa, Korona noticed the largest amount of bound water in November. The amount of free water in the leaves decreased by 4.9–5.6%. A greater increase in the amount of bound water was identifying in October by 1.6–7.8 times when compared to September, and in November by 1.1–1.9 times when compared to October.

In September 2018, precipitation fell—42.5 mm. The level of the common water content of leaves varied between 36.7 and 48.5% in September. The free water prevailed in the leaves of studied strawberry varieties in the middle of September. Its amount was in the range—from 22.5–33.2%. The level of bound water in the leaves was from 10.0–19.3% during the study period. In the first week of October, precipitation fell—14.4 mm. The water content in the leaves of strawberry varieties decreased by 1.4–13.5% over this period. In October, the content of bound water in plant leaves increased by 3.7–14.6%, compared to September. Also, in October, the highest amount of bound water was noticed in the varieties Tsaritsa and Korona. The decrease in air temperature in November (Table 1) was conducive to a decrease in the free water level by 4.3–17.8%, which varied from 9.4–24.2%. An increase of bound water (by 16.8…23.4%) was observed in the studied varieties in November,

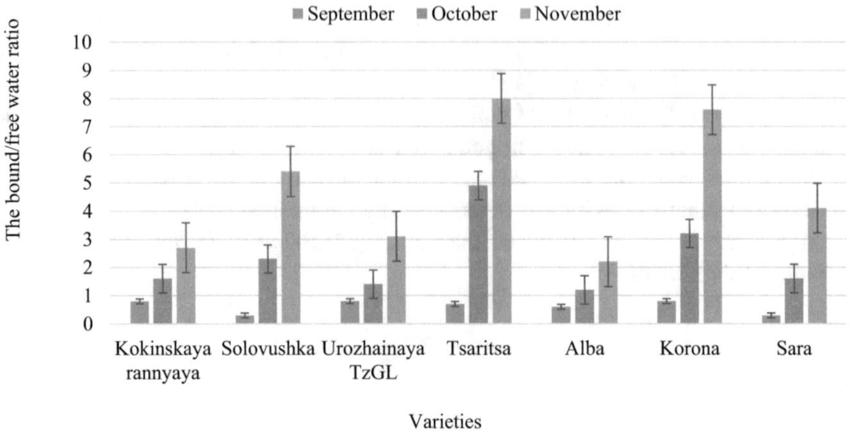

Fig. 1 The bound/free water ratio of strawberry leaves in autumn (2017–2018). *Source* Compiled by the authors

compared to September. In November (compared to October), the amount of bound water increased by 4.2–17.6% of strawberry leaves.

Earlier it was shown in [10] that winter-hardy raspberry varieties were characterized by the highest ratio of bound/free water at the end of the autumn period. In our studies, the minimum ratio of bound/free water was noted in the leaves of the studied strawberry varieties in September. In October, this indicator increased by 1.8–7.7 times. In November, the ratio of bound/free water increased by 1.6–18.0 times in the leaves compared to the previous autumn months (Table. 1). The maximum ratio of water fractions by the beginning of winter was recorded in the varieties—Solovushka, Tsaritsa, Sara, Korona, as noted earlier (Fig. 1).

In the autumn period, a close dependence was established between leave hydration ($r = 0.81$) and the ratio of bound/free water ($r = -0.97$) on the values of the minimum air temperature.

The intensity of lipid peroxidation was estimated by the accumulation of malondialdehyde, the content of hydrogen peroxide, the activity of superoxide dismutase and catalase [12, 13].

After the ambient temperature decreased, the intensity of MDA accumulation increased in all varieties. At the same time, the intensity of peroxidation of membrane lipids was significantly lower in the varieties Solovushka, Sara, Korona, and Tsaritsa, than in other genotypes. The accumulation of MDA in November compared to October increased by 11.8–23.3% in Solovushka, Sara, Korona, and Tsaritsa. In other varieties, malondialdehyde increased by 48.3–76.5%, which can indicate a more significant lipid peroxidation (Table 2).

The different levels of MDA accumulation in the studied varieties seem to be associated with a different degree of ROS formation in the cells and with a higher activity of the antioxidant defense system. For this reason, hydrogen peroxide was identified as one of the representatives of ROS.

Table 2 The accumulation of MDA of strawberry leaves in autumn (average for 2017–2018)

Varieties	MDA, microMol/g	
	October	November
Alba	8.1 ± 3.3	14.3 ± 6.8
Kokinskaya early	7.1 ± 3.0	10.6 ± 5.2
Urozhainaya TzGL	6.0 ± 2.1	8.9 ± 3.1
Sara	6.8 ± 2.8	7.9 ± 3.3
Solovushka	6.8 ± 3.0	7.6 ± 3.1
Korona	6.0 ± 2.5	7.4 ± 3.0
Tsaritsa	5.6 ± 1.9	6.7 ± 2.2

Source Compiled by the authors

Table 3 The content of hydrogen peroxide in strawberry (average for 2017–2018)

Varieties	H_2O_2, microMol/g	
	October	November
Alba	3.2 ± 0.3	19.0 ± 2.2
Kokinskaya early	3.2 ± 0.1	15.3 ± 0.5
Urozhainaya TzGL	2.6 ± 0.4	8.4 ± 1.1
Korona	2.0 ± 0.3	4.1 ± 1.2
Sara	2.0 ± 0.1	4.3 ± 0.4
Tsaritsa	1.8 ± 0.2	3.2 ± 0.5
Solovushka	1.6 ± 0.2	3.8 ± 0.4

Source Compiled by the authors

The correlation coefficient between the accumulation of MDA and the content of H_2O_2 in the leaves was: in October -0.67 and in November -0.79 (Tables 2, 3).

The high activity of antioxidant enzymes was shown for varieties with a low level of ROS and hydrogen peroxide products, except for SOD. Thus, in the Solovushka, Tsaritsa, Korona, and Sara varieties, the activity of SOD, an enzyme that recycles superoxide with the formation of hydrogen peroxide, did not significantly change in November compared to October during the two years of research, while in the other genotypes it increased by 14.6–31.4% (Fig. 2). On the one hand, this explains the different levels of hydrogen peroxide in the studied varieties.

The high activity of antioxidant enzymes—catalase, was shown for varieties Solovushka, Tsaritza, and Sara. So, the activity of catalase increased for this varieties in November 2017 by 37.7–50.5%, in comparison with October, another had an increase by 15.7–25.5%. During the experiment, we fixed low dependence ($r = -0.20$) between activity antioxidant enzyme catalase and amount of hydrogen peroxide in October. When the air temperature decreases in November (Table 1), the value of hydrogen peroxide in strawberry plants depended closely on enzyme activity $r = -0.71$. In the colder autumn period 2018 (Table 1), the coefficient of correlation between H_2O_2 and catalase was high from -0.84 to -0.82. We noted a

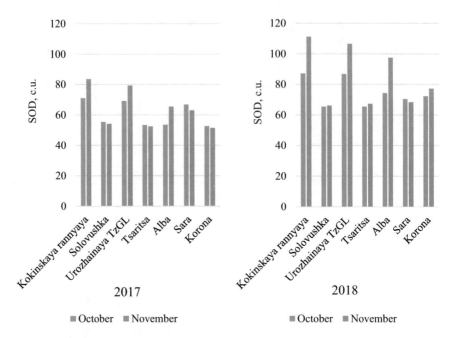

Fig. 2 The SOD activity for strawberry leaves in autumn. *Source* Compiled by the authors

significant increase in catalase activity in November 2018 for varieties Solovushka, Tsaritsa, Korona, and Sara (Fig. 3).

We noted a close correlation link between indicators of adaptability of strawberry varieties and the minimum air temperature, which were studied in the autumn period (Fig. 4).

4   Conclusion

In the autumn period, we had a study on some indicators of the adaptability of strawberry varieties of domestic and foreign breeding. The studies found that in autumn, on the background of lowering the level of hydration, for plants of strawberry, there was a characteristic increase of bound water and reduction of free water in the leaves. In late autumn, an increase in the ratio of hard-to-recover water to free water, lower levels of lipid peroxidation, and reactive oxygen species were determined in Solovushka, Tzaritsa, Sara, and Korona. All this indicates that these cultivars are characterized by a high adaptive capacity in the climatic conditions of Central Russia.

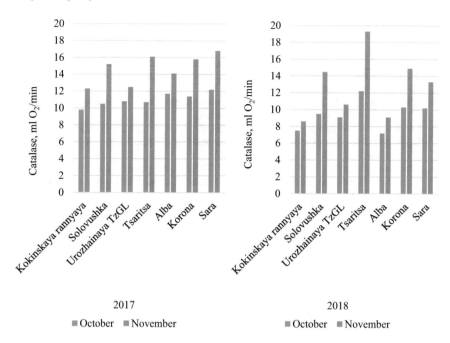

2017

■ October   ■ November

2018

■ October   ■ November

Fig. 3 The catalase activity of strawberry leaves in autumn. *Source* Compiled by the authors

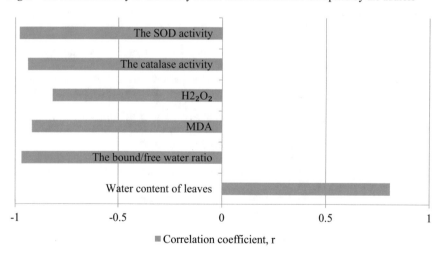

■ Correlation coefficient, r

Fig. 4 Dependence of the water regime, accumulation of MDA and H_2O_2, the activity of SOD and catalase on T_{min} in autumn. *Source* Compiled by the authors

References

1. Aaby K, Ekeberg D, Skrede G (2007) Characterization of phenolic compounds in strawberry fruits by different HPLC detectors and contribution of individual compounds to total antioxidant capacity. J Agric Food Chem 55:4395–4406
2. Aitzhanova SD (2002) Selection of strawberries in the South-Western part of the Non-Chernozem zone of Russia: PhD thesis in Agriculture
3. Cordenunci BR, Genovese MI, Nascimento JR, Aymoto Hassimoto NM, Jose dos Santos R, Lajolo FM (2005) Effect of temperature on the chemical composition and antioxidant activity of three strawberry cultivars. Food Chem 91:113–121
4. Efimova IL, Shaforostova NK, Kuznetsova AP (2008) The adaptive and productive potential of rootstocks of fruit crops in the conditions of southern gardening. Fruit berry growing in Russia, XVIII, 135–141
5. Giannopolities CN, Ries SK (1977) Superoxide dismutase. I. Occurr Higher Plant Physiol 59:309–314
6. Kumar GNM, Knowles NR (1993) Changes in lipid peroxidation and lipolitic and free radical scavenging enzyme activities during aging and sprouting of potato (Solanumtuberosum) seed-tubers. Plant Physiol 102:115–124
7. Makarova NV, Stryukova AD, Antipenko MI (2014) Comparative analysis of the chemical composition and antioxidant activity of repair and non-repair varieties of the strawberry garden. Storage Proces Agric Raw Mater 9:45–48
8. Matala V (2003) The cultivation of strawberries. Saint Petersburg
9. Ozherelieva ZE, Prudnikov PS, Zubkova MI, Krivushina DA, Knyazev SD (2019) Determination of frost resistance of strawberries under controlled conditions (methodological recommendations). Orel: VNIISPK
10. Ozherelieva ZE, Bogomolova NI (2018) Study of the water fractional composition of raspberry plants in the autumn period. Breeding Cultivar Propag Orchard Plants 5(1):89–92
11. Ozherelieva ZE, Prudnikov PS, Bogomolova NI (2016) Frost hardiness of introduced Hippophae rhamnoides L. genotypes in conditions of Central Russia. Proc Latvian Acad Sci Sect B 70, 2(701):88–95. https://doi.org/10.1515/prolas-2016-0014
12. Prudnikov PS, Ozherelieva ZE, Krivushina DA, Zubkova MI (2019) Physiological and biochemical features of the formation of resistance to hyperthermia of *FRAGARIA ANANASSA* cultivars in autumn. Bull Kursk State Agric Acad 8:118–125
13. Prudnikov PS, Ozherelieva ZE, Krivushina DA, Zubkova MI (2019) The formation of resistance to hypothermia of FRAGARIA x ANANASSA DUCH varieties of different ecological and geographical origins in the autumn period. Veg Russia 6:80–83
14. Stalnaya ID, Garishvily TG (1977) Method for the determination of malonic dialdehyde using thiobarbituric acid. Modern methods in biochemistry: under the editorship of V. N. Orehovich. Moscow: Medicine, 66–68
15. Tjurina MM, Gogoleva GA, Goloulina LK, Morozova NG, Echedi YY, Volkov FA, Arsentieva AP, Matyash NA (2002). Determination of stability of fruit and berry crops to cold season stressors in the field and controlled conditions. Moscow
16. Tretiakov NN (1990) Workshop on Plant Physiology. Agropromizdat, Moscow
17. Yakushkina NI, Bakhtenko EYu (2004) Physiology of plants. Moscow: Vlados

Comprehensive Assessment of Promising Soybean Lines of the Northern Ecotype for Cultivation in a Mixture with Corn

Inna I. Nikiforova⬤, Andrey A. Fadeev⬤, and Inga Yu. Ivanova⬤

Abstract The article presents the results of the evaluation of soybean varieties by the yield of green mass in the nursery of the competitive test. According to the results of mathematical and statistical methods, the variety type $314_{3-15-28}$ was distinguished by the yield of green mass by the beginning of bean formation, and by the wax ripeness of beans—$320_{3/1-544}$. They showed the greatest deviation from the grade of the SibNIIK 315 standard (75.98 and 94.12% relative to the standard, respectively) for the 2018–2020 years of research. The hybrid $314_{3-15-28}$ has a significant excess of the seed yield by 9.60 pcs/plant relative to the standard. The promising breeding material will be prepared for transfer to the State Export Commission for testing.

Keywords Soybean · Variety samples · Yield · Green mass · Deviation · Correlation

JEL Codes Q16

1 Introduction

Soy is a valuable forage crop. For feed purposes, use cake, meal, soy flour, green mass. The green mass of soy is readily eaten by all types of livestock, both fresh and in silage with other crops. In 100 kg of it, harvested in the flowering phase filling of beans, 90 contains up to 22 feed units and up to 3 kg of protein. There is 145–301 g of protein per one feed unit of green soybean mass [12].

I. I. Nikiforova (✉) · A. A. Fadeev · I. Yu. Ivanova
Federal Agricultural Research Center of the North-East Named After N.V. Rudnitsky, Kirov, Russia
e-mail: 8inno4ka@mail.ru

A. A. Fadeev
e-mail: chniish@mail.ru

I. Yu. Ivanova
e-mail: m35y24@yandex.ru

For the creation of productive and highly resistant forage agrocenosis and more complete use of biological factors, mixed crops (joint, compacted, dense, strip, simple, and complex grass mixtures) are increasingly used. The mixtures, thanks to the ability to select the components, give the planned quality of feed in the field. Earlier studies have shown the prospects of mixed soybean and corn crops in the main areas of their distribution. Thanks to the successful biological compatibility of these crops, the yield increases, the collection of nutrients, especially the sorghum-soy crops are superior in terms of the content of digestible protein in comparison with pure crops. In addition, soy is close to corn in its biological characteristics and cultivation methods and is the best component for a mixed agrophytocenosis [1, 5, 10].

In the southern regions, soy is used in mixed crops with cereals to obtain a protein-balanced feed, in particular with corn. For joint sowing of soybeans with corn for green fodder and silage, it is necessary to select soybean varieties with the maximum possible development of the vegetative mass, allowing achieving the highest yield according to the conditions of cultivation [2, 4].

To promote soybeans to the northern regions, new varieties are needed that meet changing environmental conditions. Therefore, the development of an early-maturing variety of the northern ecotype adapted to the conditions of a long day, well-nourished, well-leafed, with a high yield of aboveground mass for cultivation in a mixture with corn is an urgent topic in solving the problem of obtaining high-quality balanced feed in the northern regions [17].

Breeding work on the creation of new varieties of soybeans for growing in the Volga-Vyatka region in the Chuvash Research Agricultural Institute was started in 2000. Agroecological assessment of the source material contributed to the selection of biotypes that are most adapted to the conditions of Chuvashia, for further use in the selection process. Chuvashia is located between 54° and 56° s. s., and it is not a co-producing Republic, but the prospects for the production of soybeans in connection with the emergence of varieties of the northern ecotype are quite available. The thermal resources of the climate make it possible to cultivate soybeans on a production scale [6, 7].

The novelty of the work is that in the conditions of 56 °C for the first time, work is being carried out to create highly productive plastic varieties of soybeans for cultivation in a mixture with corn. Of great interest in this regard is the feed use of green soybean mass.

The research aims to create a variety of the northern ecotype with a stable yield of aboveground mass for cultivation in a mixture of corn in the middle zone of the Russian Federation.

2 Methodology

Experimental research in the nursery of competitive variety trials on loamy gray forest soils with a humus content of 5.8 in the field number 3 fodder crop rotation Chuvash Research Institute with the content of phosphorus—173 mg/kg, potassium—111 mg/kg, acidity, pH (KCI) is 5.5. At the final stage of the variety testing, four promising soybean cultivars were studied, bred in the Chuvash Research Institute of Agriculture. The standard was taken for the soybean variety SibNIIK 315, recommended for cultivation in the Volga-Vyatka region.

Field experiments were based on the method of B. A. Dospekhov [3]. The division of the plot, marking and wide-row sowing with a row spacing of 50 cm were carried out manually. The width of the track between the plots was also 50 cm to remove the edge effect. The area of the plot was 24 m^2; the repetition is threefold. Each plot was allocated accounting areas of one m^2 for phenological observation and biometric analysis of the sheaf. Sampling was carried out in the following phases: branching, flowering, the beginning of fruit formation, filling of beans, and full ripeness. When experimenting, we used the methodology of the state variety testing of crops [16], methodological guidelines for the selection of soybeans [15], and accounting for the yield of green mass [14].

The main soil treatment for soybeans was carried out with the KOS-3 unit to a soil depth of 15–17 cm. In the spring of May 05—closing of moisture with a trailed wide-reach harrow BPSh-15. Pre-sowing cultivation was carried out with the Spider-6 unit. The terrain of the area under the experiment was smooth. Sowing was carried out on May 26 by hand; the depth of seeding is 5–6 cm. The seeding rate is 450 thousand seeds of germinating seeds per hectare. During the entire growing season, double row-to-row processing of plots was carried out.

3 Results

Precocious soybean varieties of the Chuvash selection with a set of the sum of active temperatures (above 10 °C) for the growing season of 1,800–2,300 °C completes the formation of a full-fledged crop. The limiting factor in cultivation in the Volga-Vyatka region of Russia is the insufficient supply of moisture due to its uneven distribution in the phases of plant development [9].

Weather conditions in the years of research differed in the temperature regime and the amount of precipitation during the growing season. Conditions on the wet to the dry attributed 2018 (SCC = 0.68), but 2019 was moderately warm moisture deficit at the beginning of the growing season and high moisture availability in the phase of maturation culture (SCC = 1.09). The sum of active temperatures ($\sum t >$ 10 °C) in 2018 was 1,782 °C, in 2019—2,303 °C [8]. In 2020, during the period of active vegetation of plants (May–August), the average air temperature was 15.6 °C, exceeding the long-term one by 1.9 °C. Precipitation fell 325.2 mm, 37.9% of the long-term norm. The sum of the active temperatures is 2,230.6 °C. The SCC was 1.45.

Phenological observations of the growth and development of soybeans over the years of research have shown that among the tested hybrids in the group of ripeness with the weather conditions of 56 °C, there were medium-early representatives. According to the type of growth, the plants were characterized as indeterminate, i.e., an incomplete type of growth with a different bush shape (Table 1). The sign of nutting was characteristic of all the samples, to one degree or another.

For the formation of a high yield of aboveground soybean mass, such characteristics as plant height, leafiness, branching, and the number of productive nodes on the main stem are of no small importance (Table 2). In terms of plant height, all lines had growth above the standard.

Table 1 Characteristics of soybean breeding lines

Breeding line	Ripeness group	Type and shape of the bush	Pubescence of the plant	Flower coloring
SibNIIK 315—st	Early maturing	Semi-determinant Semi-spreading Curls weakly	Redhead	Purple
$29_{1/1-5kf}$	Mid-early	Indeterminate Semi-upright Curls well	Redhead	White
$314_{3-15-28}$	Mid-early	Indeterminate Semi-spreading Curls well	Sandy-red	Purple
$204_{3/2-5\,km}$	Mid-early	Indeterminate Semi-spreading Curls medium	Redhead	Purple
$320_{3/1-544}$	Mid-early	Indeterminate Semi-spreading Curls well	Redhead	Purple

Source Developed and compiled by the authors

Table 2 Morphological characteristics of soybean hybrids (2018–2020)

Breeding line	Height, cm		Quantity, pcs	
	Plants	Attaching the bottom bean	Branching	Productive nodes on the main stem
SibNIIK 315—st	60	10.7	2.3	10
$29_{1/1-5kf}$	72	10.7	3.2	11
$314_{3-15-28}$	80	10.4	2.5	13
$204_{3/2-5\,km}$	72	11.4	2.4	12
$320_{3/1-544}$	83	12.2	2.4	13

Source Developed and compiled by the authors

Our studies have established that the tested soybean hybrids had a height higher than the standard in terms of plant height; provide the height of attachment of the lower bean at the level of 10–12 cm from the soil surface, which creates conditions for reducing losses during harvesting by the combine. The highest attachment was noted in the number $320_{3/1-544}$, which is 1.5 cm higher than the standard. The hybrid $29_{1/1-5kf}$ distinguished itself by the presence of additional branches, exceeding the standard for this indicator by almost one additional stem. In terms of the number of productive nodes of the main stem, all hybrids exceeded the standard. The maximum number was noted in the numbers: $314_{3-15-28}$ and $320_{3/1-544}$, the excess of the standard variety was 18% for each hybrid.

During the experiments, the yield of the green mass was taken into account. Cleaning for the green mass was carried out in two terms: the formation of beans and wax ripeness. The yield of green mass when mowing during the period of bean formation in the studied hybrids ranged from 22.18 to 33.63 t/ha, while in the SibNIIK 315 standard this indicator was only 19.11 t/ha (Table 3). In all accounting periods, all cultivars clearly show a clear advantage in the accumulation of vegetative biomass over the standard. The largest excess of the standard for the yield of green mass at the first accounting was noted in the hybrid $314_{3-15-28}$ by an additional 14.52 t/ha relative to the standard.

In the literature, it is established that soy accumulates the main share of the green mass crop by the bean formation phase, and by the milk ripeness phase, the growth slowed down [13]. In our experience, when mowing during the period of wax ripeness, a decrease in the yield of green mass was observed in two hybrids ($29_{1/1-5kf}$ and $314_{3-15-28}$) and the standard by an average of 2.5 t/ha, which indicates the beginning of physiological aging of the plant to this phase. Therefore, the recommended harvesting time for these hybrids and the standard for green feed is bean formation.

Two hybrids with the accumulation of the maximum share of the green mass yield by the wax ripeness phase were also identified: $204_{3/2-5\ km}$ and $320_{3/1-544}$. The maximum growth rate of biomass in the range from the bean formation phase to the wax ripeness was observed in the hybrid $204_{3/2-5\ km}$—by 9.0 t/ha. Thus, the best

Table 3 Dynamics of aboveground mass accumulation by phases of development of soybean hybrids for 2018–2020 (t/ha)

Breeding line	Bean formation 11.08	Deviation ± to the standard, %	Waxy ripeness of beans 14.09	Deviation ± to the standard, %	Increase from 11.08 to 14.09
SibNIIK 315—st	19.11	–	16.67	–	−2.44
$29_{1/1-5kf}$	28.20	+47.57	25.27	+51.59	−2.93
$314_{3-15-28}$	33.63	+75.98	31.57	+89.38	−2.06
$204_{3/2-5\ km}$	22.18	+16.06	31.18	+87.04	+9.00
$320_{3/1-544}$	32.01	+67.50	32.36	+94.12	+0.35

Source Developed and compiled by the authors

Table 4 Structure of the soybean green mass yield in the bean formation phase (2018–2020)

Breeding line	Per 1 plant, pcs., g		The content of the forage, %		
	Number of beans	Number of leaves	Weight of beans	Leaf weight	Weight of the stems
SibNIIK 315—st	26	16	55	30	15
$29_{1/1-5kf}$	36	20	48	35	17
$314_{3-15-28}$	48	20	37	44	20
$204_{3/2-5\ km}$	28	16	39	41	19
$320_{3/1-544}$	42	23	28	47	25

Source Developed and compiled by the authors

harvesting period for this hybrid for the green mass is the period of wax ripeness, since the highest yield is formed by this period.

Structural analysis during the bean formation period showed that two hybrids $320_{3/1-544}$ and $314_{3-15-28}$ significantly exceeded the standard variety in the number of productive beans by 39.8 and 57.4%, respectively (Table 4). The maximum share of stems and the largest share of beans were also observed in hybrids $320_{3/1-544}$ and $314_{3-15-28}$.

The greatest contribution of leaves to the total phytomass was observed in the hybrid $320_{3/1-544}$—47.5% of the total green mass of the plant (standard—30.1%), which is more than the standard indicator by 17.4%.

One of the variable elements of soybean yield is the number of fruits and seeds per plant. The potential ability of cultivated soybean varieties to form buds, flowers, and beans is very high, but its implementation depends significantly on internal and especially external factors [11]. The results of the research showed that better indicators in the number of beans characterize the breeding number $314_{3-15-28}$ and seeds per plant than other hybrids, and in the number of seeds there is a significant excess of 9.60 pcs/plant (13.5%) relative to the standard variety with an $NSR_{0.5}$ of 5.4 pcs/plant.

4   Conclusion

As a result of the research conducted in 2018–2020, it was found that the highest phytomass at the beginning of bean formation was observed in the hybrid $314_{3-15-28}$ (+ 14.52 t/ha), and by the wax ripeness of the beans—$320_{3/1-544}$ (+15.7 t/ha). It is noted that the selection number $314_{3-15-28}$ is characterized by the best indicators for the number of beans and seeds per plant, and the number of seeds is significantly higher by 9.60 pcs/plant (13.5%). The promising breeding material will be prepared for transfer to the State Export Commission for testing.

Acknowledgements The research was carried out under the support of the Ministry of Science and Higher Education of the Russian Federation within the state assignment of the Federal Agricultural Research Center of the North-East named after N.V. Rudnitsky (theme No. 0767-2019-0097).

References

1. Bondarenko AN (2019) The efficiency of the influence of adapted cultivation technologies on the yield of leguminous crops in the Astrakhan region. Agrarian Russia 11:24–27
2. Dorokhov AS, Evdokimova OV, Bolsheva KK (2018) Overview of the world soybean market. Innov Agric 4:237–246
3. Dospekhov BA (1979) Methodology of field experience. Manuals for higher educational institutions, Moscow: Kolos, 416
4. Dronov AV, Kundik SM, Zagoruy LM (2014) Sugar sorghum in joint crops on gray forest soils of the southwest of the Central region. Innovative processes in the agro-industrial complex, pp 30–32
5. Dronov AV, Simonov VYu (2016) The efficiency of creating joint crops of fodder sorghum in the southwest of the Russian Non-Chernozem region. Combined crops of field crops in the crop rotation of the agricultural landscape, pp 34–37
6. Fadeev AA, Fadeeva MF, Vorobyova LV (2013) Ecological testing of soybean samples in the conditions of Chuvashia. Feed Prod 6:25–26
7. Fadeeva MF, Vorobyova LV (2017) Soy strategic culture in economic policy. Vladimir Farmer 1(79):27–28
8. Ivanova IYu (2020)Variety study of soft wheat in the conditions of the southern part of the Volga-Vyatka region. Agrarian Sci Euro-North-East 4(21):379–386. https://doi.org/10.30766/2072-9081.2020.21.4.379-386
9. Ivanova IYu, Fadeev AA (2020) Influence of weather conditions on soybean yield in the conditions of the Volga-Vyatka region. Legumes Cereals 4(36):93–98. https://doi.org/10.24411/2309-348X-2020-11210
10. Katysheva NB, Pomortsev AV, Dorofeev NV, Sokolova LG, Zorina SYr, Katyshev AI (2020) The comparative characteristic length of the growing season of different varieties and varieties of soybean in terms of Irkutsk region to select the contrast based on precocity. Actual problems of science in the Baikal region, pp 109–112
11. Khokhoeva NT, Kazachenko IG (2016) The nitrogen-fixing activity of soybean crops in the foothills of the Central Caucasus. Mountain Agric 3:89–94
12. Klasner GG, Gorb SS (2016) The use of soy in feed for farm animals. New Sci Problems Prosp 79:89–91
13. Krasovskaya AV, Veremey TM (2010) Comparative study of leguminous crops in Western Siberia. Proc Orenburg State Agrarian Univ 1(25–1):14–17
14. Methodological guidelines for conducting field experiments with forage crops (1987) Moscow, p 197
15. Methodological guidelines for the selection of alfalfa and soy (1977) Moscow, p 19
16. Methodology of state variety testing of crops (1983) Moscow, p 184
17. Salukova NN, Dementiev DA (2015). Symbiosis is the basis of the high productivity of soybean. Food security and sustainable development of the agro-industrial complex, 183–186

Management of Agricultural Enterprises
for Sustainable Development

System of Effective Financial Planning in the Sustainable Development of Agro-industrial Organizations

Liudmila I. Khoruzhy⬤, Yuriy N. Katkov⬤, Valeriy I. Khoruzhy⬤, and Ekaterina A. Katkova⬤

Abstract The paper develops and describes the author's system of effective financial planning in the organizations of the agro-industrial complex (AIC). The new system allows for the most effective implementation of the organizations' plans to reduce financial risks. The paper aims to develop a system of effective financial planning in the sustainable development of the AIC organizations. The authors apply the following research methods: monographic, abstract-logical, economic-mathematical, and comparison. The authors consider the elements of the planning system of agricultural formations, which is internally associated with the categories of unity, subsystem, relationship, hierarchy, and multilevel. Particular attention in the developed financial planning system is given to the Adaptive Situation Center, which operates based on artificial neural networks, allows one to respond to the possible external and internal economic influences quickly, and provides an opportunity to form a strategy for effective planning. The authors describe the realization of function in artificial neural networks and the training of neural networks. The authors present a scheme for training neural networks and a strategic map of the effective functioning of the agricultural organization, which allows linking the organization's strategic and operational objectives. The main stages of planning in the AIC organizations are discussed. The authors propose methods to improve the quality of the planning system in the AIC organizations. The introduction and implementation of

L. I. Khoruzhy · Y. N. Katkov (✉) · E. A. Katkova
Moscow Timiryazev Agricultural Academy, Russian State Agrarian University, Moscow, Russia
e-mail: kun95@yandex.ru

L. I. Khoruzhy
e-mail: dka1955@mail.ru

E. A. Katkova
e-mail: kea1459@yandex.ru

V. I. Khoruzhy
Financial University Under the Government of the Russian Federation, Moscow, Russia
e-mail: vkhoruzhiy@yandex.ru

© The Author(s), under exclusive license to Springer Nature Singapore Pte Ltd. 2022   347
E. G. Popkova and B. S. Sergi (eds.), *Sustainable Agriculture*,
Environmental Footprints and Eco-design of Products and Processes,
https://doi.org/10.1007/978-981-16-8731-0_34

the author's system of financial planning in agricultural organizations will allow organizations to function and develop steadily in current economic conditions. Moreover, it corresponds to the current economic realities, being situational and sustainable.

Keywords Sustainable development · Financial planning · Agro-industrial complex · Artificial neural networks

JEL codes Q01 · Q13 · P25

1 Introduction

The agro-industrial complex is one of the most critical sectors of the economy and is the guarantor of the social stability of the population. However, agricultural producers are forced to constantly adapt to the changes to function effectively under the influence of the rapidly changing environment.

Due to the growing competition, agricultural organizations need to constantly improve production efficiency, increase the competitiveness of products, optimize financial risks, and carry out the transition to the sustainable development of the agricultural organization. Agricultural organizations should create a financial planning system to respond to changes and maintain sustainability [12].

Modern companies are responsible for the rational use of their own (natural, labor, etc.) and society's economic resources. Therefore, the management of the agricultural organization bears the responsibility to society for the use of these resources [6]. In this regard, it is critical to assess the performance of organizations through a system of indicators in all areas of sustainable development, including the evaluation of finances, the company's social responsibility, and the environment [2].

Most researchers and economic theorists consider the company's stability in the analysis of the organization's effectiveness. In this case, the analysis of the enterprise is the study evaluating the entire activity in terms of compliance with the enterprise's goals and objectives; that is, it serves as a management element. The analysis reveals various changes in indicators reflecting the state of production, circulation of resources, the level of consumption of products (goods, services), the efficiency of resource utilization, and the quality of the end product [5].

Consequently, the organization's sustainability can be seen as finding an opportunity to improve performance by building a sound and confident financial strategy.

To maintain the image and gain an impeccable reputation in the market, the AIC organizations need to have stability [1].

Sustainability is the ability of an organization to return to its original mode of operation after a negative impact and unforeseen circumstances [5]. In the current economic environment, a system of effective planning is one of the main effective means of maintaining the sustainability of the agricultural organization, its development, and provision of its security [17].

When developing a financial planning system in agricultural organizations to ensure their sustainability, the main problem is the inability to understand the root causes of the problems creating the risks connected with unconsidered factors in implementing certain projects [4]. Another aspect challenging for agricultural organizations is the automation of accounting. The planning system of the AIC enterprises should be full-fledged and promptly respond to possible external and internal economic impacts and the predicted negative situations [8]. To achieve this effect, it is necessary to detail the available information.

2 Materials and Methods

The founder of the New Age concept was M. Strong. In his report at the Stockholm Conference in 1972, he put forward the idea of measuring economic development against the idea of a balance between ecological and economic conditions [20]. In the same report, he first called for the global community and the transition from economic development to eco-financial development. The concept of eco-financial development was short-lived and evolved into the concept of sustainable development.

The concept of sustainable development was based on the following provisions: environmental security and the rights of society to it; the likelihood of environmental restoration; optimization and prevention of the warming of non-renewable natural resources; prevention of the loss of the human gene pool [11].

Based on the provisions of the sustainable development concept, many scholars, depending on their professional affiliation, attempted to formulate the conditions and provide a clear definition of sustainable development [16]. However, the demand for sustainable development introduced by John Hartwick in the 1970s became world-famous.

Hartwick's rule is that sustainable development can only be achieved if all rent from natural resources, defined as the difference between the market price of the resource and the marginal cost of its extraction, is invested in the reproducible capital [15].

Thus, global reviews of secondary literature have established that sustainable development consists of the following components: social responsibility, ecology, and economy. The ecology focuses on the impact of business on the environment. Social responsibility includes workforce diversity, working conditions, and professional ethics. The economy focuses on long-term financial development and effective planning.

The paper aims to develop a system of effective financial planning in the sustainable development of the AIC organizations.

The research methods include monographic, abstract-logical, economic-mathematical, and method of comparison. The solution of the indicated problem is based on experimental data and known theoretical provisions in the field of sustainable development of organizations.

3   Results

It is necessary to consider several areas to build a system of effective financial planning in the AIC organizations.

1.   Mission and strategy of the organization.

The formation of a strategy and mission in an AIC organization ensures the future success of the business. Thus, it is necessary to achieve a rational distribution of resources available to the organization [19]. The mission and strategy of AIC organizations should consist of operational, tactical, and strategic development.

2.   Control over the receipt and use of financial resources of the organization.

The features of ratified budgets are the basis for the development of budgetary requirements and the limitation of revenues, and the use of financial resources. The responsibility centers must set specific limits on the receipt and expenditure of financial resources [10]. Receipts of financial resources are necessary only to cover expenditures. In turn, restrictions on using financial resources are mandatory in planning and monitoring the implementation of plans.

The control over the receipt and use of financial resources in the organization includes the following:

- Cost control;
- Performance control;
- Cash flow control.

The procedure for monitoring the receipt and use of financial resources should be reflected in the provisions of the planning and budgeting system and job descriptions [14].

3.   Implementation of automated financial planning systems.

It is proposed to introduce an automated planning system to improve the planning and financial security of agricultural organizations. Planning involves developing a financial plan for the organization, which reflects the main activities aimed at achieving the goals.

The main goal of financial planning is to increase the company's value, improve its competitiveness and production efficiency, and ensure financial stability [13].

4.   A strategy for effective flexible planning.

To build a system of effective financial planning, it is necessary to develop a strategy for effective flexible planning allowing agricultural organizations to gain additional profits. The strategy should involve the continuous collection of contacts of the target audience to maximize the benefits [18].

5.   Development of personnel competence.

Training and qualification of personnel of agricultural organizations become increasingly popular in building a system of effective financial planning [7].

The development of personnel competence positively impacts strategic planning through better decision-making and consistency of common action. When constructing a primary goal, an employee of an agricultural organization must have personal training needs [9].

The development of personnel competence in building a system of financial planning of agricultural organizations allows the following:

- Identify problem areas and make adjustments to the plan for the further development of the organization;
- Examine the current situation;
- Identify deviations from the plan;
- Choose priority areas for strategic planning;
- Develop effective plans for further development;
- Evaluate the results and propose relevant measures [3].

Planning in the AIC organizations should be carried out in the following stages:

- Definition of the organization's aim, mission, and strategy;
- Assessment of external and internal factors;
- Consideration of the ways of organizing activities and appointing responsible persons;
- Provision of an accessible plan;
- Definition of a strategy for effective flexible planning;
- Development of a system of plans;
- Control over the implementation of plans.

Figure 1 shows the system of effective financial planning for agricultural organizations.

4 Discussion

Let us briefly describe the algorithm of the developed system. A specially created department is the basis for the financial planning in the AIC organizations. The formation of such a department is a prerequisite for the sustainable development of organizations since agro-industrial holdings and complexes have a complex structure and numerous information and financial flows.

The department of financial planning performs a full range of work, from collecting and evaluating financial information to forming a system of plans and monitoring their implementation.

Based on the assessment of external and internal factors, the department of financial planning determines the mission and strategy of the agricultural organization.

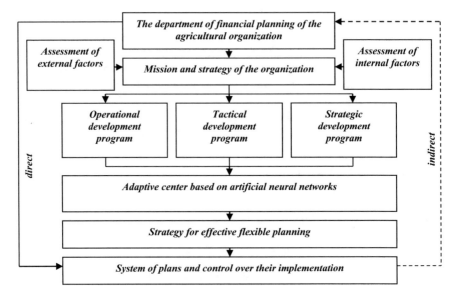

Fig. 1 The system of effective financial planning in the AIC organizations. *Source* Compiled by the authors

The programs for the operational, tactical, and strategic development of the organization are created to implement the developed mission and strategy. The Adaptive-Situation Center, which operates based on artificial neural networks, occupies a particularly important place in the presented financial planning system. This center allows the financial planning system to respond to possible external and internal economic impacts quickly. Moreover, artificial neural networks allow for an anticipatory response to predicted negative situations that may arise within the organization or come from the external environment.

Artificial neural networks are self-learning subsystems that simulate the human brain. Their operation is based on a computational model with many uncomplicated processes interacting with each other.

The functioning of artificial neural networks is shown in Fig. 2.

To reconstruct the learning process, it is first necessary to obtain the available information for the network and present the paradigm of the external environment with the functioning neural network.

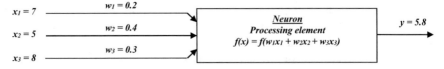

Fig. 2 The realization of a function in artificial neural networks. *Source* [15]

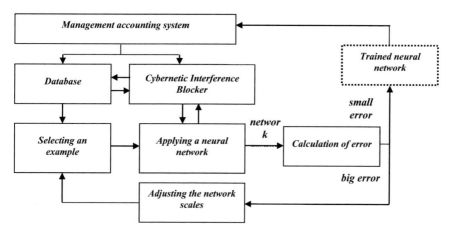

Fig. 3 The mechanism for training a neural network. *Source* Compiled by the authors

The model of the external environment with a functioning neural network describes the theory of learning. This model has a tremendous ability to penetrate the content of learning at all levels and engage the mechanisms of continuous learning. The training procedure involves the use of certain interrelated algorithms. Figure 3 presents the learning mechanism of the neural network.

The mechanism for training a neural network has some peculiarities. Thus, a strategic goal is defined. The goal is embodied in the form of a business project with the formed budgets and plans (in other words, a database is formed). The artificial neural network will be used to build a predictive model with parameters maximally close to the target.

Next, we consider the strategic map of the productive life of an agricultural organization. This map allows us to link the strategic objectives and operational objectives of the organization (Table1).

The strategic map of the productive life of an agricultural organization consists of the following elements: capital, consumers, management, and employees. After developing a strategic map and highlighting the main elements, it is necessary to form a database and run the training (Tables 2 and 3).

The adaptive-situational center allows to form a strategy for effective planning, which is not static, is in constant motion depending on the evolving situation, and allows for flexible planning.

The last element of financial planning is a system of plans and control over their implementation. This element employs direct communication to connect to the department of financial planning, developing a system of plans. Feedback between these elements allows one to monitor the implementation of plans and make the necessary adjustments to the target planning indicators.

Table 1 Strategic map of the productive life of an agricultural organization

	Roadmap	Indicator	Plan
Capital	Increasing business value	Net profit	320% increase
	Increasing sales growth	Increasing sales in the local market	385% increase
	Increasing the organization's profitability	Production profitability	6.8% increase
Consumers	Enhancing the organization's reputation	Consumer loyalty	56% increase
	Increasing customer satisfaction with the product range	Number of secondary purchases	38% increase
	Service and service quality	Customer loyalty indicator	31% increase
Management	Creating an original product of an agricultural organization	Share of indicators of the introduction of unique products	18% increase
	Stock management	Business activity of commodity stocks	38% increase
Employees	Developing competencies of employees	Compliance of employees with qualification requirements, %	51% increase

Source Compiled by the authors

5 Conclusion

Our research presents the results of significant transformations and proposals for financial planning in agricultural organizations. The authors developed a system of effective financial planning in the sustainable development of the AIC organizations. Additionally, the authors presented a neural network learning process in which a goal is translated into a business project with formed budgets and plans. Moreover, the authors built a strategic map of the effective functioning of the agricultural organization. This map allows highlighting the main cause–effect relationships, the application of which will make it possible to achieve the objectives.

Thus, the goal of our research has been achieved. The considered developments can be reflected in accounting and analytical systems, as well as in planning systems used in agricultural organizations. At this stage, research in financial planning in the sustainable development of the AIC organizations is not over and requires further development in this direction.

Thus, the presented financial planning system allows agricultural organizations to implement their plans most effectively since it is adaptive, situational, and allows to minimize the possible financial risks arising in economic activity. The introduction and implementation of the developed system in the agricultural organization allow organizations to develop and operate steadily in a dynamically changing economic environment.

Table 2 Indicators of the productive activity of an agricultural organization

Year	Capital		Consumers		Management		Employees	Capitalization, mln. RUB
	Sales volume, mln. RUB	Production profitability, %	Number of second purchases, %	Loyalty indicator, %	Indicator of the introduction of unique products, %	Business activity of commodity stocks, days	Qualified personnel, %	
Fact								
2017	385	5.8	27	16	79	90	34	63
2018	677	8.1	34	20	68	77	45	67
2019	880	6.8	50	32	92	74	54	74
2020	1755	8.8	63	49	43	70	86	99
Plan								
2021	3229	9.4	74	56	59	70	108	124
2022	3564	10.4	77	68	47	67	121	538
2023	5346	12.8	85	76	61	52	140	835
2024	11,340	15.7	92	81	63	45	153	1391
2025	14,040	18.5	97	80	70	36	162	1620

Source Compiled by the authors

Table 3 Sampling for the mechanism of learning of neural network

Production profitability, %	Loyalty indicator, %	Business activity of commodity stocks, days	Qualified staff, %	Capitalization, mln. RUB
Fact				
5.8	16	90	34	**63**
8.1	20	77	45	**67**
6.8	32	74	54	**74**
8.8	49	70	86	**99**
Plan				
9.4	56	70	108	**124**
10.4	68	67	121	**538**
12.8	76	52	140	**835**
15.7	81	45	153	**1391**
18.5	80	36	162	**1620**
Capital	Consumers	Management	Employees	–

Source Compiled by the authors

References

1. Andreeva L, Mirgorodskaya E (2004) A look at systemic competitiveness as the dominant factor of sustainable economic development. Economist 1:81–88
2. Baranenko SP, Shemetov VV (2004) Strategic stability of the enterprise. ZAO Tsentrpoligraf, Moscow, Russia
3. Bautin VM, Katkova EA, Dzhikiya MK, Zaruk NF, Ukolova AV (2018) Knowledge management in the system of ensuring the economic security for agricultural organizations. Astra Salvensis, VI(Special Issue), 891–898
4. Bogomolov VA (2009) Economic security: study guide for university students studying in the field of economics and management, 2nd edn. YuNITI-DANA, Moscow, Russia
5. Bowersox DJ, Kloss DJ (2005) Logistics: an integrated supply chain process, 2nd edn. (Trans. from English). Olymp-Business CJSC, Moscow, Russia (Original work published 1996)
6. Bulavko VG, Gusakov VG, Nikitenko PG (2009) Economic security: theory, methodology, practice. Pravo i ekonomika, Minsk, Belarus
7. Christensen PO, Feltham GA, Sabac F (2003) Dynamic incentives and responsibility accounting: a comment. J Account Econ 35(3):423–436
8. Davis RV, McLaughlin LP (2009) Breaking down boundaries. Strat Financ 4:46–53. https://sfmagazine.com/wp-content/uploads/sfarchive/2009/04/Breaking-Down-Boundaries.pdf
9. Druker P (2001) The essential drucker: selections from the management works of Peter F. Drucker. HarperBusiness, New York, NY
10. Euske KJ, Riccaboni A (1999) Stability to profitability: managing interdependencies to meet a new environment. Acc Organ Soc 24(5–6):463–481. https://doi.org/10.1016/S0361-3682(99)00020-3
11. Goncharenko LP, Akulinina FV (2015) Economic security: textbook for universities. Yurayt, Moscow, Russia
12. Hammer M, Champy J (1993) Reengineering the corporation: a manifesto for business revolution. Harper Business, New York, NY

13. Harvard Business Review (2014) The chart that organized the 20th century. https://store.hbr. org/product/the-chart-that-organized-the-20th-century/F1409Z
14. Jensen MC (2001) Corporate budgeting is broken—let's fix it. Harv Bus Rev. https://hbr.org/ 2001/11/corporate-budgeting-is-broken-lets-fix-it
15. Katkov YuN (2013) The use of artificial neural networks in accounting managerial accounting. Controlling 1(47):34–40
16. Khoruzhiy LI (2011) Registration and analytical system of branch of agriculture: theoretical and practical problems of development. RIO BGU, Bryansk, Russia
17. Khoruzhy LI, Katkov YN, Khoruzhiy VI, Dzhikiya KA, Stepanenko EI (2018) Current approaches to assessing and enhancing the efficiency of managerial decisions in agrarian organizations. Astra Salvensis, VI(Special Issue):835–845
18. Masaaki I (2006) Gemba kaydzen: way to cost-cutting and improvement of quality. Alpina Biznes Buks, Moscow, Russia
19. Verschoor CC (2012) Responsibility reporting is getting more attention. Strat Financ 12:14–15. https://sfmagazine.com/wp-content/uploads/sfarchive/2012/12/ETHICS-Responsibility-Reporting-Is-Getting-More-Attention.pdf
20. Voronkov NA (2006) Ecology: general, social, applied. Textbook for university students. Agar, Moscow, Russia

Improving the Price Mechanism in Milk Production and Processing

Olga A. Stolyarova⒟, Lyubov B. Vinnichek⒟, and Yulia V. Reshetkina⒟

Abstract One of the priority tasks for the Russian Federation is to ensure the country's food security since the production of high-quality domestic dairy products directly depends on it. The paper aims to identify and disclose the significant problems of low efficiency of milk production by agricultural manufacturers. The research object is the dairy subcomplex of the Russian Federation and the Penza Region. The methods of the research are abstract-logical, monographic, economic-statistical, and analysis. The authors present the main reasons hindering the effective development of dairy cattle breeding. The authors substantiate the idea that trade organizations unreasonably raise retail prices of finished dairy products to obtain the highest profits. The authors are convinced that it is necessary to develop a methodology for accounting for produced and sold milk at the national level. Additionally, the authors propose the methodology for calculating the purchase price of milk, considering its quality, which will stimulate manufacturers to produce higher quality products. According to the research results, the authors conclude that the improvement of the industry's efficiency requires the improvement of the pricing mechanism and the development of its concept, contributing to the settlement of incomes of milk manufacturers and processors.

Keywords Milk and dairy products · Pricing · Import substitution · Price parity coefficient

JEL codes Q11 · Q13 · Q17

O. A. Stolyarova (✉) · Y. V. Reshetkina
Penza State Agrarian University, Penza, Russia
e-mail: stolyarova.o.a@pgau.ru

Y. V. Reshetkina
e-mail: reshetkina.y.v@pgau.ru

L. B. Vinnichek
St. Petersburg State Agrarian University, St. Petersburg-Pushkin, Russia
e-mail: l_vinnichek@mail.ru

© The Author(s), under exclusive license to Springer Nature Singapore Pte Ltd. 2022   359
E. G. Popkova and B. S. Sergi (eds.), *Sustainable Agriculture*,
Environmental Footprints and Eco-design of Products and Processes,
https://doi.org/10.1007/978-981-16-8731-0_35

1   Introduction

Pricing is an essential element of the economic relations between partners in producing, processing, and selling milk and dairy products. It largely depends on the production cost, based on the production direction of the enterprise [9, p. 490]. The milk and dairy products market is one of the main links in the food industry since milk is an indispensable food product. As an economic space of relations between owners, the market for milk and dairy products is developing under the influence of further separation of social labor, consumer income, and an innovative renewal of productive forces [4, p. 143].

Nowadays, the market for milk and dairy products sees significant price fluctuations, which indicates the seasonality of prices for raw milk (Fig. 1). Thus, the average price for raw milk in the Penza Region increased by 23679.4 rubles in 2000–2020 and equaled 26535.9 rubles per one ton in 2020. In Russia, the average price for raw milk equaled 25860.9 rubles per ton (25.9 rubles per kg). The average price for raw milk in July 2020 was 24.8 rubles/kg, in December—27.1 rubles/kg, and in March 2021—27.3 rubles/kg. Significant growth in raw milk prices in Russia is associated with the desire of agricultural manufacturers to increase profitability by modernization of production facilities and modernization of the production process.

Agricultural producers, processing enterprises, and trade organizations are the parties having economic interests in the dairy industry and the main subjects of production relations. The importance of the agricultural sector is explained by the fact that without milk production, the very existence of other participants in the dairy industry and the existence of the domestic dairy market is questionable. Nevertheless, milk is a perishable product and is subject to further processing, which creates the preconditions for establishing close ties with the processing industry and determines the need for their proportional development and mutual accommodation [6, p. 133].

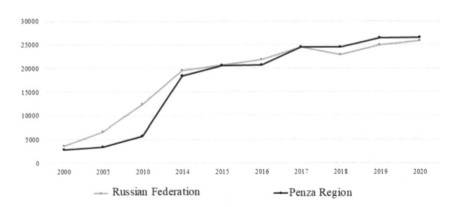

Fig. 1 The average producer price for raw milk, rubles per 1 ton. *Source* Compiled by the authors based on [5, 7]

The maximum level of prices for raw milk in October 2020 is observed in the Yamal-Nenets Autonomous Area—77.3 rubles per kg; the minimum price level in the Republic of Dagestan—19.2 rubles per kg. In the Tula Region, Trans-Baikal Territory, Moscow, Primorye Territory, and the Republic of Sakha, the price of raw milk is higher than 30 rubles per kg. In the Kemerovo Region, the Republic of Mordovia, the Republic of Mari El, and the Kirov Region, the price of raw milk averages 23.3 rubles per kg (Fig. 2).

The formation of prices affects the indicators of the economic efficiency of milk production. In 2019, milk production in the Penza Region with the available production facilities and resources was profitable, with cost-effectiveness equal to 37.9% (Table 1).

Agricultural producers suffer the most from the unfairness of pricing since the level of milk prices is significantly undervalued, and the prices in retail stores exceed them by 3–5 times and are constantly changing. The price of one ton of milk sold by agricultural organizations of the Penza Region was 25865.9 rubles in 2019, 5670 rubles in 2005, and 20298.8 rubles in 2015. The total cost price was 1874.5 rubles, 5161 rubles, and 1659.9 rubles, respectively. The level of cost profitability was 9.9% in 2005, 22.3% in 2015, and 37.9% in 2019. The consumer price index for milk and dairy products has been decreasing over the past years, namely:

- In 2003, it was 114% compared to last year;
- By 2005, it reached 110.4% compared to last year;
- By 2019, it was only 103.1% compared to last year.

When forming prices, the difficulty lies in the fact that the price is a conjunctural category depending on political, economic, social, psychological, and other factors. The impact of these factors on market development is different. Thus, the introduction of sanctions in 2014 affected Russia's foreign trade relations with other

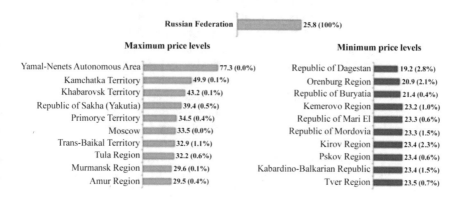

Fig. 2 Rating of the subjects of the Russian Federation by the level of prices for raw milk in October 2020, rubles per 1 kg (share of the subject of the Russian Federation in the all-Russian volume of milk production in January–September 2020, %). *Source* Compiled by AC MilkNews based on the data of the Federal State Statistics Service and National Union of Milk Producers (Soyuzmoloko) [1]

Table 1 Indicators of milk production efficiency in agricultural organizations of the Penza Region

Indicator	2000	2005	2015	2016	2017	2018	2019
Milk yield per average annual cow, kg	1683	2344	4676	5153	5820	6571	7499
Gross milk yield, thousand tons	168.3	161.1	155.4	155.6	158.2	171.9	172.6
Total milk sales, thousand tons	139.9	133.3	144.2	140.1	147.1	154.6	154.2
Marketability rate, %	83.1	82.7	92.8	90.0	93.0	89.9	89.3
Total cost of 1 cwt of milk, rubles	323	5161	1659.9	1682.2	1767.5	1819.4	1874.5
Profit (loss) from the sale of 1 cwt of milk, rubles	-31.6	50.9	369.98	392.6	751.4	646.6	712.1
Level of profitability (unprofitability) of costs %	-9.8	9.9	22.3	23.3	42.5	35.5	37.9

Source Compiled and calculated by the author

countries, which led to a decrease in exports of dairy products by 7% and imports by 27.3% compared to the previous year. From the economic point of view, the price is determined by the cost factor. From the psychological point of view, its level may depend on the behavior of buyers. Modern dairy industry enterprises operate in a competitive market and are interdependent [2, p. 169].

In current conditions, the development of competitive dairy cattle breeding is impossible without an effective organizational and economic mechanism capable of triggering and developing the fundamental factors affecting the functioning of the industry. Continuity of the technological production process in the dairy industry is conditioned by the rational organization of receiving raw milk, producing finished dairy products, and bringing them to the end consumer. Organizations and individuals engaged in retail trade often overestimate the cost of goods for obtaining the highest income, which contradicts the law of demand—when prices for products increase, the demand for them will decrease. To increase demand for products, retail outlets use a system of discounts. Each of the economic entities of the dairy industry, connected by a technological chain, is interested in maximizing profits, which is impossible without efficient production and sale of finished dairy products.

Let us consider the formation of retail prices for dairy products (Fig. 3).

Wholesale and selling prices should be within the limits on which the efficiency of dairy processing enterprises will depend. Wholesale prices are formed by processing enterprises based on the incurred costs of production, production capacity utilization, and material incentives for the staff.

In March 2021, retail prices for dairy products increased by 0.5% compared to February 2020 and were 3.3% higher than in March 2020. The average consumer price for pasteurized 2.5–3.2% whole milk in the Russian Federation was 57.7 rubles in 2019. In 2020, this price was 59.5 rubles, which was 6.3 rubles and 8.1 rubles higher than in 2016. In 2020, the average consumer price of hard and soft rennet cheeses rose by 142.6 rubles compared to 2016 and by 52.3 rubles compared to

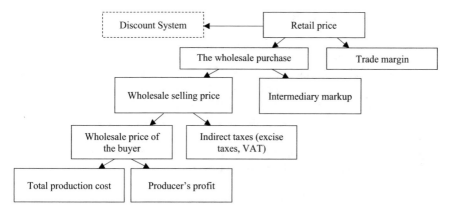

Fig. 3 The formation of prices for dairy products. *Source* Compiled by the authors based on research materials

2019. The average consumer price of fat cottage cheese from 2016 till 2020 was 315.2 rubles.

Development of the concept to improve the pricing mechanism in the dairy industry will increase the efficiency of milk and dairy products and all activities in general. The main provisions of this concept include the following:

1. Creating fair and reasonable types of prices depending on the channels of sales of products: contract prices, export prices, moving and sliding prices, market (free), wholesale, procurement, etc.;
2. Finalization of technical regulations for milk, providing for the reduction of requirements for somatic cells, and limiting the use of tropical oils for the production of dairy products, which will allow increasing the purchase price of raw milk from agricultural producers;
3. Compliance with the content of micronutrients and food additives and the limits of permissible deviations of nutritional value indicators in the labeling of the finished product in accordance with technical regulations in the production of milk and dairy products;
4. Pricing in milk production and processing is influenced by such factors as the quality of raw milk and ready-made final products, supply and demand in the dairy market, its capacity and presence of competitors, state regulation, the profitability of consumers, the necessary amount of resources, and costs;
5. Formation of all types of prices at each stage of the technological chain in the production of milk and dairy products should be determined by the market price of the finished dairy products, which depends on the material, financial, general production, general economic, and other necessary costs for its production and market conditions.

2   Materials and Methods

In our opinion, the methodology used to account for the milk produced and sold in the country and its constituent entities is still imperfect. The method of accounting for milk supplied to processing plants is based on the following main parameters characterizing the quality of milk: protein, fat content, acidity, density, somatic cells, and bacterial contamination. Any of these indicators may vary depending on the quality of milk produced and delivered.

In accordance with the quality indicators, we propose to calculate the price of 1 kg of milk according to the following formula:

$$Pm = (P \times Pp \times F \times Pf) \times Cq, \tag{1}$$

where

   Pm—price for milk, rubles/kg;
   P—presence of protein in milk, %,
   Pp—price of milk protein, rubles/kg;
   F—presence of fat in milk, %;
   Pf—price of milk fat, rubles/kg;
   Cq—coefficient of quality.

For the purpose of calculations, we chose the largest enterprise in the Penza Region, OJSC Milk Plant "Penzensky," as a research object. Thus, the price of 1 kg of milk at OJSC Milk Plant "Penzensky" will be 27.72 rubles against the current price of 26.54 rubles:

$$Pm = (0.032 \times 294.75 + 0.046 \times 185) \times 1.3 = 23.32 \text{ rubles.}$$
$$\Delta P = 27.72 - 26.54 = 1.18 \text{ rubles.}$$

The purchase price of 1 kg of raw milk should be formed from the base price and markup to it. The markups depend on the parameters characterizing the quality of milk supplied to milk processing plants, which exceed the basic values (fat content, grade, acidity inhibitors, etc.). The base price for milk from agricultural organizations is determined based on the prevailing average market price for second-grade milk. As a result, the purchase price of milk (based on the offset weight) may significantly exceed the base price if high-quality milk is produced. In our opinion, to encourage manufacturers to produce the milk of the highest grade, its markup can be at least 15% of the base price since the Penza Region sells only 25% of milk of the highest grade. Nowadays, there is a 10% markup to the base price for the highest-grade milk. Such a payment system for milk should consider not only the interests of buyers (processors of raw milk) but also the interests of suppliers (agricultural producers). The processing enterprises of the dairy industry will get higher quality raw milk with increased fat and protein content. Respectively, the output of 1 ton of raw milk will increase, the quality of the final dairy products will improve, the competitiveness of the final dairy products will increase, and customer demand for quality dairy products will be satisfied.

In our opinion, the payment for sold milk should also depend on the dry matter content of milk, which affects the dry matter content of the finished product. Secondary dairy resources (buttermilk, skim, and others) are created at dairy industry enterprises, and, therefore, their prices do not stimulate agricultural manufacturers to increase the dry matter content of milk for food production. However, many companies in the country use skimmed milk powder as the main raw material to reduce the cost of the finished product. For example, LLC "Siberian Dairy Products Plant" produces a dairy product with milk fat substitute "Sweet Cheese" with vanilla and raisins, "Glazed Cheese" with raisins and dried apricots, "Cream Cheese," etc., which allows using milk fat substitutes, sugar, and flavorings.

Based on the skim milk powder estimate, the following payment options for milk are possible:

(1) Additional payment for skimmed milk powder exceeding 8.2% – for each 0.1% at a rate of 2% to the base price;
(2) Reduced payment for milk in case of deviation from the base figures for skimmed milk powder below 8.1%—for every 0.1% in the amount of 2.5% of the base price.

This will contribute to the development of the organizational and economic mechanism in the production and processing of milk at the enterprises of the dairy industry (by means of pricing) and increase in the finished dairy products from one ton of raw milk. In the current conditions, the processing enterprises do not consider the grade of milk when calculating the cost of final dairy products. Since the grade of milk is known, the quality of produced dairy products depends on the quality of raw milk. When calculating the unit cost of dairy products, it is necessary to determine the cost of skim milk, which is a secondary product that is further used as raw material, since the largest share in the cost of production is taken by the cost of raw materials.

At fat content of raw milk 3.6–3.8%, the cost of skimmed milk at OJSC Milk Plant "Penzensky" is 58%. The following formula should determine the cost of the normalized mixture:

$$Cn = NAm \times CMa + NQc \times CCa, \tag{2}$$

where

Cn—the cost of the normalized mixture, thousand rubles;
NAm—the amount of milk necessary to produce 1 ton of mixture, tons;
CMa—the cost of milk of actual fat content, thousand rubles;
NQc—the quantity of cream necessary to produce 1 ton of mixture, tons;
CCa—the cost of cream of actual fat content, thousand rubles.

$$Cn = 1.3 \times 26.54 + 0.56 \times 64.25 = 70.48 \text{ thousand rubles.}$$

In our opinion, the cost of cream of actual fat will depend on the experimental amount of milk of actual fat and skimmed milk.

$$CCa = 5.5\ CMa - 4.5Ps, \tag{3}$$

where

5.5—experimental amount of milk of actual fat content required to produce one ton of cream, t;
4.5—experimental skim milk yield, t;
Ps—skim milk price, rubles.

$$CCa = 5.5 \times 26.54 - 4.5 \times 15.39 = 76.72\ \text{thousand rubles.}$$

In our opinion, the coefficient assessing the cost of skimmed milk from the cost of milk, for example, at OJSC Milk Plant "Penzensky" should be 0.58 (calculated in accordance with the volume of milk processing and its fat content). Then, Ps $= 0.58 \times$ CMa.

Currently, resource-saving technologies for new dairy products aimed at saving dairy raw materials and increasing production profitability have been developed. Russia is experiencing high growth rates in the production of specialty fats, including milk fat substitutes. The capacity of the Russian specialty fat market varies from 800 thousand tons to 1 million tons per year. The production of ice cream with vegetable fat accounts for 8%, milk products—9%, and spreads—10% of the total volume [8, p. 48].

In our opinion, to assess the economic efficiency of the production of new types of dairy products, it is necessary to calculate such an economic category as milk capacity. This category would characterize the rate of raw milk with a basic fat content of 3.4% for the production of 1 ton of products and would allow conducting a comparative assessment of fermented dairy products.

$$Mc = Smb/Wp, \tag{4}$$

where

Smb—standard of raw milk with a basic fat content of 3.4%;
Wp—product weight, kg.

$$Mc = 1000/325 = 3.1$$

The introduction of new technologies will reduce the milk capacity of production. For dairy industry enterprises, innovations will help reduce the cost of production of final dairy products and increase the profitability of its production. On the other hand, in our opinion, it will affect the health of the population and, above all, children. Therefore, when developing recipes for dairy products and spreads, it is necessary to follow the recommendations of the World Health Organization and "Norms of physiological needs in energy and nutrients for different population groups of the Russian Federation."

In the production and sale of dairy products, it is necessary to be based on the equivalence of exchange between the subjects of the technological chain. It is reasonable to use the coefficient of price parity for dairy cattle breeding and processing industry products. In our opinion, the coefficient of parity (Kp) should reflect the ratio of procurement prices for milk in farms of all categories and wholesale prices for dairy products sold:

$$Cp = Pw/Pp, \tag{5}$$

where

Pw—wholesale selling price of 1 ton of dairy products, rubles;
Pp—purchase price of 1 ton of milk in farms of all categories, rubles.

$$Cp = 45570/26374 = 1.72$$

Based on the coefficient of price parity, we can determine the purchase prices of milk for each dairy processor.

$$Pp = (Pwa/Cr) \times C, \tag{6}$$

where

Pwa—actual wholesale price of one ton of dairy products, rubles;
Cr—ratio of actual wholesale and purchase prices.

$$Pp = (48360/1.85) \times 1.72 = 44962 \text{ rubles.}$$

Based on the price parity coefficient, it is necessary to substantiate the normative levels of profitability of producing the corresponding types of products for agricultural organizations and dairy enterprises. The normative level of profitability should provide not only simple but also expanded reproduction.

3 Results

Based on our calculations, we can conclude that the coefficient of parity relative to processing enterprises increases from 1.7 to 1.85, reflecting the unprofitability of milk production in agricultural organizations. Consequently, the retail trade markup should be formed taking into account the milk prices of producers and processing organizations and the average level of profitability, which will allow to equalize the income of agricultural organizations and dairy industry enterprises and increase consumer demand for dairy products.

4 Discussion

Given a significant degree of uncertainty in the domestic market of dairy products, pricing strategy should be flexible and adapted to changes in market conditions, which will promptly make changes in the product range and pricing policy, predict the effects of these measures and their impact on the financial performance of the enterprise, assess the level of the enterprise's competitiveness, take strategic decisions in accordance with changes in the market situation, and form a modern system of relations between producers and consumers [2, p. 169].

With the growth of purchase prices for raw milk in the dairy industry, there is an increase in the price of ready-made dairy products and a reduction of the commodity market in this segment. It is possible to increase the economic interest of agricultural producers in the production of raw materials for industrial processing by applying a contractual system that can satisfy the interests of all subjects of relations on a parity basis.

The most important features of the developed market of dairy products are as the following framework [10, p. 324]:

- Satisfaction of consumer demand for milk and dairy products;
- State influence on pricing, which would guarantee at least a minimum profitability from the sale of products;
- Availability of regulations.

5 Conclusion

The rise in food prices, including milk, is an integral part of the import substitution. The dairy market can be called as an example of a successful import substitution policy since many Russian products and products of new supplier countries have appeared in this market, replacing the prohibited expensive imported products [3, p. 621]. The solution to the problem of pricing for imported and Russian dairy products should be systemic in nature. Moreover, it should not depend on a single case of increasing the purchase price of raw milk and affect the development and improvement of organizational and economic relations between the stakeholders and the fair distribution of profits between them.

To regulate and balance prices in the dairy industry, it is necessary to improve the organizational and economic mechanism of relations between agricultural producers, processing enterprises, and trade organizations. In recent years, Russia has been under conditions of economic instability, which leads to a significant increase in food prices and a decrease in people's incomes. For this reason, consumers are forced to search for and purchase food products that do not always meet quality standards at lower prices, which leads to a reduction in work capacity and deterioration of the health of the country's population.

References

1. AC MilkNews (2020, December 7) Rating: regions with the highest and lowest raw milk prices. https://milknews.ru/analitika-rinka-moloka/reitingi/reting-ceny-syroe-moloko-regiony.html
2. Belolipov RP, Konovalova SN (2019) Pricing in the milk market: problems and ways to solve them. J Voronezh State Agrarian Univ 3(62) 168–175. http://vestnik.vsau.ru/wp-content/upl oads/2019/11/168-175.pdf
3. Berendeeva EV (2019) Transformation of the Russian food market: income and substitution effects. Econ J High Sch Econ 23(4):605–623. https://doi.org/10.17323/1813-8691-2019-23-4-605-623
4. Dozorova TA, Bannikova EV (2012) Regulation of the milk market and dairy cattle breeding. J Ulyanovsk State Agricult Acad 4:143–146
5. Federal State Statistics Service of the Russian Federation (Rosstat) (2000–2020) Agriculture, hunting, and forestry. https://rosstat.gov.ru/enterprise_economy
6. Reshetkina YV (2020) Features of development of organizational and economic relations to ensure food security in the sphere of production and processing of milk (On the materials of the Penza Region) (Dissertation for the Degree of Candidate of Economic Sciences). Russian State Agrarian University—Moscow Timiryazev Agricultural Academy, Moscow, Russia
7. Rosstat Regional Office of Penza Region (2000–2020) Average producer prices of agricultural products. https://pnz.gks.ru/ofstatistics https://pnz.gks.ru/storage/mediabank/jz7G8cXj/TD_CX1.XLS
8. Stepanova LI (2010) Substitutes for milk fat SDS and SOUZ—a guarantee of the quality of your products. Dairy Indus 10:48–49
9. Vinnichek LB, Reshetkina JV, Stolyarova OA (2021) Pricing of milk and dairy products by manufacturers and trade organizations. In: Bogoviz, AV (ed) The challenge of sustainability in agricultural systems. Springer, Cham, Switzerland, pp 489–496. https://doi.org/10.1007/978-3-030-73097-0_55
10. Vinnichek LB, Stolyarova YuV, Stolyarova OA (2019) The development of milk market in the conditions of ensuring food security in Russia and Germany. Adv Econ Bus Manag Res 79:322–325. https://doi.org/10.2991/iscfec-19.2019.91

Peculiarities and Prospects of the Application of the Unified Agricultural Tax

Lidiya A. Ovsyanko⬤, Natalia I. Pyzhikova⬤, Georgy N. Kutsuri⬤, Kristina V. Chepeleva⬤, and Tatiana A. Borodina⬤

Abstract The relevance of the research topic is determined by the need to choose an effective taxation system by agricultural producers, which would contribute to improving their financial results and performance. The paper aims to identify the features and prospects of applying the unified agricultural tax (UAT) by business entities in Russia. The authors use abstract-logical, economic-statistical, monographic, calculation, and constructive research methods. Moreover, the authors use the results of their theoretical and practical research. The paper summarizes the main positive and negative sides of using the UAT by agricultural producers, which should be the basis for developing and improving the taxation system for economic entities in the agricultural sector. The analysis for 2015–2019 revealed that the UAT remains sufficiently demanded by agricultural producers and its share in the structure of payments under special tax regimes was 66.3% in 2019. In terms of federal districts, the largest share of tax payments is registered in the Far Eastern (28.3%), Southern (22.0%), and Northwestern Federal Districts (21.1%). The proposed grouping of federal districts by the share of UAT payers in their total number allowed the authors to evaluate the impact of the UAT on the efficiency of economic entities. In the Volga and Southern Federal Districts, where 22 and 27.4% of all UAT payers are concentrated, the tax burden averages 5.2%.

L. A. Ovsyanko (✉) · N. I. Pyzhikova · K. V. Chepeleva · T. A. Borodina
Krasnoyarsk State Agrarian University, Krasnoyarsk, Russia
e-mail: lidiya-ovs@mail.ru

N. I. Pyzhikova
e-mail: pyzhikova@kgau.ru

K. V. Chepeleva
e-mail: kristychepeleva@mail.ru

T. A. Borodina
e-mail: rigik25@mail.ru

G. N. Kutsuri
Financial University Under the Government of the Russian Federation, Moscow, Russia
e-mail: nimageo@mail.ru

Keywords Russian Federation · Unified agricultural tax · Tax burden · Efficiency

JEL codes E62 · H25

1 Introduction

Due to the specific features of the functioning of the subjects of the Agro-Industrial Complex (AIC), the funds they invest in production are not always returned in full. This fact leads to a decrease in efficiency and a significant reduction in the number of agricultural organizations. Currently, the government is actively involved in supporting agricultural production through a system of direct and indirect measures affecting the results of the financial and economic activities of economic entities. Indirect measures in the form of preferential taxation systems act as an essential supplement to state financing of business entities.

Balakin cites the following vital tasks of any government [1]:

- Building an optimal ratio of taxes and fees;
- Effective implementation of the functions of taxes, including through minimizing the costs of the entire tax system;
- Achieving fairness and transparency in taxation;
- Management of tax relations at a sufficiently high level, etc.

At the same time, it is critical to respect the principle of proportionality (i.e., the balance between the taxpayer's interests and the state budget).

According to Buklanov, "the crisis phenomena of the past decade have demonstrated a decrease in the effectiveness of the current fiscal mechanism" [2]. Therefore, the issue of ensuring the country's food security remains relevant, which means primarily supporting the subjects of the AIC through preferential tax regimes. Although agricultural producers do not have an absolute tax exemption, they have always had tax benefits such as reduced taxable base and reduced tax rates [3].

According to the Federal Law "On the development of agriculture" [4], one of the measures of the national agrarian policy is the application of special tax regimes for agricultural producers. Along with other activities, this measure should help to achieve the indicators provided in the Food Security Doctrine of the Russian Federation [5]. Thus, within the framework of the State program of agricultural development and regulation of markets of agricultural products, raw materials, and food [6], it is planned to increase the volume of the country's tax expenditures to 84002.3 million rubles (by 18.8%) from 2020 to 2023. These measures are confirmed by the constant improvement of the existing preferential tax regimes. Simultaneously, choosing an effective tax regime for a particular agricultural producer is a strategically important aspect.

A significant contribution to studying the current state and solution of the taxation issues of agricultural producers has been made by such scientists as Kataev,

Merkulova, Moiseeva, Pestryakova, Starkova, Sushentsova, and others. Their works served as the scientific basis for our research [3, 7–11].

2 Methodology

The authors use the abstract-logical method to reveal the peculiarities of the functioning of business entities in AIC and the use of a special tax regime (unified agricultural tax) by agricultural producers.

The statistical method of research allowed the authors to identify the dynamics of the main indicators characterizing tax payments of economic entities in agriculture, forestry, fishing, and fish farming.

The authors use the monographic method to assess the application of the unified agricultural tax (UAT) in Russia and its federal districts.

The application of the calculation and constructive method allowed us to determine the impact of preferential taxation on the efficiency of activities of the subjects of AIC based on the conducted grouping of federal districts by the specific weight of UAT payers in their total number.

3 Results

Several scholars correctly point out that one of the distinctive features of the functioning of agriculture in Russia is its taxation under a general or special regime [12].

Simultaneously, the multivariant system of taxation increases its efficiency but involves the solution of the following dilemmas [8]:

- Which of the proposed system is the easiest and most convenient to use;
- Which of the proposed system provides the greatest opportunity to minimize tax payments to the budget.

The taxation conditions for business entities continue to improve due to changes in external and internal economic conditions. Chapter 26.1 of the Tax Code of the Russian Federation "Taxation System for Agricultural Producers (UAT)" was adopted in 2004 [13]. In this case, "a special tax regime can be applied by those enterprises or individual entrepreneurs whose income from the sale of manufactured or agricultural products, including primary processing products made from agricultural raw materials of own production, is not <70% of the total income from the sale of goods (works, services)" [13]. Many researchers attribute this fact to the disadvantages of the UAT.

According to Pestryakova, "this system is effective for agricultural producers who have high stock and labor intensity of production, but low profitability of economic activity" [11].

Starting from 2019, UAT payers must pay VAT if their income exceeds 100 million rubles for 2018, 90 million rubles for 2019, 80 million rubles for 2020, 70 million rubles for 2021, and 60 million rubles for 2022 [13, 14].

Researchers in the field of taxation note the following. "Exempting taxpayers from paying VAT reduces the number of possible buyers of agricultural products because of the impossibility of presenting this tax for deduction" [3]. "It should be considered that even with the exemption from VAT, UAT payers pay VAT in a stealthy form" [15].

Moiseeva notes that "for the correct calculation and payment of VAT, producers must comply with various requirements, primarily technical ones" [9].

The ambiguity of the application of the UAT is emphasized in the works of several authors. Thus, they indicate the following main advantages [7, 10, 16, 17]:

- Reducing the tax burden by replacing the set of taxes on one tax;
- Convenient payment terms;
- Simpler accounting.

The use of the UAT is not beneficial to all categories of agricultural producers. Consequently, the decision to switch from the general taxation system to the UAT requires a thorough analysis. Agricultural production in Russia is carried out in almost every region, considering natural and climatic conditions, resource endowment, etc. The number of organizations and enterprises operating in agriculture, forestry, hunting, fishing, and fish farming from 2015 to 2019 decreased from 146,822 to 102,915 units. However, the share of profitable organizations in the reporting year was 71.4%, which is 4% higher than in 2015. Simultaneously, the profitability of production of such entities decreased from 20.7 to 17.2%. Among federal districts, a large share of AIC organizations and enterprises accounts for the Volga (18.4%), Siberian (12.5%), and Southern Federal Districts (11.2%) [18–23].

For 2015–2019, there was an upward trend in state budget revenues in the form of tax revenues. In the reporting year, the consolidated budget of Russia received 2610.6 billion rubles in taxes and fees, which is 1.64 times higher than in 2015. Over the study period, the share of such revenues attributable to the study area changed insignificantly. Its minimum value was recorded in 2018 (0.62%) and the maximum in 2016 (0.77%) (Fig. 1).

In the structure of all tax revenues to the consolidated budget of the Russian Federation, the greater part is accounted for by business entities in the field of mining (more than 30%) [24].

During the study period, tax payments of economic entities in agriculture, forestry, fisheries, and fish farming increased by 53% and amounted to 152411.1 million rubles in the reporting year. The value of federal taxes and fees increased by 54.1% and amounted to 107,743 million rubles, which is 70.7% of the total. In the aggregate of federal taxes and fees, personal and corporate income tax account for a significant percentage (87% and 18.2%, respectively) in 2019.

The amount of taxes paid by entities applying special tax regimes also increased by 11664.7 million rubles, or almost twice, which amounted to 15.5% of the total.

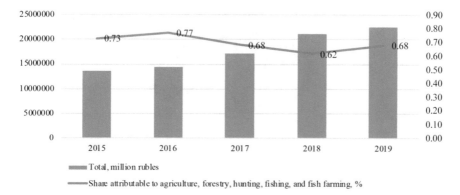

Fig. 1 Receipts of taxes and fees in the consolidated budget of the Russian Federation. *Source* Compiled by the authors based on [24]

The amount of regional taxes increased by 32.2%. Local taxes and fees decreased by 7.1% (Table 1).

Figure 2 presents the structure of tax revenues of economic entities of the considered sphere in the context of federal districts of Russia. In 2019, the largest tax revenues were registered in the Southern Federal District (19.55%), the Northwest Federal District (17.03%), and the Central Federal District (16.98%). This figure depends on the turnover of this sphere in each of the regions.

In the total amount of taxes under special tax regimes, the UAT accounted for a large share—61.7% in 2015 and 66.3% in 2019. The value of the calculated UAT more than doubled over the study period and amounted to 15,714.9 million rubles (Fig. 3).

Indicators of the use of the UAT in Russia show that the number of taxpayers of this tax decreased by 3677 (3.7%). This situation can be explained by the general trend of

Table 1 Composition and structure of tax payments of economic entities in agriculture, forestry, fisheries, and fish farming, million rubles

Taxes and fees	2015	2016	2017	2018	2019
Federal	69,913.1	76,666	81,428.1	87,450.2	107,743.0
% of total	70.2	72.3	69.7	67.1	70.7
Regional	11,638.3	11,792.3	12,417.1	16,705.1	15,385.9
% of total	11.7	11.1	10.6	12.8	10.1
Local	6006.4	5273.6	5074.3	5600.2	5578.5
% of total	6.0	5.0	4.3	4.3	3.7
Special tax regimes	12,039.0	12,259.2	17,887.0	20,586.4	23,703.7
% of total	12.1	11.6	15.4	15.8	15.5
Total	99,596.8	105,991.1	116,806.5	130,341.9	152,411.1

Source Compiled by the authors based on[24]

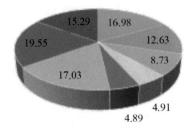

- Central Federal District   ▪ Volga Federal District   ▪ Siberian Federal District
- Ural Federal District   ▪ North Caucasus Federal District ▪ Northwestern Federal District
- Southern Federal District   ▪ Far Eastern Federal District

Fig. 2 The structure of tax revenues to the budget from economic entities in agriculture, forestry, hunting, fishing, and fish farming by federal districts of Russia in 2019, %. *Source* Compiled by the authors based on [24]

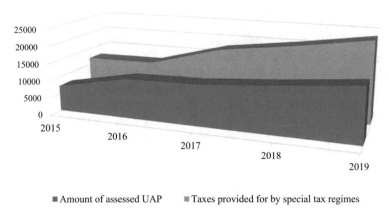

▪ Amount of assessed UAP   ▪ Taxes provided for by special tax regimes

Fig. 3 Dynamics of calculated taxes in connection with the application of special tax regimes in the Russian Federation, million rubles. *Source* Compiled by the authors [24]

a decrease in agricultural producers. Simultaneously, the calculated tax allows us to conclude that the profit of business entities, applying a special tax regime, increased. On average, the UAT per organization amounted to 166.1 thousand rubles in 2019, which is 2.2 times higher than in 2015.

On the whole, the share of the UAT in the consolidated budget of each constituent entity of Russia is insignificant. However, it saw a slight increase from 0.11% in 2015 to 0.16% in 2019. Virtually the entire amount of the tax is credited to the local budget of the subjects (i.e., the tax is local) [25].

The application of the UAT exempts agricultural organizations from paying corporate income tax and corporate property tax (individual entrepreneurs—from paying personal income tax and personal property tax). This fact significantly reduces the tax payments of the industry to the budget and, accordingly, the tax burden on business

Table 2 Impact of the application of UAP on the efficiency of economic entities in the federal districts of Russia

Indicator	Grouping of federal districts by the share of UAT payers in their total number, %			
	Up to 10	From 10.1 to 20	More than 20	Average
Number of federal districts in the group	3	3	2	3
Share of UAT payers in the total number of taxpayers on average in one federal district, %	4.2	12.6	24.7	12.5
Turnover of subjects on average per federal district, billion rubles	290.0	518.4	496.5	427.3
Transferred to the consolidated budget of taxes and fees on average per federal district, billion rubles	18.9	15.5	24.5	19.05
Share of the UAT in the total amount of taxes and fees transferred to the budget on average per federal district, %	11.6	6.3	9.1	9.0
Tax burden on average per federal district, %	6.2	3.7	5.2	5.0
Federal districts	Northwestern Ural Far Eastern	Siberian North Caucasus Central	Volga Southern	x

Source Compiled by the authors based on [24]

entities. Next, let us group federal districts of Russia by the share of single agricultural taxpayers of each district in the total number of economic entities applying the UAT to assess the effectiveness of the preferential treatment (Table 2).

Thus, all federal districts of Russia were divided into three groups. The third group includes the Volga and Southern Federal Districts with a large number of UAT payers, which amounts to 20,880 and 25,940 units of agricultural producers, respectively. This figure is higher than the average share of UAT payers in their total number in all districts by almost two times. The share of UAT, among other taxes and fees of this group, is slightly above average and amounts to 9.1%. Additionally, the tax burden in this group is only 2% higher than the average value. The first group of federal districts (the Northwestern, Ural, and Far Eastern Federal Districts) has less effective indicators. Thus, with a smaller turnover of 290 billion rubles per federal district, the tax burden in these districts is higher than in other groups—6.2%.

Consequently, the application of UAT creates more favorable conditions for the functioning of agricultural producers. Additionally, new opportunities for differentiation of tax rates depending on several conditions make this regime more attractive.

4 Conclusion

The effective functioning of the tax system of any country should be reduced to the implementation of the fiscal function, the primary function of taxes. Simultaneously, fiscal policy should be differentiated and consider the peculiarities of business entities, in particular agricultural producers. Therefore, the choice of the most appropriate system of taxation is usually made for a short period.

In 2015–2019, tax revenues to the budget system of Russia increased by more than 1.5 times and amounted to 22510.6 billion rubles. During the study, the share of taxes and fees attributable to the AIC organizations averaged 0.7%. In the reporting year, tax payments of economic entities in the sphere of agriculture, forestry, fishing, and fish farming amounted to 152.4 billion rubles, which is 53% higher than in 2015. Most of them (70.7%) are federal taxes and fees.

Among agricultural producers, the most popular special tax regime is the unified agricultural tax. In the reporting year, the number of UAT payers amounted to 94,633. The amount of calculated tax is 15714.9 million rubles, which is 2.1 times more than in 2015. Almost the entire amount of the tax is credited to the local budget.

Based on the analytical grouping of federal districts in terms of the share of UAT payers in their total number, the authors revealed the following:

- In the third group of federal districts (Volga and Southern Federal Districts), with a higher concentration of UAT payers (over 20%), the tax burden on business entities in agriculture is 5.2%;
- In the second group of federal districts (Siberian, North Caucasus, and Central Federal Districts), with the concentration of UAT payers from 10.1 to 20%, the tax burden on business entities in agriculture is 3.7%.
- In the first group of federal districts (Northwestern, Ural, and Far Eastern Federal Districts), with a smaller concentration of UAT payers (up to 10%), the tax burden on business entities in agriculture is higher than average and equals 6.2%.

Based on such an enlarged analysis, the authors can note a positive impact of UAT on the efficiency of agricultural producers. At the same time, the results indicate that additional research is needed to assess the effectiveness of the application of preferential treatment by each business entity. Given the changes in legislation, agricultural producers have new opportunities to apply the unified agricultural tax.

Acknowledgements The project "Efficiency of application of different types of taxation systems by agricultural organizations in the Krasnoyarsk Territory" was supported by the Krasnoyarsk Regional Science Foundation (application code: 2021020407246).

References

1. Balakin RV (2020) Profitability and risk of the tax system of the Russian Federation and the factors determining them (Dissertation of Candidate of Economics). Plekhanov Russian University of Economics, Moscow, Russia
2. Buklanov DA (2020) Sustainability of tax revenues of the Russian Federation: Assessment of the current state and directions for regulation (Dissertation of Candidate of Economics). North-Caucasus Federal University, Stavropol, Russia
3. Starkova OY (2021) Unified agricultural tax: prospects for application. Econ Agricult Russia 1:22–24
4. Russian Federation (2006) Federal Law "On the development of agriculture" (December 29, 2006 No. 264-FZ, as amended October 15, 2020). Moscow, Russia. http://www.consultant.ru
5. President of the Russian Federation (2020) Decree "On approval of the Food Security Doctrine of the Russian Federation" (January 21, 2020 No. 20). Moscow, Russia. http://www.consultant.ru
6. Government of the Russian Federation (2012) Decree "On the state program of agricultural development and regulation of markets of agricultural products, raw materials, and food" (July 14, 2012 No. 717, as amended April 6, 2021). Moscow, Russia. http://www.consultant.ru
7. Kataev VI, Sushentsova SS (2018) Justification of the choice of an effective system of taxation in the farmer sector. Econ Agricult Russia 9:2–10
8. Merkulova EY, Logvin DD (2017) Problems of choosing a tax regime for small businesses. Socio-Econ Process Phenomena 2:74–81
9. Moiseeva OA (2019) Taxation of agricultural producers. Econ Agricult Russia 6:37–43
10. Moiseeva OA (2020) Tax planning tools for small and medium-sized businesses in the agricultural sector. Econ Agricult Russia 7:2–7
11. Pestryakova TP (2007) The system of tax planning of an agricultural enterprise. Econ Anal: Theory Pract 5(86):51–57
12. Kazmin AG, Orobinskaya IV (2013) Taxation of agricultural producers in Russia. Modern Econ: Prob Trends Perspect 9:56–72
13. Russian Federation (2000) Tax Code of the Russian Federation (Part Two) (August 5, 2000 No. 146-FZ, as amended June 11, 2021). Moscow, Russia. http://www.consultant.ru
14. Russian Federation (1998b) Tax Code of the Russian Federation (Part One) (July 31, 1998 No. 146-FZ, as amended April 20, 2021). Moscow, Russia. http://www.consultant.ru
15. Borlakova TM (2019) Actual problems of taxation of agricultural enterprises. Eurasian Sci J 6(11):1–8
16. Ivanova OE, Sidorkina MY (2014) Methods for determining the tax burden of agricultural producers. Naukovedenie 5(24):1–13
17. Ivanova YN (2012) Development of methods of tax regulation of agricultural enterprises (Dissertation of Candidate of Economics). Tomsk, Russia
18. Alexandrova OA, Bugakova NS, Voronina IV, Gokhberg LM, Grigoriev LM, Zubarevich NV, … Kharlamova IV (2020) In: Okladnikov SM (ed) Regions of Russia. Socio-economic indicators: statistical collection. Moscow, Russia: Rosstat.
19. Baranov EF, Bezborodova TS, Bobylev SN, Bugakova NS, Vagan IS, Gokhberg LM, … Shapoval IN (2020) In: Malkov PV (ed) Russian statistical yearbook. Moscow, Russia: Rosstat.
20. Baranov EF, Bugakova NS, Gelvanovsky MI, Gokhberg LM, Egorenko SN, Elizarov VV, … Khoroshilov AV (2016) In: Surinov AE (ed) Russian statistical yearbook. Moscow, Russia: Rosstat.
21. Bobkov VV, Zasko VN, Kirillova GN, Konovalova OA, Krivenets AN, Kruglova MV, … Nechaev AV (2020) In: Shapoval IN, Afonin MM (eds) Finance of Russia, 2020: Statistical collection. Moscow, Russia: Rosstat.
22. Bugakova NS, Gokhberg LM, Grigoryev LM, Zhitkov VB, Zubarevich NV, Klimanov VV, … Kharlamova IV (2016) In: Egorenko SN (ed) Regions of Russia. Socio-economic indicators: Statistical collection. Moscow, Russia: Rosstat.

23. Zinchenko AP, Ivchenko VN, Kiselev SV, Klevakina MP, Novokshchenova EI, Obychaiko EE, ... Shashlova NV (2019) In: Laikam KE (ed) Agriculture in Russia. Moscow, Russia: Rosstat.
24. Federal Tax Service of the Russian Federation (2020) Consolidated tax passport of the Russian Federation as of January 1, 2020. https://analytic.nalog.ru/portal/index.ru-RU.htm
25. Russian Federation. (1998a). Budget Code of the Russian Federation (July 31, 1998 No. 145-FZ, as amended April 30, 2021). Moscow, Russia. http://www.consultant.ru

Biologization of Spring Wheat Cultivation with Application of Sugar Beet Waste to the Soil

Irina V. Gefke⦿, **Larisa M. Lysenko, and Olga V. Bychkova**⦿

Abstract The paper considers the main aspects of the development of sugar beet production in Russia. The authors provide the dynamics of exports of sugar beet pulp as a by-product of sugar production. It is concluded that most of the by-products of sugar beet production are not subsequently used in the economic turnover, which creates certain environmental risks. Additionally, the authors determine the efficiency of ecologization of grain production in the Altay Territory when applying waste from beet sugar production to the soil.

Keywords Sugar beet pulp · By-products · Ecologization of production · Grain yield · Production profitability

JEL codes Q53 · Q55 · Q150

1 Introduction

The research was conducted to reveal the influence of introducing wastes from beet sugar production on the efficiency of grain production in the Altay Territory since the current state of the material-intensive sugar industry is accompanied by a significant excess of raw materials and products used in production over the volume of products manufactured. As a result, the sugar industry generates significant amounts of production wastes which are a source of environmental pollution. This production waste includes molasses, sugar beet pulp (desugared sugar beet chips), and filtration

I. V. Gefke (✉) · O. V. Bychkova
Altai State Agricultural University, Barnaul, Russia
e-mail: ivgefke@mail.ru

O. V. Bychkova
e-mail: olga4ka_asu@mail.ru

L. M. Lysenko
The Altai Branch of the Russian Presidential Academy of National Economy and Public Administration, Barnaul, Russia
e-mail: vip.lml.2014@mail.ru

sludge (defecate). The generated production waste occupies about 5.4, 83.0, and 8.0% of the total weight of processed sugar beet, respectively. Sugar beet pulp (the main by-product of sugar beet production) is used for cattle feed in fresh form (not more than 35.0–40.0%) by agricultural enterprises within a radius of 100 km from sugar factories. It is also used as raw material for biogas plants. About 30.0% of sugar beet pulp is dried and used for various purposes or exported. In 2019–2020, beet sugar factories produced about 4760.9–6207.2 thousand tons of sugar beet pulp, with about 1347.1–1444.7 thousand tons (23.32–28.30% of total production) for export. The rest of the sugar beet pulp goes sour in the storage, forming unclaimed agricultural and industrial pulp water with low nutritional value. All waste products from the sugar industry can be used by the chemical industry (as raw materials for producing alcohol, glycerin, yeast, acetone, etc.) and agriculture (as fertilizers). The use of waste from beet sugar production in cultivating crops is one of the directions of biological farming and obtaining environmentally cleaner products. The use of sugar beet pulp acquires particular relevance in the context of reducing the number of cattle in large agricultural enterprises in the territories of sugar factories, which leads to the unsystematic removal of sugar beet pulp to the areas adjacent to the sugar factories and contamination of soil and water with oily acid entering the groundwater and adjacent water bodies.

2 Methodology

The sources of information on the volumes of mineral and organic fertilizers applied by Russian agricultural organizations were the EMISS database of the Federal State Statistics Service (Rosstat). The information on the export volumes of sugar beet pulp was obtained from the database of the Federal Customs Service of Russia. The efficiency of grain production with the introduction of sugar beet waste as fertilizer was calculated in normative-technological maps, taking into account the most common technology of growing grain crops in the beet-growing regions of the Altay Territory. In particular, the increase in yield was directly reflected in transportation costs and the cost of grain processing at the mechanized thrashing floor. When calculating the normative-technological map, the authors also considered additional production processes related to weed control operations (harrowing) and the application of sugar beet pulp and defecate to the soil. The joint application of defecate and sugar beet pulp is a prerequisite for neutralizing the adverse effects of the separate application of wastes from the sugar industry in the soil. As confirmed by the results of many studies, soils were subjected to oxidation when applying sugar beet pulp and alkalinization when applying defecate. Thus, A. B. Basov notes that the joint application of sugar beet pulp (5 tons/ha) and defecate (5 tons/ha) allows increasing the yield of spring wheat on average by 2.2 c/ha compared to the control with the accumulation of organic matter in the soil, increasing its mobile forms of potassium and phosphorus, and reducing acidity [1]. Applying pulp and defecate under winter wheat stabilizes soil acidity and increases yield by 5.8 c/ha (20.2%) compared to the technology without organic fertilizers. These results are at the level of yield with the joint application of 40 tons/ha of manure and $N_{60}P_{60}K_{60}$ [8]. Gurin and Gneusheva revealed that the application of sugar beet pulp as a separate fertilizer and in combination with defecate promotes active growth and an increase in the mass of weed plants, which implies measures to control weeds [6]. In other countries, studies on the use of the by-products of the sugar industry are mainly aimed at assessing the effectiveness of their use in feeding cattle [7, 9] and poultry [3], as well as their use as a raw material in the production of fuel ethanol [13], cellular proteins, organic acids, biologically important secondary metabolites, enzymes, prebiotic oligosaccharides, and other valuable products [2].

3 Results

The Altay Territory is one of the sugar beet-growing regions of Russia and the only region cultivating sugar beet beyond the Urals. The region's share was about 2.1–2.5% of the total area of sugar beet cultivation in Russia (decrease by 1.7–2.1% compared to 1990). In 2018–2020, the main areas of cultivating sugar beet were concentrated in the Central Federal District (53.5–54.4%), Volga Federal District (19.5–20.5%), and Southern Federal District (20.0–20.1%) (Table 1).

The reduction of sown area of sugar beet in the Altay Territory in 1990–2020 is due to the bankruptcy of three large sugar factories situated in the Aleisky (Aleisky sugar factory, halting of production—2008), Biysky (Biysky sugar factory, halting

Table 1 Distribution of sown areas under sugar beet by federal districts of Russia, thousand hectares

Federal district		1990	2010	2018	2019	2020
Central	Total	880.7	620.1	604.2	612.7	503.4
	% of total	60.3	53.5	53.6	53.5	54.4
	% of 1990	100.0	70.4	68.6	69.6	57.2
Southern	Total	256.3	269.0	225.6	229.4	186.3
	% of total	17.5	23.2	20.0	20.0	20.1
	% of 1990	100.0	104.9	88.0	89.5	72.7
Volga	Total	262.1	254.6	231.4	233.7	180.6
	% of total	17.9	22.0	20.5	20.4	19.5
	% of 1990	100.0	97.1	88.3	89.2	68.9
Siberian	Total	61.3	15.9	23.4	27.5	23.5
	% of total	4.2	1.4	2.1	2.4	2.5
	% of 1990	100.0	25.9	38.1	44.8	38.4
Others	Total	–	−0.2	42.1	41.7	32.1
	% of total	–	–	3.7	3.6	3.5

Source Compiled by the authors based on [5]

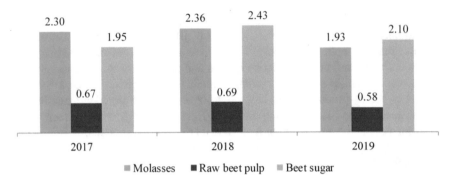

Fig. 1 Specific weight of the Altay Territory in the production of main and by-products of the sugar industry in Russia, %. *Source* Compiled by the authors based on [5]

of production—2007), and Bystroystoksky (Bystroystok sugar factory, halting of production—1997) districts. As a result, agricultural producers in the surrounding rural areas withdrew sugar beets from crop rotation since transportation to the only functioning sugar factory (OJSC "Cheremnovsky Sugar Factory") was unprofitable and led to a decrease in the quality of tubers. In 2004, there were 154 beet growers in the Altay Territory. In 2019–2020, beet was grown in only six farms under the general management of LLC "Agrofirm 'Cheremnovskaya',," part of the Group of Companies "Dominant" [10]. Overall, between 2017 and 2019, OJSC "Cheremnovsky Sugar Factory" produced 31.9–39.0 thousand tons of molasses (1.93–2.30% of the molasses production of all Russian sugar mills) and 48.3–49.0 thousand tons of sugar beet pulp (0.58–0.67% of the production volume in Russia) (Fig. 1).

In terms of exports of molasses and dried sugar beet pulp, Russia has become a world leader since 2010. In terms of its share in export revenues, exports of molasses and dried sugar beet pulp are at the level of sugar exports. For example, in 2017, exports of sugar from Russia totaled $281 million, and exports of pulp and molasses totaled $134 million and $54 million, respectively. In 2018, sugar exports in value terms were lower than the combined sales of pulp and molasses—$183 million and $240 million, respectively. In 2018–2020, 29.8–48.2 thousand tons of sugar beet pulp were exported, representing 2.33–3.33% of total exports of these products in Russia in physical terms and 2.52–3.37% in value terms. The export of sugar beet pulp by OJSC "Cheremnovsky Sugar Factory" as the only producer of the sugar industry is significantly ahead of the average for Russia. The ratio of sugar beet pulp exports in the Altay Territory compared to the Russian average for 2018–2020 is 1.43 for natural indicators and 1.34 for value indicators (Table 2).

There are six rural areas within a radius of up to 100 km from the OJSC "Cheremnovsky Sugar Factory." For 2010–2020, only the Shelabolikhinsky district increased the number of cattle in agricultural enterprises. This indicator increased by 10.6% and amounted to 15,156 heads. In other rural areas, the number of cattle significantly decreased (the total number of cattle in these areas was 24,683 heads):

- in the Talmensky district—by 13.2%;

Table 2 Volume of exports of Russian sugar beet pulp in 2018–2020

Indicators			2018	2019	2020	
					Total	% to 2018
Export volume, thousand tons	Russia—total		1279.02	1347.10	1444.71	112.95
	including the Altay Territory	Total	29.78	32.55	48.16	161.75
		% in all-Russian data	2.33	2.42	3.33	x
Export volume, thousand dollars	Russia—total		186.15	200.59	219.01	117.65
	including the Altay Territory	Total	4.69	5.09	7.38	157.32
		% in all-Russian data	2.52	2.54	3.37	x

Source Compiled by the authors based on [4]

- in the Pavlovsky District—by 24.8%;
- in the Rebrihinsky district—by 29.4%;
- in the Topchikhinsky district—by 38.1%;
- in the Kalmansky district—by 43.5%.

This decrease reduces the potential need of agricultural producers to feed for livestock. Only the Talmensky district has some potential, which is currently building a cowshed for 6000 cows (LLC "Altai-niva," the village of Berezovka) [11].

Under these conditions, with limited opportunities to export by-products of sugar production, environmental risks of waste accumulation in the form of sugar beet pulp and molasses in the surrounding areas of OJSC "Cheremnovsky Sugar Factory" increase. In 2018, an administrative case was initiated against the plant for improper use of production waste. A significant risk caused by the activities of OJSC "Cheremnovsky Sugar Factory" is the pollution of water sources, primarily the freshwater lake Anisimovo (the area of the water surface is 1.63 km^2), located at a distance of less than one km east of the village of Cheremnoye (Fig. 2).

Additional sources of organic waste in the cultivation of crops are clearly demanded in the agriculure of the Altay Territory. In 2000–2020, the amount of organic fertilizer applied per ha did not exceed 0.2 tons, which is 5–9 times lower than the average in Russia (0.9–1.3 tons/ha). Although the number of mineral fertilizers increased from 1.6 to 24.5 kg/ha over the same period, it remains 3.1–16.8 times lower than the Russian average (20.5–75.6 kg/ha) (Table 3).

In 2016–2020, the share of spring wheat in the structure of sown areas in the Altay Territory was quite high and exceeded 34.0–39.5% (1747.8–2126.8 thousand ha). In the calculations, it was considered that the most effective of the options for applying waste from sugar production is a joint application of sugar beet pulp and defecate in the ratio of 1:1 with the introduction to the soil before sowing and embedding the fertilizer to a depth of 22–25 cm. In this case, the yield increases by 25.0%, and protein content—by 1.2% compared to the control (with the introduction of manure

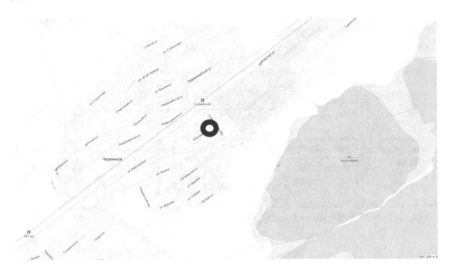

Fig. 2 Location of the "Cheremnovsky Sugar Factory" in relation to Lake Anisimovo. *Source* [12]

Table 3 Amount of applied mineral and organic fertilizers per one hectare of crops in Russia

Indicators		2000	2005	2010	2018	2019	2020	
							Total	% to 2000
Mineral fertilizers, kg	Russia	20.5	28.6	41.4	60.5	65.9	75.6	368.9
	Siberian Federal District	4.5	7.6	9.9	17.6	18.2	27.7	616.0
	Altay Territory	1.6	1.7	3.2	11.1	14.6	24.5	1533.8
Organic fertilizers, tons	Russia	1.0	0.9	1.0	1.3	1.3	1.2	122.0
	Siberian Federal District	0.3	0.4	0.5	0.7	0.7	0.6	210.0
	Altay Territory	0.2	0.1	0.2	0.2	0.2	0.2	115.0

Source Compiled by the authors based on [5]

as an organic fertilizer in the amount of 40 tons/ha but without the introduction of sugar beet pulp).

Standard costs determined by technology charts in the variant with the combined introduction of sugar beet pulp and defecate amounted to 1167.4 thousand rubles per 100 ha, which is higher than the variant with the introduction of manure only by 0.67%, and per one ton of grain—lower by 19.6%. In absolute terms, per 100 ha of spring wheat crops, alternative application of organic fertilizers increased electricity costs by 25.0%, fuel and petroleum products by 0.47%, transportation by 0.64%, and depreciation and repairs of fixed assets by 0.94%. At the same time, seed costs remained unchanged, and fertilizer costs decreased from 30.0 thousand rubles to 22.5 thousand rubles (25.0%) (Table 4).

Table 4 Standard costs of growing spring wheat with different variants of organic fertilizers, RUB per 100 ha

Costs	Fact (application of manure)			Project (application of sugar beet pulp and defecate)		
	Total	% of total		Total	% of total	% of fact (control)
Remuneration with deductions	84,782	7.31		91,190	7.81	107.56
Material costs—total	359,886	31.03		356,297	30.52	99.00
Including fuel and petroleum products	197,966	17.07		198,896	17.04	100.47
Seeds	120,000	10.35		120,000	10.28	100.00
Fertilizers	30,000	2.59		22,500	1.93	75.00
Electricity	11,921	1.03		14,901	1.28	125.00
Depreciation and repairs	95,369	8.22		96,270	8.25	100.94
Transportation work	525,071	45.28		528,413	45.26	100.64
Total costs	1,159,619	100.00		1,167,387	100.00	100.67
Including per one hectare	11,596	X		11,674	x	100.67
Per one ton of grain	9810	X		7890	x	80.44

Source Compiled by the authors

If we consider the need to use part of the obtained grain for seeds in subsequent years (seeding rate 0.2 t/ha), then the commercial production on the control is 1.1 t/ha. The commercial production with the joint application of pulp and defecate is 1.43 t/ha. When selling grain at a price of 10,229 RUB/t (the average price in the Altay Territory in the first half of 2021), alternative application of organic fertilizers can increase the profitability of production from 13.95 to 38.38% (24.43%), which is 6.69% higher than the profitability of the production of spring wheat on average in the region (Fig. 3).

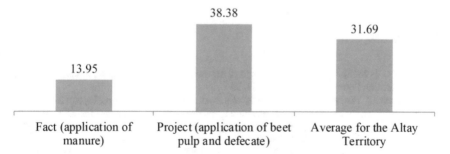

Fig. 3 Profitability of the production of spring wheat grain with alternative variants of organic fertilizers, %. *Source* Compiled by the authors

Due to high transportation costs, the use of waste from beet sugar production as organic fertilizer can be recommended to agricultural producers only if they are located near the sugar factories. In any case, it is necessary to assess the agrochemical composition of the fields, which varies depending on the prevailing soils, crop rotations used, and other factors that significantly differ within the region, municipalities, and economic entities. The use of waste from sugar production will solve the urgent problem of their disposal and create conditions for the preservation and improvement of the fertility of agricultural land in municipalities.

4 Conclusion

In the current management conditions, the use of by-products of the Russian sugar industry is a rather urgent problem. Before 1990, this problem was not on the agenda since all waste products were used to feed livestock in agricultural enterprises. In subsequent years, a sharp decrease in the number of cattle led to the fact that the sugar beet pulp and molasses (as the main types of by-products in sugar production) became completely unclaimed by agricultural producers. In these conditions, significant volumes of production waste are accumulated, disturbing the ecological balance in places where sugar factories are located. The gradual development of exports of sugar beet pulp and molasses solves this problem only partially. In this case, the use of sugar beet pulp and molasses as organic fertilizers allows solving problems of sugar producers (the accumulation of industrial waste), society (pollution of the soil, water), and agricultural producers (preservation and improvement of soil fertility and increasing crop yields, especially grain and leguminous crops). The application of sugar beet pulp and molasses should be combined. The research allows us to evaluate the high efficiency of alternative fertilization of spring wheat crops in the Altay Territory. Costs per hectare of crops increase only by 0.67%, while the cost of grain decreases by 19.6%, and the profitability of grain production increases from 13.95 to 38.38% (24.43%) compared to the traditional technology of growing wheat involving the introduction of manure as an organic fertilizer.

References

1. Basov UV (2016) Agroecological aspects of the use of sugar production waste in spring wheat. Agricult Eng Power Supply 4–2(13):23–30
2. Berlowska J, Binczarski M, Dziugan P, Wilkowska A, Kregiel D, Witońska IA (2018) Sugar beet pulp as a source of valuable biotechnological products. In: Advances in biotechnology for food industry. Lodz University of Technology, Lodz, Poland, pp 359–392. https://doi.org/10.1016/B978-0-12-811443-8.00013-X
3. Berrocoso JD, García-Ruiz A, Page G, Jaworski NW (2020) The effect of added oat hulls or sugar beet pulp to diets containing rapidly or slowly digestible protein sources on broiler

growth performance from 0 to 36 days of age. Poult Sci 99–12:6859–6866. https://doi.org/10.1016/j.psj.2020.09.004

4. Federal Customs Service of the Russian Federation (2021) Customs statistics of foreign trade of the Russian Federation. http://stat.customs.gov.ru/analysis. Accessed 8 June 2021

5. Federal State Statistics Service (2021) Unified interdepartmental information and statistical system. https://fedstat.ru/organizations/. Accessed 8 June 2021

6. Gurin AG, Gneusheva VV (2018) The change of agrochemical soil features and its biological activity when using wastes of sugar production in spring wheat. Bull Agrarian Sci 1(70):3–7. https://doi.org/10.15217/issn2587-666X.2018.1.3

7. Heydaria M, Ghorbania GR, Sadeghi-Sefidmazgia A, Rafieeb H, Ahamdia F, Saeidya H (2020) Beet pulp substituted for corn silage and barley grain in diets fed to dairy cows in the summer months: feed intake, total-tract digestibility, and milk production. Animal 15–1:100063. https://doi.org/10.1016/j.animal.2020.100063

8. Koltsova OM, Stekolnikova NV, Zhitin YuI (2018) Sugar beet production wastes and their use in agriculture. Vestnik Voronezh State Agrarian Univ 4(59):52–58. https://doi.org/10.17238/issn2071-2243.2018.4.52

9. Münnich M, Klevenhusen F, Zebeli Q (2018) Feeding of molassed sugar beet pulp instead of maize enhances net food production of high-producing Simmental cows without impairing metabolic health. Anim Feed Sci Technol 241:75–83. https://doi.org/10.1016/j.anifeedsci.2018.04.018

10. Vorobyov SP, Trotskovsky AI, Vorobyova VV (2019) Economic efficiency of the integrated formations functioning in the regional agriculture. IOP Conf Ser: Earth Environ Sci 274:012031. https://doi.org/10.1088/1755-1315/274/1/012031

11. Vorobyov SP, Vorobyova VV, Chernykh AA (2021) Profitability modeling in agricultural organizations. In: Bogoviz AV (ed) Complex systems: innovation and sustainability in the digital age. Springer, Cham, Switzerland, pp 435–442. https://doi.org/10.1007/978-3-030-58823-6_48

12. Yandex Maps (n.d.) Location of the "Cheremnovsky Sugar Factory" (53.176026, 83.236071). https://yandex.ru/maps/-/CCUibKcQOC. Accessed 8 June 2021

13. Zheng Y, Yu C, Cheng Y-S, Lee CM, Simmons C, Dooley TM, … Gheyns JV (2012) Integrating sugar beet pulp storage, hydrolysis and fermentation for fuel ethanol production. Appl Energy 93:168–175. https://doi.org/10.1016/J.APENERGY.2011.12.084

Study of the Ratio of Heat and Electrical Energy Expended in Microwave-Convective Drying of Grain

Dmitry A. Budnikov⬤

Abstract Nowadays, the design of the power supply of agro-industrial enterprises should be carried out considering the required installed capacity and the peculiarities of production. Thus, livestock farming involves waste, the disposal of which incurs costs. Simultaneously, preparation of feed (including drying of feed grain) is associated with significant energy costs. Thus, the availability of biogas equipment will allow synthesizing utilization technologies by processing waste into biogas and supplying energy to the equipment to carry out drying. Simultaneously, it is necessary to pay attention to technologies with reduced energy consumption for technological processes, for example, microwave-convective or infrared-convective drying of grain. These technologies have reduced energy consumption for moisture extraction, but the installed capacity of the equipment is higher than in traditional technologies. This paper aims to investigate the ratio of heat and electrical energy expended during microwave-convective drying and choose possible sources of renewable energy for implementing technological operations. The fact that grain drying is mainly carried out in the harvesting period before placing it in storage allows us to consider energy equipment as a source of thermal energy in the cold season when drying is not required.

Keywords Grain drying · Microwave field · Energy costs

JEL code Q220

D. A. Budnikov (✉)
Federal State Budgetary Scientific Institution "Federal Scientific Agroengineering Center VIM", Moscow, Russia
e-mail: dimm13@inbox.ru

1   Introduction

1.1   *Energy Costs of Heat Drying*

The power supply of agricultural production facilities is associated with the study of technological processes and operations of a given farm. This is typical both for the design of new equipment and the reconstruction of existing plants and production lines. Implemented projects must be based on the energy balance of the introduced technologies of production and energy generation [15]. Synthesis of various existing generation and waste disposal technologies allows for a significant reduction of the cost of the final product. Drying can be considered an example of one of the energy-intensive processes in agricultural production since the energy intensity of agricultural production is primarily related to the cultivation of crops. The gross harvest of grains in the world is currently at 2 billion tons per year. At the same time, several processes require significant energy costs to process the resulting crop. One of the mandatory and energy-intensive processes is drying of grain to bring it to the normal amount of moisture [6, 8, 13, 14]. As one of the main technological stages of grain processing, drying is primarily characteristic of regions marked with unsatisfactory weather conditions. Although the leading grain producers(by volume) are countries with mild conditions, drying is applied to about 30–40% of the gross grain harvest.

In some regions, the use of grain dryers is required only once every 2–3 years. Simultaneously, the energy intensity of drying grain is at a very high level and is 3.5–14 MJ per evaporation of one kilogram of moisture [12, 13]. It is necessary to consider that many developers, such as Petkus (Germany), Cimbria (Denmark), Dozagrant (Russia), AVG (Russia), Stela Laxhuber GMBH (Germany), Tornum (Sweden), Altinbilek (Turkey), and provide data only on the flow rate of heat drying agent or indicate the basic capacity and do not provide data for a detailed analysis of the energy intensity of drying. This is mainly due to the influence of weather conditions under which the grain is processed and the state of the harvested crop. It should also be noted that grain harvesting and drying are mainly carried out in the warm season, which allows us to consider the sources of energy idle at this time of the year as sources of heat.

The existing and currently developing drying technologies (e.g., the use of infrared radiation and microwave fields) have lower energy costs for drying than the classic heat drying methods, amounting to 2.8–4.5 MJ for evaporation of one kilogram of moisture [12]. However, their implementation increases the share of electrical energy in the total balance of energy costs. There are works devoted to using renewable energy sources (e.g., solar energy) to prepare the drying agent and reduce the cost of grain drying [2, 9]. The existing and the plants under creation often allow for both types of energy [1, 4–7, 10, 11], which allows to consider them as energy units for grain drying technology. Additionally, these units can be used to provide heat and power for technological operations at the plant. Such operations may include decontamination and preparation of fodder, as well as the provision of daily needs.

Thus, the paper aims to evaluate and analyze the energy costs of drying grain and determine the possibility of using renewable energy sources. The following tasks are carried out:

- Experimental evaluation of the balance of electrical and heat energy during drying of wheat grain in the microwave-convective drying unit;
- Selection of possible energy installations based on renewable energy sources for the process of drying of grain.

1.2 Technologies of Intensification of Grain Drying

The improvement of grain drying equipment, aimed at increasing its productivity, leads to an increase in capital costs in most cases. In this case, the costs may be caused by the purchase of new equipment, reconstruction of existing lines, or implementation of new automation systems to implement control algorithms. Given that the primary goal of agricultural enterprises is to make profits, the development and implementation of energy-efficient technologies and equipment is a promising area for researchers. It is necessary to remember about the need to ensure the quality of the crop.

The main directions of increasing the productivity of grain drying equipment are increasing the capacity (mainly by increasing the number and geometric dimensions of apparatuses) and the use of factors intensifying the process of heat-mass transfer. As proved by the results of various studies, the use of electrophysical factors can lead to a reduction in specific energy consumption for the drying of grain [3, 12, 14]. Factors can affect the properties of the drying agent, such as the use of ozone-air mixtures or directly on the material to be treated, such as the use of thermal fields (infrared, high frequency, microwave) as well as an acoustic influence (ultrasound). These factors have both advantages and disadvantages that should be considered when designing equipment.

Currently, the work of many researchers is aimed both at intensifying the transfer of heat and moisture in the process of drying and reducing energy costs for drying. The electrophysical factors influencing the course of drying are actively considered. Many of the factors (e.g., ultrasound) have a minimal application due to the peculiarities of the penetration of the influencing factor into the material. Other factors (e.g., infrared and microwave) involve direct heating of the material. Therefore, it is necessary to develop specialized modes of operation tied to the current moisture content of the material due to the peculiarities of propagation of the above factors in the layer of the treated material, the possible high energy densities dissipated in the material, and high intensities of heating the grain of high humidity.

For agricultural producers, the choice of the performance criteria of equipment (high-intensity drying or low energy intensity drying) is related to the cost of the final product, which is associated with the cost and availability of energy resources required for technological processes and the provision of safety of the product during its processing and storage.

Table 1 Grain drying technologies

Technology	Energy intensity, MJ/kg of moisture	Production capacity, t/h	Execution	Advantages
Heat drying	5–14	4–250	Stationary; mobile	High output range
Ultrasound	4.0–6.0	0.01–0.1	Laboratory	High intensity of the moisture transfer
Ozone	3.5–5.5	0.1–5.0	Containerized; stationary; mobile	Low power consumption; disinfection
Aeroions	3.3–5.0	0.1–1.0	Stationary; laboratory	Low energy intensity
Radiofrequency drying	4.5–6.5	0.1–1.0	Laboratory	Decontamination; increased shelf life of the final product
Infrared radiation	5.0–9.0	0.1–10.0	Stationary; laboratory	High intensity of heating

Source Compiled by the author

In terms of determining the direction of the study of grain drying equipment, it is worth comparing the use of electrical technology. Table 1 shows some indicators of drying technologies.

These intensifications of heat and moisture transfer require additional energy capacities. However, the balance of thermal and electric capacity can change significantly. Thus, drying in the mine dryer with a capacity of five tons per hour uses a heat generator with a thermal capacity of 440–500 kW with installed electrical power equipment of 12 kW. In our case, we will consider the unit for microwave-convective drying of grain, which allows intensifying heat and moisture transfer and carrying out the preparation of the drying agent and direct heating of the grain mass.

2   Materials and Methods

2.1   Laboratory Unit

To conduct experimental studies, the author used a laboratory unit, the process flow diagram of which is shown in Fig. 1. This unit contains six sources of microwave field (magnetrons—2.45 GHz, power—700 W). Magnetrons are powered through individual transformers. The operation mode of each magnetron is controlled by a programmable relay PR-114, which allows implementing not only continuous and pulsed modes of operation of the microwave field sources but also different operation modes of magnetrons by level. This feature is vital for accounting for the moisture

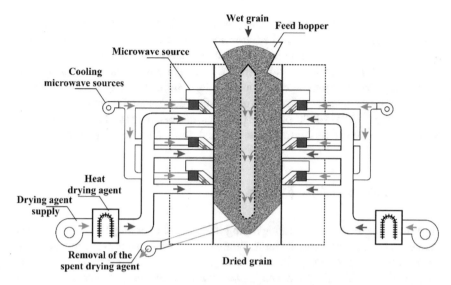

Fig. 1 Technological scheme of laboratory microwave-convective dryer of grain. *Source* Compiled by the author

loss that occurs during the movement of the grain layer along the grain pipeline. Since the moisture content in the grain decreases during drying, the depth of penetration of the microwave field and the uniformity of its heating increases. During this research, the author used modes with the magnetrons working equally on all levels.

The transformers and magnetrons were cooled in groups; the air thus heated was supplied into the drying zone, thus reducing energy losses with the exhaust air. Individual protection of magnetrons and transformers against overheating is provided to prevent emergencies. The drying agent was supplied centrally, and the drying agent was prepared in the block of flame heaters. In real conditions, the block of heat preparation of the drying agent can be implemented using electric heaters, various furnaces, recovery units, etc.

The technological process and energy consumption are controlled through a SCADA system with logging and visual presentation to the operator. In addition to the power consumption, the current temperatures of the grain layer in the drying zone are displayed to prevent disruption of the technological process. The connection is made through a common counter, which records the total costs of the technological process, which is used to estimate the specific energy intensity of the drying process or other heat treatment of the grain. The energy consumption of electrophysical sources is also evaluated in the laboratory unit utilizing an additional electricity meter. The electricity meter powers the required positions of consumers. This allows us to estimate the energy consumption for the drying process and the influence of electrophysical factors on the reduction of the specific energy intensity, as well as the share of energy consumed by them.

The presented laboratory unit allows us to estimate the energy intensity of grain processing and the intensity of heat transfer using electrophysical factors. In this case, the influence of a single factor or a combination of factors can be realized. The control of equipment operation modes (e.g., periods of magnetron activation, the delivery rate of drying agent, required temperature of drying agent, control of process parameters, and accumulation of statistical data) is carried out via the SCADA system. The MODBUS-RTU protocol provides data exchange and data logging. The details of the unit and its control system were described in previous works [3].

2.2 Conditions of the Experiment

The microwave power cycle during drying (the ratio of operating time of microwave sources to drying time) and the temperature of the drying agent were considered as factors influencing the energy intensity of drying. The magnetron activation time was set through the SCADA system and implemented through programmable relays. The cycle was 20 s. Thus, at a fraction of microwave power 0.25, the magnetron worked for 5 s, then turned off for 15 s, and the cycle was repeated. The parameters of the experiment are presented with the results in Table 2.

The current moisture content was controlled by an express method using a portable moisture meter Fauna-M and by sampling with subsequent measurement in laboratory conditions. It is worth noting that the readings of the moisture meter immediately after treatment with the microwave field do not correspond to reality due to the change in the form of bonding of liquid with the dry matter of grain. The equalization of the readings of moisture meter and indicators obtained in the laboratory occurs only after 30–40 min of treatment with a microwave field. During the unit's operation, the temperature of the grain layer in the immediate vicinity of the apertures of microwave power sources was monitored to avoid overheating of the grain above the temperatures regulated by the technological requirements. The control was carried out by fiber-optic temperature sensors OSMT-313.

Table 2 Parameters of the experiment

Mode	1	2	3	4	5	6	7	8	9
Microwave power supply cycle, p.u	0	0	0	0.25	0.25	0.25	0.5	0.5	0.5
Temperature of drying agent, °C	20	30	470	20	30	40	20	30	40

Source Compiled by the author

2.3 Considered Energy Sources

We considered solar and wind energy sources and biogas plants as sources of renewable energy typical for agricultural enterprises and the regions in which they are located [13, 14]. However, in the case of wind power plants, an additional stage of preconversion of electrical energy into thermal energy would be required to provide the preparation of the drying agent, which could negatively impact the final energy balance.

Solar energy should be considered the primary energy source for regions with a sufficient number of sunny days and the required intensity of solar radiation [2, 9]. Researchers have noted the high efficiency of solar energy for the preparation of the drying agent of various products, including grains of various crops. The application of these sources may be particularly interesting for cases where the crop is harvested at a moisture content close to conditionality. This allows one to carry out drying over a long period (several days) without additional costs to intensify the process. However, in many cases, drying will need to be performed on a tight schedule, which will somewhat limit the use of this type of energy.

Livestock farms engaged in harvesting grain for fodder can consider processing into biogas with its subsequent conversion into heat and electricity [4, 7]. Since the accumulated waste of livestock products must be utilized, biogas synthesis has a high proportion of relevance for livestock farms.

3 Results

The author used the taken drying curves for the studied modes as input data for the analysis. The processing of the obtained data considered the standard parametric model and the free model set during processing. The quality of the approximation was evaluated graphically using various suitability criteria. The specified criteria include R-square and Adjusted R-square. The dependence of the change in grain moisture (W, %) on the drying time (t, min) can be represented by the following dependence:

$$W = a \cdot e - b \cdot t + c, \tag{1}$$

where:

a, b, c—the coefficients of model a, %; b, min-1; c, %.

Table 3 shows the data obtained from the statistical processing of the drying curves.

The author also determined the confidence intervals for the model and confidence bands for the approximation and the data.

Table 3 Results of statistical processing of drying curves

Mode	a	b	C	R-square	Adj R-square
1	3.438	0.0911	13.9	0.9819	0.9747
2	3.769	0.07893	13.48	0.9836	0.9754
3	3.293	0.09729	13.57	0.9996	0.9996
4	3.035	0.05973	14.07	0.9911	0.9889
5	3.013	0.04815	13.79	0.9864	0.9825
6	3.332	0.07748	13.81	0.9745	0.966
7	2.959	0.02602	13.95	0.9879	0.9855
8	2.99	0.04361	13.96	0.9727	0.9666
9	2.963	0.05021	13.92	0.9667	0.9584

Source Compiled by the author

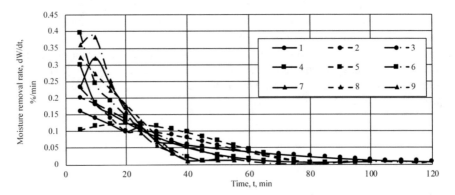

Fig. 2 Moisture removal rate during wheat drying in laboratory tests of the unit of microwave-convective treatment of the grain. *Source* Compiled by the author

At the next stages, the author obtained diagrams of moisture removal rate (Figs. 2 and 3), determined the energy consumption for moisture removal, and calculated the ratio of energy power (electrical and thermal), which are required for the operation of the unit in the considered modes of its operation. The results of these calculations are presented in Table 4. These results of determining the ratio of electricity and heat costs, together with data on available heat sources derived from renewable energy sources, will assess the possibility of energy supply of these units with different types of energy. It is also possible to determine the composition of the equipment and the required level of overlap of their capacities.

It is worth noting that the time for drying using microwave power is reduced by 2–3 times, which allows one to intensify the processing of the crop and reduce losses in case of untimely processing.

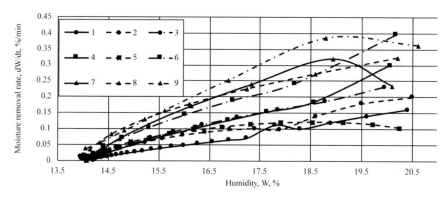

Fig. 3 Moisture removal rate at current moisture content in laboratory tests of the microwave-convective grain processing module. *Source* Compiled by the author

Table 4 Parameters and results of the experiment

Mode	1	2	3	4	5	6	7	8	9
Drying costs, MJ/kg of evaporated moisture	6.17	6.8	8.57	6.1	7.59	4.64	4.36	3.74	3.4
Heat power consumed, %	40	57	67	28	43	53	21	35	45
Electrical power consumed, %	60	43	33	72	57	47	79	65	55
Of which microwave	0	0	0	31	24	20	47	39	33
Of which ventilation	60	43	33	41	33	27	32	26	22

Source Compiled by the author

4 Discussion

The obtained results suggest that the correct choice of operating modes of equipment for the initial and current moisture content of the material can significantly affect the speed of drying and the energy consumption for its implementation. It is also worth noting that the introduction of sources of electrophysical effects, on the one hand, increases the installed capacity of the unit and, on the other hand, leads to a reduction in specific energy consumption due to the intensification of drying.

The ratio of electric and thermal energy costs suggests the possibility of using solar energy sources for the energy supply of these units. In this case, thermal energy will be used to prepare the drying agent, and electrical energy will be used to power microwave power sources and drive conveying devices. In this case, it is possible to implement stationary plants of relatively small capacity (less than three tons per hour). In this case, the electrical power will be 8–15 kW.

Mobile units based on commercially available pick-up attachments can also be implemented. In this case, it is possible to operate the mobile unit on the grain tank. Its power will be supplied by batteries or by cable from the feeding unit.

Another option is the use of biogas plants. Unlike solar installations, these units are independent of weather conditions. They can be used as a thermal energy source in the heating period and for the energy supply of grain drying equipment in the warm period. In this case, the unit will be relevant primarily for livestock enterprises. The use of biogas plants, in this case, will reduce the cost of the final product, emissions into the atmosphere, and the cost of processing and disposal of waste production.

In addition to the process of microwave-convective drying, the considered energy sources can be used during disinfection, micronization of grain, provision of daily needs of the enterprise, etc., which should be considered at the stage of design or reconstruction of specific farms.

5 Conclusion

Based on the results, the author can make the following conclusions:

- It is advisable to generate energy at the facilities of agricultural production, considering the full list of technological and domestic processes and the ratio of electric and thermal power needed;
- The share of electrical energy spent on the implementation of the technological process varies from 30 to 60% when conducting drying with the use of electrophysical influences;
- For the processes of drying crops, it is advisable to use solar thermoelectric plants and generators operating on biogas (considering the local generation of biogas from industrial waste).

The results allow choosing an available energy source for post-harvest processing based on the expected properties of the grain and the available resources of a particular enterprise.

The practical significance of this research is that the necessity of using electric and thermal energy in the operation of microwave convective drying of grain is proved. The ratio of electric and thermal energy allows for the use of local sources of energy generation.

Further research can be aimed at experimental testing of the combined operation of these energy sources with stationary and mobile units of microwave-convective drying of grain, as well as on the development of mobile devices of microwave convective treatment of the grain.

References

1. Appels L, Baeyens J, Degreve J, Dewil R (2008) Principles and potential of the anaerobic digestion of waste-activated sludge. Prog Energy Combust Sci 34(6):755–781
2. Baniasadi E, Ranjbar S, Boostanipour O (2017) Experimental investigation of the performance of a mixed-mode solar dryer with thermal energy storage. Renew Energy 112:143–150
3. Budnikov DA, Vasilyev AN (2020) Development of a laboratory unit for assessing the energy intensity of grain drying using microwave. Adv Intell Syst Comput 1072:93–100
4. Gladchenko MA, Kovalev DA, Kovalev AA, Litti YuV, Nozhevnikova AN (2017) Methane production by anaerobic digestion of organic waste from vegetable processing facilities. Appl Biochem Microbiol 53(2):242–249
5. Gusarov VA, Godzhaev ZA (2018) Development of low-power gas turbine plants for use at industrial facilities. J Mach Manuf Reliab 47(6):500–506
6. Izmaylov AY, Dorokhov AS, Vershinin VS, Gusarov VA, Majorov VA, Saginov LD (2019) Window frame installed photovoltaic module for feeding of low power devices. Int J Renew Energy Res 9(1):187–193
7. Kovalev AA, Kovalev DA, Grigoriev VS (2020) Energy efficiency of pretreatment of digester synthetic substrate in a vortex layer apparatus. Eng Technol Syst 30(1):92–110
8. Kupfer K (1999) Drying oven method, infrared- and microwave drying as reference methods for determination of material moisture. Tech Mess 66(6):227–237
9. Mathew AA, Venugopal T (2021) Solar power drying system: a comprehensive assessment on types, trends, performance and economic evaluation. Int J Amb Energy 42(1):96–119
10. Nesterenkov P, Kharchenko V (2019) Thermo physical principles of cogeneration technology with concentration of solar radiation. Adv Intell Syst Comput 866:117–128
11. Varfolomeev SD, Efremenko EN, Krylova LP (2010) Biofuels. Russ Chem Rev 79(6):491–509
12. Vasilyev AN, Dorokhov AS, Budnikov DA, Vasilyev AA (2020) Trends in the use of the microwave field in the technological processes of drying and disinfection of grain. AMA-Agricult Mech Asia Africa Latin America 51(3):63–68
13. Vasiliev AN, Goryachkina VP, Budnikov DA (2020) Research methodology for microwave-convective processing of grain. Int J Energy Optim Eng 9(2):1–11
14. Wang Y, Wang S, Zhang L, Gao M, Tang J (2011) Review of dielectric drying of foods and agricultural products. Int J Agric Biol Eng 4(1):1–19
15. Wang XF, Ma PP, Wang ZN (2017) Combining SAO semantic analysis and morphology analysis to identify technology opportunities. Scientometrics 111(1):3–24

Clustering of Agribusiness and Its Role in Forming the New Architecture of Rural Areas (Case Study of the Republic of Bashkortostan)

Vilyur Ya. Akhmetov⑩, Rinat F. Gataullin⑩, Razit N. Galikeev⑩, Salima Sh. Aslaeva⑩, and Ramil M. Sadykov⑩

Abstract The authors analyze international and Russian experiences of the formation and development of agricultural clusters. Additionally, the authors identify priority areas of agribusiness clustering in relation to each municipal district of the Trans-Ural part of the Republic of Bashkortostan through the development of vertical and horizontal integration—the development of agricultural holdings and cooperatives of various forms. The meat, dairy, and grain clusters were identified as priority areas. Tourist and recreational, scientific and educational, and ethnical product clusters are also promising in the Republic of Bashkortostan. The authors substantiate the need to adjust federal, regional, and municipal regulations, programs for developing the agro-industrial complex, small and medium enterprises, tourism, and socioeconomic development of rural areas. The authors provide proposals for the formation of agricultural clusters. From the authors' point of view, the multiplier effect allows agricultural business clustering to contribute to the formation of a new architecture of rural areas and improve the quality of life by organizing new enterprises, diversifying the rural economy, creating new jobs, and improving transport communications, engineering, and social infrastructure.

Keywords Agro-industrial complex · Rural economy · Integration · Cooperation · Cluster · Ethnic cluster · Ethnic products · Region · Republic of Bashkortostan

V. Ya. Akhmetov (✉) · R. F. Gataullin · R. N. Galikeev · S. Sh. Aslaeva · R. M. Sadykov
Institute of Socio-Economic Research, Ufa Federal Research Center of the Russian Academy of Sciences, Ufa, Russia
e-mail: willi76@mail.ru

R. F. Gataullin
e-mail: gataullin.r2011@yandex.ru

R. N. Galikeev
e-mail: razitg@inbox.ru

S. Sh. Aslaeva
e-mail: salima2006a@mail.ru

R. M. Sadykov
e-mail: sadikovrm@mail.ru

JEL codes Q13 · Q18

1 Introduction

In recent years, the Government of the Russian Federation has consecutively adopted several programs for the development of rural areas:

- "Social Development of the Village" in 2003–2013 [1];
- "Sustainable development of rural areas for 2014–2017 and until 2020" in 2013 [2];
- "Integrated Rural Development for 2020–2025" [3].

According to these and other federal programs for the effective socioeconomic development of rural areas, regional and municipal programs are also being developed at the level of the territorial entities of Russia. Thus, the Republic of Bashkortostan by the Decree of the Government of the Republic of Bashkortostan No. 728 (December 12, 2019, as amended December 8, 2020) approved the state program "Integrated development of rural areas of the Republic of Bashkortostan" [4]. Moreover, in 2010–2011, the Institute of Socio-Economic Research of Ufa Federal Research Center of the Russian Academy of Sciences (ISER UFRC RAS), with the participation of the authors of this research, developed two programs for the development of depressed rural areas in the Northern-Eastern and Southern-Eastern districts of Trans-Ural part of the Republic of Bashkortostan for 2011–2015 (later extended until 2020, and then until 2024).

As a result of the research, the authors substantiate the expediency of adjusting the subregional medium-term comprehensive program of economic development of Trans-Ural part of the Republic of Bashkortostan and including promising projects for the development of various agricultural clusters—new "growth points" for particular municipal districts and cities of the subregion.

The feasibility of clustering of agricultural business in the Republic of Bashkortostan is explained by the following reasons:

1. The Republic has enormous agricultural potential. There are more than 7 million hectares of land, of which about 5 million hectares is arable land. The region steadily holds a leading position in the production of major agricultural products in Russia.

2. The Republic of Bashkortostan should gradually transform from a supplier of agricultural raw materials into a producer of final food products, including deep processing. So far, the food and processing industry is a weak link in the region's economy, especially in its South-Eastern districts.

3. The lack of a proper supply of agricultural products from local producers leads to the penetration of enterprises from other subjects of Russia to the regional market and the promotion of their agricultural and ethnical brands. Therefore, a comprehensive approach to developing local and regional brands, including state and regional support, is vital.

The formation and development of agricultural clusters are studied in many works of international and Russian authors, including Ketels [5], Nguyen [6], Morozov, Konakov [7], Dorzhieva [8], Kostenko [9], and others. However, this issue is not well developed in relation to the conditions of specific rural areas of Russia.

2 Materials and Methods

During the research, the authors analyzed scientific articles and monographs of scholars leading in the field of agricultural and ethnic economics. The authors applied a wide range of general scientific and special research methods. Moreover, the authors used statistical data on the regions of Russia, including the Republic of Bashkortostan, and the results of socioeconomic research of the ISER UFRC RAS.

The research aims to identify priority agricultural clusters for the Republic of Bashkortostan.

The authors solve the following tasks:

- To analyze international and Russian experience of forming and developing agricultural clusters;
- To allocate priority areas of clustering of agricultural business in relation to each municipal district of the Trans-Ural part of the Republic of Bashkortostan;
- Consider the prospects for the formation of meat, dairy, grain, tourist, recreational, scientific, educational, and ethnical clusters in the South-Eastern districts of Bashkortostan;
- To substantiate the need to adjust federal, regional, and municipal legal acts and programs for the development of the agro-industrial complex (AIC), small and medium enterprises, tourism, and social and economic spheres of rural areas, as well as to include proposals for the formation of agricultural clusters in them;
- To identify the role of agricultural business clustering in the formation of a new architecture of rural areas and improvement of the quality of life by organizing new enterprises, diversifying the rural economy, creating new jobs, and improving transport communications, engineering, and social infrastructure.

3 Results

The formation of a new architecture of the regional economic space in the Republic of Bashkortostan implies a comprehensive progressive development of certain sectors of the rural economy, the creation of new enterprises in rural areas and jobs, and the development of rural production and social infrastructure to improve the quality of life. Therefore, continuing federal programs, the regional, subregional, and local authorities develop concepts and programs for rural development, implying various funding sources. An example of one of the subregional programs in the Republic of Bashkortostan is the "Midterm Comprehensive Program of Economic Development of Trans-Urals for 2011–2015" [10], which has been repeatedly edited and extended. This program is aimed at effective socioeconomic development and the formation of a new architecture of depressed South-Eastern rural areas, the total area of which exceeds 1/3 of the region.

From our point of view, for a more comprehensive socioeconomic development of rural areas of the Trans-Ural part of the Republic of Bashkortostan, it is advisable to include various areas of diversification and clustering of the rural economy and non-agricultural industries (tourism, folk crafts and trades, catering system, etc.) as new "points of growth" in this program. Since some districts of the Trans-Ural part are populated by more than 90% of the indigenous Bashkir population, who have preserved many national traditions, customs, and production technologies of unique ethnic products and services, it is very promising to organize not only agricultural but also ethnical clusters, including the development of gastronomic and ethnographic types of tourism and production according to international standards "Halal."

In the author's opinion, many economic and social problems of rural areas (including areas of the Trans-Ural Republic of Bashkortostan) can be solved by the multiplier effect provided by the formation and development of clusters in the following areas:

- Dairy and beef cattle breeding;
- Horse breeding and production of kumis;
- Fish farming;
- Poultry farming;
- Ethnographic, rural, health, and event tourism;
- Production of organic (eco) products, collection, and processing of wild plants.

During the formation of agricultural clusters, it is advisable to pay attention to the best international (American, Chinese, Canadian, German, and English) and Russian experience (Kaluga, Samara, and Leningrad Regions, the Republic of Tatarstan, etc.).

Adapting the definition of M. Porter, the founder of the cluster theory, to our problems, we can conclude that agricultural clusters are networks of agricultural, processing, tourist, recreational, and other enterprises, as well as scientific and educational institutions of the region, which are located in a particular area and work together to address the production, processing, and marketing of products and services, including through the introduction of innovation [11].

Agricultural clusters will unite producers of agricultural products, processors, logistics centers, and retailers, which can help small local producers get into large retail chains.

The clustering of agricultural business can play a key role in forming a new architecture of rural areas of the Republic of Bashkortostan. It is necessary to reasonably combine market mechanisms with the use of federal and regional target programs for accelerated socioeconomic development of territories, implying not only issues of organizing new enterprises but also issues of modernizing the existing and constructing new utilities, transport, and social facilities (schools, hospitals, cultural and recreational centers, etc.).

It is essential to form an effective "government-business-science" tandem to solve the problems mentioned above. In this tandem, the staff of ISER UFRC RAS can provide all possible assistance in the construction of the optimal architecture of the economic space of the Republic of Bashkortostan through scientific support and consulting support of projects for the development of rural areas.

4 Discussion

The works of Barlukova [12], Voronov [13], Gryadov [14], Khukhrin [15], Tokhchukov [16] are of particular interest when dealing with the agrarian economy and the need to diversify it.

In the Republic of Bashkortostan, these issues are presented in the works of Barlybaev et al. [6, 8], Galikeev [17], Gataulin et al. [18], Kuzyashev [19].

Nevertheless, this issue is still insufficiently developed in relation to specific rural areas of the studied region and other regions of Russia.

5 Conclusion

The lack of consistency and complexity in managing the formation and development of regional and subregional agricultural clusters is the primary reason for the weak pace of their formation in the rural regions of Russia (including the Republic of Bashkortostan). For the large-scale and stable development of agricultural clusters in the Republic of Bashkortostan, it is advisable to develop the following:

1. "Scientific and methodological concept for developing agricultural and ethnical clusters in the Republic of Bashkortostan for the period up to 2030." This scientific development will act as a logical continuation of various programs to develop agriculture, small and medium-sized rural enterprises, ethnical and ecotourism, etc. In particular, the concept will identify specific actions to increase employment of the local population and algorithms and mechanisms for the organization of certain areas of environmental (including ethnographic) tourism;
2. Strategies of socioeconomic development of rural areas, long-term plans, and programs for the development of agricultural and ethnical clusters for each territorial unit of the Republic of Bashkortostan and the whole region;
3. Business portfolio for rural investors in various areas of agricultural business and social development in rural areas.

Also, to develop entrepreneurial, legal, and financial literacy of the rural population and intensify agricultural and ethnical business, it is advisable to hold courses, trainings, foresight sessions, forums, and scientific and practical conferences on various areas of organization of different forms of business in rural areas.

Acknowledgements This study was performed within the framework of the state task No. 075-00504-21-00 of the Ufa Federal Research Centre of the Russian Academy of Sciences for 2021.

References

1. Government of the Republic of Bashkortostan (2011) Mid-term comprehensive program of economic development of Trans-Urals for 2011–2015. Ekonomika, Moscow
2. Government of the Republic of Bashkortostan (2019) On approval of the state program "Integrated development of rural areas of the Republic of Bashkortostan" and on amendments to some decisions of the Government of the Republic of Bashkortostan (as amended April 7, 2021). Ufa, Republic of Bashkortostan. Retrieved from https://docs.cntd.ru/document/561 663813 (Accessed 20 January 2020)
3. Government of the Russian Federation (2002) Resolution "On the federal targeted program 'Social development of the village until 2013'" (December 3, 2002 No. 858, as amended July 15, 2013). Moscow, Russia. Retrieved from http://www.consultant.ru/document/cons_doc_LAW_64705/ (Accessed 20 January 2020)
4. Government of the Russian Federation (2013) Resolution "On the federal target program 'Sustainable development of rural areas for 2014–2017 and for the period up to 2020'" (June 15, 2013 No. 598). Moscow, Russia. Retrieved from http://base.garant.ru/70419016/#ixzz5z 5mgstWr (Accessed 20 January 2020)
5. Government of the Russian Federation (2019) Resolution "On approval of the state program of the Russian Federation 'Integrated development of rural areas' and on amendments to some acts of the Government of the Russian Federation" (May 31, 2019 No. 696). Moscow, Russia. Retrieved from http://static.government.ru/media/files/aNtAARsD8scrvdi zD7rZAw0FaFjnA79v.pdf (Accessed 20 January 2020)
6. Akhmetov VY, Lisica AV (2014) Territorial Branding and its role in increasing the investment attractiveness of depressive sub-regions (Based on the example of Trans-Ural Region of the Republic of Bashkortostan). Sci Rev 6:289–294
7. Barlukova AV (2010) Classification status of ethnic tourism. Bulletin of Baikal State University 4(72):106–108
8. Barlybaev AA, Akhmetov VY, Nasyrov GM (2009) Tourism as a factor in the diversification of the rural economy. Problemy Prognozirovaniya 6(117):105–111
9. Dorzhieva EV (2012) Formation and development of competitive agro-industrial clusters at the mesolevel of the economy. Publishing house of the St. Petersburg University of Management and Economics, St. Petersburg, Russia
10. Galikeev RN, Gataullin RF, Akhmetov VY (2017) Problems of improving the land-use mechanism in the region. Fundamen Res 10–2:323–327
11. Gryadov SI (2009) Agro-industrial cluster: Problems and prospects for development. Bull Altay State Agrarian University 4(54):74–79
12. Ketels S (2003) The development of the cluster concept – present experiences and further developments
13. Khukhrin AS, Primak AA, Pekhutova EA (2008) Agroindustrial clusters: the Russian model. Econ Agricult Process Enterprises 7:30–34
14. Konakov MA, Morozov NM (2009) Small scale agro-industrial clusters. Econ Agricult Process Enterprises 2:30–33
15. Kostenko OV (2016) Agro-industrial clusters in the economic policy of Russian regions. economics: yesterday. Today, Tomorrow 5:55–68
16. Nguyen ND, Martin J (2015) Implementing clusters for economic development in emerging economies: the case of Luong bamboo sector in Thanh Hoa Province, Vietnam. In: Proceedings of the 18th Toulon-Verona conference: excellence in services. Palermo, Italy: University of Palermo, pp 355–372

17. Porter ME (1990) The competitive advantage of nations. Free Press, New York
18. Repushevskaya OA, Nasretdinova ZT, Kuzyashev AN, Beschastnova NV, Shamshovich DA (2021) The role of credit cooperatives in financing the real sector of the economy. In: Bogoviz AV, Suglobov AE, Maloletko AN, Kaurova OV, Lobova SV (eds), Frontier information technology and systems research in cooperative economics, pp 3–11. https://doi.org/10.1007/978-3-030-57831-2_1
19. Tazhitdinov IA, Gainanov DA, Akhmetov VY (2017) Priority areas of agribusiness clustering in the South-Eastern and North-Eastern regions of the Republic of Bashkortostan. Manag Econ Syst: Sci Electron J 9(103). Retrieved from http://www.uecs.ru/regionalnaya-ekonomika/item/4535-2017-09-20-08-24-38 (Accessed 20 January 2020)

Assessment of Efficiency and Production Risks in Crop Production Innovative Development

Tatiana N. Kostyuchenko⦾, Darya O. Gracheva⦾, Natalia N. Telnova⦾, Alexander V. Tenishchev⦾, and Marina B. Cheremnykh⦾

Abstract The purpose of the article is to offer a comparative evaluation methodology of the economic efficiency of field crop production in risk farming conditions. This technique can be used as the basis of risk management tools formation, the need of which is increasing in the context of the innovative development of the agricultural sector. The authors propose to use an integral indicator to assess the economic efficiency of crop production, taking into account the differences in the profitability and riskiness of the arable use in the cultivation of various crops. The use of this indicator is advisable to optimize the structure of arable areas of agricultural enterprises, making decisions on the innovative production technologies implementation. It is shown that the use of traditional indicators of economic efficiency for a comparative assessment of various crops production—the level of profitability and (or) the amount of profit per 1 hectare of arable land—in the zones of unstable farming should be supplemented with indicators reflecting the degree of uncertainty and variability of production results. The proposed method is advisable to use in the optimizing of the sown areas structure and the improving of the machine and tractor fleet. As well as changing the sort composition of seed, planning the use of innovative cultivation technologies. The obtained results are required to use in creating storage capacities in the regions and crop production processing, determining the practicability of state investment support of agricultural production, etc.

T. N. Kostyuchenko (✉) · D. O. Gracheva · N. N. Telnova · A. V. Tenishchev · M. B. Cheremnykh
Stavropol State Agrarian University, Stavropol, Russia
e-mail: kostuchenkotn@mail.ru

D. O. Gracheva
e-mail: pochtadg@mail.ru

N. N. Telnova
e-mail: telnatnik@mail.ru

A. V. Tenishchev
e-mail: nepier@mail.ru

M. B. Cheremnykh
e-mail: Che.212@mail.ru

Keywords Crop production · Risks · Methods · Economic efficiency · Innovative development

JEL Codes O22 · O33 · Q12

1   Introduction

Effective management in the sphere of agricultural production in modern conditions is only based on a systematic analysis of activity results. These results are both influenced by objective factors (these primarily include weather conditions) and subjective factors that are provided by the agricultural enterprise itself. Among the subjective factors to improve efficiency and competitiveness, the use of innovative technologies is becoming increasingly important. However, the innovative activity of agricultural producers is still insufficient. Therefore, according to NRUHSE [1], the aggregate level of innovative activity in agriculture is only 3.7%, with an average value for the Russian economy at 8.5%.

The main reason—the high cost of investment in innovative development and the significant payback period, which increases the riskiness of entrepreneurship. Thus, in agriculture, industrial production risks have a multiplier effect on the risks inherent in innovation [2]. Therefore, for effective innovative development management in crop production can be proposed a two-step analysis:

1.   a systematic analysis of the comparative efficiency of production by crops, taking into account the existing cultivation risks in a particular enterprise or area;
2.   an analysis of the innovative technologies' effectiveness use.

The following provisions determine methodological approaches to the first step implementation. The annual change in weather and climatic conditions is the reason for the high indicator variability of crop yields. This is especially typical for regions of insufficient and unstable moistening, in which agricultural producers are having difficulties in ensuring the financial sustainability of their activities. The significant non-additive effect and stochasticity of weather and climatic conditions explain the high riskiness of agribusiness, the complexity of effective planning, as well as the adoption of timely management decisions aimed at losses reducing due to high/low temperatures or insufficient/excessive precipitation in the context of crop growth phases.

Therefore, to assess the economic (technical) efficiency of agricultural production in Russia, as in the other countries, stochastic frontier analysis (SFA) is increasingly being used, which allows identifying its key factors for a specific period. Over the past decades, more and more attention among these factors has been given to risks that play a decisive role in agribusiness. Risk management strategies should be based on their quantitative assessment. Among the main factors, Peltonen-Sainio et al. [3] consider the crop production risks associated with climate change, Sannikova et al. [4], Stulec et al. [5]—with weather conditions, Tiedemann and Latacz-Lohmann

[6]—with land use, labor, and capital. Novickytė [7] includes insurance, weather derivatives, contract agriculture, a transformational adaptation of agribusiness to new conditions, product diversification.

Russian scientists have also paid considerable attention to assessing the weather conditions that influence crop production results. Ivagno et al. [8] propose to use the statistical parameters of the dynamic series of crop yields. Siptits [9] focuses on climate risk reduction based on adaptive activity management both a single commodity producer and agri-food systems, Chepurko [10]—measuring the zonal-sectoral risk in agricultural production based on aligning the dynamics series using the linear programming method.

Thus, there is still no unified approach to assessing the influence of weather and climatic risks on crop cultivation efficiency. Moreover, most of the proposed methods remain methodologically rather complicated and, therefore, are recommended to use at the macro-management level, regional consulting services, and the largest insurance companies. According to Sannikova et al. [4], assessment algorithms are needed to develop the adaptive strategies for individual agricultural producers, consisting, first of all, in the sown area distribution between cultivated crops, their varieties, and cultivation technologies.

2 Methodology

In most studies, the authors assess the agricultural business risk based on the study of the crop yields dynamic series, which is one of the main indicators of production efficiency. However, our studies demonstrate that in Russian reality, the economic impact of both yield growth and reduction is largely offset by sales price changes (Fig. 1).

It should be noted, that deflated prices were used in the calculations to obtain more reliable results of this comparison. As a result, it was found that the correlation coefficient between the analyzed indicators for the period from 2011 to 2019 was at (-0.70), which confirms the presence of a close inversely proportional relationship.

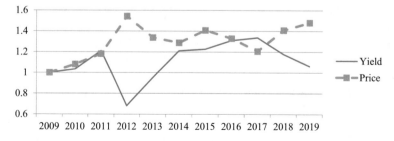

Fig. 1 Trends changes in average grain yield and average grain prices among the enterprises of the Stavropol Territory. *Source* Compiled by the authors based on [7]

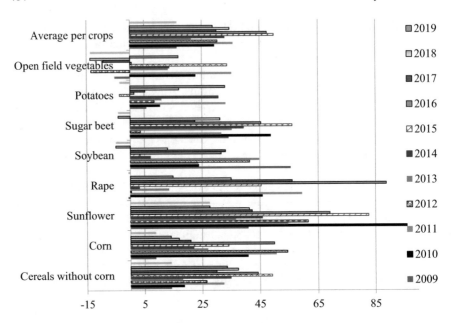

Fig. 2 The level of profitability of production of crop products grown by enterprises in the Stavropol Territory. *Source* Compiled by the authors based on [7]

Therefore, in further calculations, we used economic efficiency indicators, reflecting the production results changes and the situation on the crop product markets.

The most common indicator for assessing the agricultural enterprise's economic efficiency is the level of profitability, calculated both for individual types of products and for the whole enterprise. These average data of agricultural enterprises in the Stavropol Territory are presented in Fig. 2.

The data analysis shows that in general for field crop cultivation and across cultures, there are significant differences in profitability by year—from (-13%) to (+ 96%), these reflect the extreme volatility of the results of crops cultivation. At the level of separate enterprises, volatility can be even more significant.

It can be concluded that the efficiency of invested resources has changed since the level of profitability is calculated as the ratio of profit to sales cost. But, at the same time, it is impossible to assess the optimal merging structure of the sown area. It is so because it remains a high level of future results uncertainty especially in production conditions in the insufficient moisture zone and the case of low regulated product prices.

Assessing the efficiency of land use, the profit indicator per 1 hectare of the arable area is of greater interest (Fig. 3).

Analysis of the data for the period from 2009 to 2019 shows that the largest profit (on average per year) was received by enterprises from the cultivation of the most labor-intensive products—potatoes and sugar beets—29.35 and 15.69 thousand rubles per hectare with year volatility from -8.45 to 72.0 rubles per hectare. Their

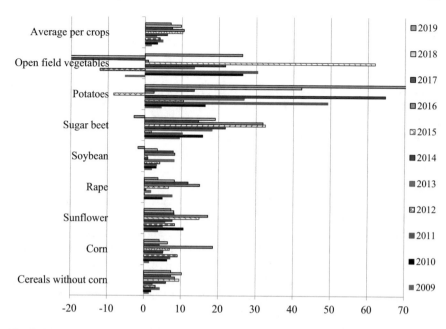

Fig. 3 The amount of profit per 1 hectare of crop areas, received by the enterprises of the Stavropol Territory. *Source* Compiled by the authors based on [7]

share in the structure of crop areas is not more than 0.2%. The second place is occupied by open field vegetables and sunflowers, which sales ensure the profit from -20.79 to + 61.94 rubles per hectare, providing an average annual rate of 11.35 and 8.85 rubles per hectare. The share of its crops does not exceed 11%. The third place is taken by corn and grain crops with profit from 7.22 and 5.75 thousand rubles per hectare in average annual terms.

However, the cultivation of crops provides the possibility to obtain the profit from 1 hectare with varying probability. In our opinion, to measure the risks of achieving the indicated average annual results, it is advisable to use the coefficient of variation, which is the ratio of the standard deviation to the expected value of the useful result—in our case, profit per 1 hectare.

3 Results

The crop with the lowest risk among the considered crops is the one whose profit per 1 hectare has the lowest value of the variation coefficient. It is considered that if the variation index is less than 33%, the set of numbers is indiscrete. In the opposite case, it is usually characterized as discrete and, therefore, the production of this crop can be considered riskier. The data analysis, presented in Fig. 3, shows that any crop production cannot ensure such indiscrete value. Even the average variation coefficient of field cultivation for the analyzed 11 years exceeds 47%.

The group of the least risky crops of the Stavropol Territory enterprises includes crops with the variation coefficient of profitability per 1 hectare up to 66%: such as sunflower (43.8%), cereals (53.0%), corn (61.1%), and sugar beet (min—vertical shading of shapes in Fig. 4). The group of medium risky crops (horizontal shading of shapes) includes soybeans, rapeseed, and potatoes. High risky is the cultivation of field vegetables (max—solid color shapes).

According to some authors, only the negative deviation from the average level of the indicator should be considered in risk assessment. In our opinion, unsustainable agricultural business is characterized by the full volatility amplitude of output indicators. Moreover, we determined that in further calculations it is necessary to take into account changes of only one of represented production efficiency indicators of different crops because the correlation between their variation coefficients was 0.994.

For an integral assessment of efficient arable land use, we propose the methodology shown in Table 1. In this case, the difference in the amount of profit per 1 hectare of the cultivated area can be assessed through the points of profit growth, taking the unit value of this indicator for the least profitable crops. "The growth score" of variation coefficient can estimate the difference in risk cropping rate.

In this case, it is advisable to calculate the score of the integral assessment of the efficiency of the use of arable land as the ratio of the first indicator to the second.

As a result of the calculations, it was found that the highest efficiency of arable area cultivation is achieved in the high-profit and medium-risk products such as potatoes and sugar beets (integral evaluation scores—3.81 and 2.90). The lowest one—high-risk cultivation of field vegetables and soybeans, that generates the smallest income per 1 hectare (integral assessment scores—0.64 and 0.52).

At the same time, it is not correct to conclude that the cultivation of rapeseed, vegetables, and soybeans is not economically feasible, because analysis should be complemented by market research prognosis for relevant products. However, it is quite reasonable to use the proposed approach in the process of perspective forming of arable areas structure.

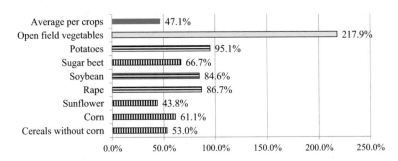

Fig. 4 The profit variation coefficient per 1 hectare of crop areas for the period from 2009 to 2020, received by the Stavropol Territory enterprises. *Source* Compiled by the authors based on [7]

Table 1 Integral assessment of the efficiency of the use of arable land in the cultivation of various crops

Agricultural crop (risk group)	Score		
	Profit growth per 1 ha	Growth of coefficient of variation	Integral assessment
Potatoes	8.3	2.17	3.81
Sugarbeet	4.4	1.52	2.90
Sunflower (min)	2.5	1.00	2.49
Corn (min)	2.0	1.39	1.46
Cerealswithoutcorn (min)	1.6	1.21	1.34
Rape	1.3	1.98	0.66
Openfieldvegetables (max)	3.2	4.97	0.64
Soybean	1.0	1.93	0.52

Source Compiled by the authors

Using this approach to study the innovation effectiveness is required to consider the additional risks in various innovative technologies practice applying. Most often, the riskiness of innovation is recommended to be carried out using expert assessment, in particular, the assessment algorithm "An Intellectual analysis technology" by Cedano and Granados [11]. In the publications of Russian scientists, the assessment of improving technology efficiency in crop production provides taking into account the size of investments, volume, cost, and product price without risks [12] or using the project analysis methodology with risk assessment [13]. The proposal to take into account the unsustainable results of crop cultivation in the process of innovation effectiveness assessment is an actual way to improve the previously offered methods. That would provide more reliable results and effective management of innovation activity in agriculture.

4 Conclusion

Planning the structure of crops areas (their sort and (or) hybrid composition, other innovative changes) for a comparative assessment of crop production economic efficiency (profitability and (or) profit per 1 hectare of arable area) in unsustainable farming zones should be supplemented by the variation coefficient as a degree of their uncertainty and variability.

This method can be used as the first step in determining ways of innovative development in agribusiness. Decision-making level of innovative technologies validity may be increased in case expert assessment combination with results of comparative crops production efficiency according to current risks of their cultivation in a particular enterprise or farming area.

References

1. Bannikova N, Telnova N, Markarova V (2021). Innovation activity in agriculture and the issues of its assessment. Western Balkan J Agricult Econ Rural Develop 3(1)
2. Bannikova NV, Telnova NN, Izmalkov SA, Kostyuchenko TN et al (2019) Economic assessment of the production of organic products of vegetable and fruit growing in the Stavropol Territory: monograph. Stavropol. p 376
3. Cedano KG, Granados AH (2021) Defining strategies to improve the success of technology transfer efforts: an integrated tool for risk assessment
4. Chepurko VV (2019) Methodical approach to assessing the zonal-sectoral risk of agricultural production. Scientific Notes of the Crimean Engineering and Pedagogical University 1(63):206–213
5. Ivanyo YM, Petrova SA, Polkovskaya MN (2018) Probabilistic assessment of recurrence of droughts and determination of risks of agricultural production. Bull Irkutsk State Tech Univ 22(4), 73–82
6. Kostyuchenko T, Telnova N, Orel Y, Izmalkov S, Baicherova A (2020) Forecasting the efficiency of technological development by the example of crop research. In: Growth poles of the global economy: emergence, changes and future perspectives. Lecture notes in networks and systems, Plekhanov Russian University of Economics. Luxembourg, p 835–846
7. Novickytė L (2019) Risk in agriculture: an overview of the theoretical insights and recent development trends during last decade – a review. Agricultural Economics, Czech 65:435–444
8. NRUHSE (2019) Indicators of innovative activity: statistical digest. National Research University, Higher School of Economics, Moscow, Russia, p 376
9. Peltonen-Sainio P, Venäläinen A, Mäkelä HM, Pirinen P, Laapas M, Jauhiainen L et al (2016) Harmfulness of weather events and the adaptive capacity of farmers at high latitudes of Europe. Climate Res 67:221–240
10. Sannikova MO, Providonova NV, Sharonova EV (2020) The influence of technical efficiency and weather risk on crop production in Russian agriculture. In: IOP conference series: materials science and engineering, vol 753, (7), 5 March, 072023
11. Siptits SO (2019) Models of adaptive behavior of agriculture as a tool to reduce the climatic risks of the development of the agricultural sector of the economy. Sci Works Free Econ Soc Russia 216(2):227–235
12. Stulec I, Petljak K, Bakovic T (2016) Effectiveness of weather derivatives as a hedge against the weather risk in agriculture. Agricult Econ (Czech Republic) 62(8):356–362
13. Tiedemann T, Latacz-Lohmann U (2013) Production risk and technical efficiency in organic and conventional agriculture - the case of arable farms in Germany. J Agricult Econ 64(1):73–96.

Peculiarities of Organizational and Economic Interaction of Organizations in the System of Product Subcomplexes

Lidia A. Golovina◉ and Olga V. Logacheva◉

Abstract The production and commercial activities of organizations within the modern food subcomplexes are becoming increasingly complex, which leads to the need to find ingenious measures to ensure their survival and sustainable development. The accelerating development of competition and market relations in the country's economy and the development and application of the project-based approach on the part of the authorities have determined the need to change the organizational and economic conditions in the system of interaction of organizations of food subcomplexes. The paper highlights the critical conditions ensuring the balance of interaction between organizations of food subcomplexes in the context of their categories (entering, fund-forming, functionally related, and interacting with the product subcomplex) using the motives that determine the development of the system of economic relations. The authors prove that positive changes in the system of food subcomplexes have led to the provision of the country's food security and the transformation of the model of economic interaction from import-substituting to export-oriented.

Keywords Interaction of organizations · Food subcomplex · Efficiency · Sustainable development · Project-based approach · Digital transformation

JEL codes Q01 · Q13 · Q16

L. A. Golovina (✉) · O. V. Logacheva
Federal Research Center for Agrarian Economy and Social Development of Rural Areas – All-Russian Research Institute of Agricultural Economics, Moscow, Russia
e-mail: golovina.lidia@yandex.ru

O. V. Logacheva
e-mail: ro22ashka@mail.ru

© The Author(s), under exclusive license to Springer Nature Singapore Pte Ltd. 2022
E. G. Popkova and B. S. Sergi (eds.), *Sustainable Agriculture*,
Environmental Footprints and Eco-design of Products and Processes,
https://doi.org/10.1007/978-981-16-8731-0_41

1   Introduction

A distinctive feature of the Russian economy is the presence of developed inter-industry complexes. The universality of their functioning lies in the organic interaction of industries and spheres integrated by economic relations to perform national economic tasks based on accelerated development of scientific and technological advances, reduction of the time necessary to promote products from production to sale, and the optimal use of production resources and sustainable development. In this case, the critical factor in ensuring effectiveness is a project-based approach to management on the part of the government.

During the transition of the Russian economy to a capitalist mode of economic management (until 2006), the state administration gave preference to the development of the oil and gas and defense complexes. Over the past decade, the government switched its focus to the agro-industrial complex (AIC). The structure of AIC has several food subcomplexes, among which the most importance is given to grain, sugar beet, potato, fruits and vegetables, meat, dairy, fat and oil, etc. [2]. Each subcomplex, taking into account its specifics, combines the interaction of the relevant specialized agricultural organizations, enterprises specializing in the manufacturing of production means and technological upgrading, research institutions, procurement, and enterprises specializing in the storage and transportation of raw materials and processing and marketing of finished products. In this case, the primary target function of food subcomplexes is to satisfy the population's needs in the relevant types of products. The efficiency of the activity of these subcomplexes is mainly due to the specific conditions of functioning of particular industries [6].

2   Materials and Methods

The methodological basis of the research is the analysis of the economic activities of organizations of certain product subcomplexes. The analysis is based on calculations in various directions and mechanisms of agribusiness. The research uses the methods of statistical analysis, monographic and abstract-logical methods, and method of rating.

3   Results

Currently, food subcomplexes are the leading producer of GDP in some Russian regions. In Russia, they include more than 80 industries, which account for about one-third of GDP, fixed production assets, and the number of employees. The complex solution of problems in implementing measures of agrarian policy stipulated by state projects allowed the agribusiness to get a serious impetus for the development of agricultural organizations and organizations of food and biological profile. Moreover, it positively influenced fund-raising organizations, organizations specializing in agricultural machinery, organizations producing chemical fertilizers and plant protection products, organizations constructing agricultural facilities and engineering infrastructure, and organizations providing veterinary and sanitary-epidemiological services. The high level of subsidies attracted major investors to the agricultural market; there was a steady growth in production indicators of economically interrelated segments of the AIC [10].

Effective interaction between government agencies and business entities is achieved through the step-by-step implementation of the following project management principles:

- Stability (provided through the implementation of the most relevant areas and state support to agricultural production);
- Consistency (consists in the effective combination of measures to support agricultural production with effective regulation of markets and development of rural areas);
- Co-financing (consists of reasonable co-financing of agricultural production from the federal and regional budgets);
- Public–private partnership (combining the efforts of government and business to achieve the set goals) [12].

Figure 1 presents the multistage configuration of directions for developing the agrarian sector on the project basis for 2006–2019.

I. Priority national project "Development of the agro-industrial complex"	II. State Program of Agricultural Development and Regulation of Markets for Agricultural Products, Raw Materials, and Food for 2008-2012
3 directions: - Accelerated development of cattle breeding; - Stimulating the development of small forms of farming in the AIC; - Providing affordable housing for young professionals in rural areas.	**5 directions:** - Sustainable development of rural areas; - Creation of general conditions for the functioning of agriculture; - Development of priority agricultural sub-sectors; - Achieving the financial sustainability of agriculture; - Regulation of the market for agricultural products, raw materials, and food.
III. The current state program of agricultural development and regulation of markets of agricultural products, raw materials and food	IV. State program "Integrated development of rural areas"
9 directions, the main ones are: - Development of AIC industries; - Providing conditions for the development of AIC; - Export of agricultural products; - Sustainable development of rural areas; - Scientific and technical support for the development of the sectors of AIC; - Development of land reclamation of agricultural lands in Russia; - Ensuring general conditions for the functioning of the agroindustrial complex.	**5 directions:** - Creating conditions for affordable and comfortable housing for the rural population; - Development of the labor market in rural areas; - Creation and development of infrastructure in rural areas; Analytical, normative, and methodological support of integrated development of rural areas; - Ensuring the implementation of the state program of the Russian Federation "Integrated development of rural areas."

Fig. 1 Projects for the development of the AIC implemented in 2006–2020. *Source* Compiled by the authors

The nodal vector of economic interaction in the system of implementation of these directions is large-scale integration and a non-ordinary approach to the formation of innovative knowledge.

The complexity of the problem lies in the regulation of the production and commercial relations, as well as in ensuring a connection of agricultural organizations with fund-forming and functionally linking food subcomplex, especially with those delivering products to the consumer. The implementation of measures of state support based on the project-based approach contributed to the construction of an intersectoral vertical, which ensures the interaction of all actors at all levels of government and accountability for achieving the established results [3].

It is not easy to consider the effectiveness of the used measures in combination. However, if we evaluate such indicators as the net profit rate and the return on funds of organizations of certain industries, it is essential to note that agricultural organizations are the leaders (Fig. 2).

Thus, from 2013 to 2015, the rate of net profit in agricultural organizations increased from 6.1 to 10.1%. Nevertheless, its value decreased to 7.6% by 2019. In organizations specializing in the production of food products, the rate of return

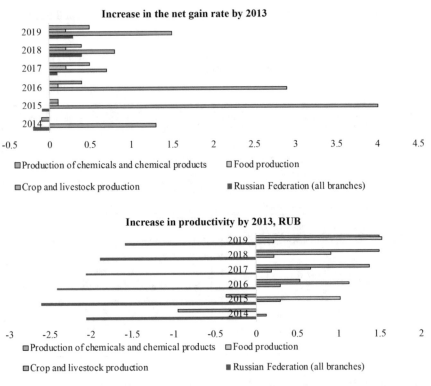

Fig. 2 The rate of increase (decrease) of net profit and productivity of funds in the Russian Federation and individual organizations of the AIC to the level of 2013. *Source* Compiled by the authors based on [8]

for the studied period was in the range of 1.4–1.6%. In organizations producing chemicals and chemical products, the rate of return increased from 1.9 to 2.4%. The decrease of the net profit rate in agricultural organizations since 2016 can be explained by strong fluctuations in the exchange rate of the ruble to foreign currencies, which significantly affected the final result. For the totality of organizations in all sectors of the Russian economy, the rate of net profit did not exceed 2.2%. In some years, its growth rate had negative values against the level of 2013.

One of the characteristics of the stable state of the agricultural sector of the economy is the sustainable development of all its food subcomplexes. This development depends primarily on the stable state of the economic environment and the outcome of an active and effective response to internal and external changes. The key indicator of sustainability is the investment coverage ratio (long-term financial independence ratio), which gives an idea of what share of assets is financed from sustainable sources (long-term liabilities and equity). Deviation from the normative values (0.7–0.9) indicates the inability to repay current accounts or insufficient use of sources of debt financing. In the agricultural sector, the investment coverage ratio is at the upper limit of the normative value (Fig. 3).

The value of the investment coverage ratio in organizations producing food and chemicals is worse than in agricultural organizations due to the low share of equity and long-term borrowed capital.

In our view, the balance in the consistency of the actions of organizations within the product subcomplex is ensured by the observance of conditions determining the development of the system of organizational and economic interaction. One of the most critical principles in scientific methodology is the principle of determination. The methodology of this principle allows designating a view of the relationship of all existing phenomena and processes, which, in economic research, is understood as "the objective conditionality of the state of the economy and trends in the development of its certain processes" [11]. Figure 4 shows the most significant determinants of economic interaction between organizations of product subcomplexes in current functioning conditions.

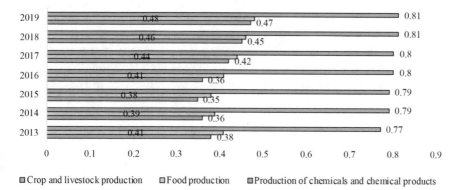

Fig. 3 Dynamics of change in the coefficient of investment coverage in certain AIC organizations for 2013–2019. *Source* Compiled by the authors based on [8]

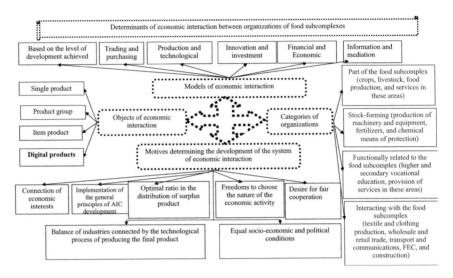

Fig. 4 Determinants of economic interaction between organizations of food subcomplexes. *Source* Compiled by the authors

New challenges in the economic interaction of organizations of food subcomplexes (lower prices, changes in consumer demand, natural and biological phenomena, and state support) increase the importance of two basic postulates of connecting their interests:

(1) Adherence to the principles of mutual economic benefit;
(2) In the process of interaction, the economic interest of a particular organization is realized regardless of the economic interest of the other.

The first case is marked with the most favorable conditions for the functioning of each party, mutual control over the activities, and the alignment of interests. In the second case, as a rule, there is a one-sided relationship of interests, causing unequal interaction between the parties. In turn, this inevitably sets the stage for the subordination of the economic interest of one subject to the other, weakening the desire to motivate high economic results in each of the parties [13].

The most important vector of modern economic interaction of organizations in the system of food subcomplexes is to ensure the country's food security.

Food security is determined by the level of self-sufficiency in basic agricultural products of domestic production. According to the estimates of the Ministry of Agriculture of Russia, in 2019, the thresholds of food independence (self-sufficiency) of the Russian Federation have been achieved or exceeded: grain—155.5%; sugar—125.4%; vegetable oil—175.9%; meat and meat products—96.7%. Self-sufficiency in the following products remains below the threshold of the Food Security Doctrine: milk and dairy products—84.4%; vegetables and gourds—88.4%; fruits and berries—39.5%. Self-sufficiency in potatoes remained at the level of last year—94.9% [1].

Fig. 5 Dynamics of imports and exports of agricultural products for 2013–2019, $ billion. *Source* Compiled by the authors [5]

The successful implementation of the federal project "Export of Agro-Industrial Products" changed the model of economic interaction from import-substituting to export-oriented. Russia has entered the international market for agricultural products and foodstuffs as the largest exporter. The agro-industrial complex of the country has taken the leading place among the non-resource exports [7]. Figure 5 shows the dynamics of imports and exports of agricultural products for 2013–2019.

Even though in 2019 the gap between exports and imports in value terms is $4.3 billion, the value of imports decreased by 31%, and the value of exports increased by 52% compared to the level of 2013.

4 Conclusion

It should be noted that the large-scale transformation of production processes has deeply affected the system of functioning of food subcomplexes. The accelerated development of the latest technologies in food subcomplexes is directly linked to implementing the departmental project "Digital Agriculture," which was successfully launched in the agricultural sector. Innovative business models created on end-to-end information and communication platforms provide a close and direct connection between the consumer and the manufacturer, excluding the most marginal segments from the turnover—retail.

Thus, the modern agro-industrial complex of Russia claims to be the main site for demonstrating the technological revolution's results: the robotic technology of Industry 4.0 translates almost all agricultural machinery into uncrewed mode. The Internet of Things and the Internet of Everything unite the entire production chain in a single ecosystem: from creating new fertilizers and animal and plant species to the production of functional products that dramatically improve human properties [4].

The digitalization potential in the system of economic interaction of organizations of product subcomplexes is aimed primarily at reducing target conflicts in production and management. Positive results are achieved by applying innovative technological methods by the highly specialized and targeted use of certain resources to ensure economic growth. At the same time, the multifaceted application of digital transformation provides the basis for sustainable development in the case of a simultaneous increase in productivity in the production process [9]. An incomplete list of factors that will affect the agricultural sector in Russia and the world in the future includes new information, bio- and nanotechnologies, significant changes in the value-added distribution chain, changing consumer preferences and attitudes, and climate change [4].

Acknowledgements The paper was prepared within the framework of the state task of the Ministry of Science and Higher Education of the Russian Federation No. 0569-2019-0045 "To develop the conceptual framework and improve the mechanisms of economic interaction between organizations of agro-industrial complex in terms of structural and technological transformation," stage 3—2021 "To improve the mechanisms of economic relations of organizations of agro-industrial complex in the system of inter-branch interaction."

The authors of this research would like to express their gratitude to the Acting Director of Federal Scientific Center of Legumes and Groat Crops, Doctor of Economics, Professor of the Russian Academy of Sciences, Andrey Aleksandrovich Polukhin, for his advice during this study.

References

1. Government of the Russian Federation (2019) National report on the progress and results of the state program of agricultural development and regulation of markets of agricultural products, raw materials, and food in 2019. Retrieved from https://mcx.gov.ru/upload/iblock/98a/98af7d 467b718d07d5f138d4fe96eb6d.pdf (Accessed 15 February 2021)
2. Altukhov AI (2019) Modern internal and external threats to development of the agrarian sphere of economy. Econ Agricult Russia 12:2–10. https://doi.org/10.32651/1912-2
3. Bespakhotnyy GV (2018) Program-target planning and project management in agriculture. Models, Syst, Netw Econ, Eng, Nat Soc 2(26):3–15
4. Chulok A (2019) The agro-industrial complex of the future. A look at agriculture through the prism of big data analysis. Agroinvestor. Retrieved from https://issek.hse.ru/news/248112278. html (Accessed 17 February 2021)
5. Federal Customs Service of the Russian Federation (nd) Official website. Retrieved from https://customs.gov.ru/ (Accessed 25 February 2021)
6. Golovina LA, Logacheva OV (2019) Reproductive situation in agricultural organizations: perspectives from the Orel region. In: IOP conference series: earth and environmental science, vol 274, pp 012018. https://doi.org/10.1088/1755-1315/274/1/012018
7. Golovina LA, Logacheva OV (2021) Financial aspects in the system of economic interaction of grain product subcomplex organizations. Ration Remuner Labor Agricult 2:20–28
8. Handbook of Industry Financial Indicators (2020) Official website. Retrieved from https://www.testfirm.ru (Accessed 10 February 2021)
9. Lyasnikov NV, Romanova YuA (2019) Global challenges and threats of agrarian sector development in Russia. Food Policy Secur 6(2):85–96. https://doi.org/10.18334/ppib.6.2.41386
10. Panin AV (2018) On priorities in the system of scientific and technological development of agricultural production. In: Semenova EI, Rodionova OA, Polukhin AA, Savkin VI, Ponitkina NB, Golovina LA, Logacheva OV (eds), Parity relations in the agricultural sector of the economy: Scientific and practical support and implementation mechanisms, vol 1, pp 271–277. Moscow, Russia

11. Panin AV, Lidinfa EP, Baranova SV, Efremov IA (2020) Strategy of socio-economic development and formation of regional personnel policy. Bull Agrarian Sci 3(84):116–126. https://doi.org/10.17238/issn2587-666X.2020.3.116
12. Poddueva IS (2015) Assessment of the effectiveness of the implementation of the target program "Development of agriculture and regulation of markets of agricultural products, raw materials, and food for 2013–2020 years" on the example of a municipal district of the Novosibirsk Region. Molodoy Uchenyy, 21(101), 434–438. Retrieved from https://moluch.ru/archive/101/22927/ (Accessed 9 January 2021)
13. Polukhin AA, Yusipova AB (2019) Method of economic justification of projects of development of reserves of increase of efficiency of the agricultural organizations. J Econ Entrepreneurship 9(110):1031–1035

Stock Analysis as an Element of Financial Flows Optimization in Enterprises of the Agricultural Sector of Economics

Anna A. Babich⬤, Anna A. Ter-Grigor'yants⬤, Elena S. Mezentseva⬤, and Tatiana A. Kulagovskaya⬤

Abstract <u>Purpose</u>: At the current stage of development of Russian economics, the problem of efficient management of financial flows, which make a significant impact on the availability of financial resources in the companies is one of the major issues for companies of the agricultural sector of economics. Management of financial flows is a key aspect in activities of companies of the agricultural sector as it regulates all spheres and directions of functioning of each economic agent. The main purpose of the research is to develop models and methods of financial flows management taking into account analysis of economic agents' financial resources immobilized into stock. The article deals with methodological instruments required for financial flows management of agricultural entities and based on available methods of stock analysis and optimization. Correct and efficient organization of cash flows ensures reasonable and smooth economic activities and serves as the prerequisite for achieving sound total financial performance results by enterprises of the agricultural sector in present-day conditions. <u>Design/methodology/approach</u>: Methodological framework of the research includes general scientific and specific cognitive methods. Solutions to research issues were specified using a set of additional methods: economic and statistical methods, e.g., comparison, grouping, methods used in the system and functional analysis, analytical modeling, systematicity, integrity, logical modeling, etc. <u>Findings</u>: Results of the research show that adaptive application of developed methods of financial flow optimization shall help to increase the efficiency of activities carried out by economic agents of the agricultural sector under an unstable market environment.

A. A. Babich · A. A. Ter-Grigor'yants (✉) · E. S. Mezentseva · T. A. Kulagovskaya
North Caucasus Federal University, Stavropol, Russia
e-mail: ann_ter@mail.ru

A. A. Babich
e-mail: delightful@bk.ru

E. S. Mezentseva
e-mail: mez.elen_@mail.ru

T. A. Kulagovskaya
e-mail: kulagovskaya@mail.ru

Keywords Stock · Financial flows · Efficiency · Management · Working capital · Analysis

JEL Codes B26 · Q14

1 Introduction

Financial resources and their equivalents are an important element in the system of economic entities' resources turnover and form the basis for the increase in their financial stability and financial solvency. Currently, methods of efficient management of financial flows are applied as a major economic entities' bankruptcy prevention measure [1].

Development of the efficient financial flow management system of economic entities of the agricultural sector plays an important role in reaching financial stability and sustainable development of agricultural companies as a prerequisite for ensuring competition of companies in the agricultural sector in Russian and global markets in the context of limited available financial resources. Analysis of company financial flows means analysis of a set of figures, formation of financial flows within the company detection of major tendencies, and patterns to find means for the subsequent increase in efficiency of financial flows management.

It is necessary to point that the tendency to evaluate the performance of economic entities and the efficiency of invested capital with the help of financial flows has been traced lately in world practice. Analysis of financial flows of agricultural sector enterprises means analysis of a set of figures, formation of financial flows within the enterprise detection of major tendencies, and patterns to find means for the subsequent increase in efficiency of financial flows management.

In their turn, financial flows in cooperation with material flows serve as a controlled element and shall be subject to major laws of the economic system. Financial flows management is carried out based on the application of certain models and methods selected by the company as basic models and methods due to market conditions, specific features of company activities, possibility to regulate them by way of processes that are appropriate from any point of view. These flows are also a controlled element and shall be governed by uniform laws of the economic system.

Such coordination characterized by integral operational principles in any activities of the economic agent helps to reach synergies and increase the efficiency of company performance.

Currently, influence on material flows through management of financial resource flows is a promising approach that helps to target the financial aspect of company performance.

Thus, activities of agricultural entities depend mainly on the territorial organization of the agricultural production, its seasonal patterns, biological nature of resources in use, and output products. Due to the fact that stock is one of the least liquid assets, special attention shall be paid to the rationalization and restructuring of its' management system. Proper management and optimization of commodity stock of an agricultural company shall help to find working capital concealed within the company.

The task of analysis of financial flow movement is to evaluate the ability of the company to accumulate financial resources in the amount and within the period required for covering planned expenditure. Analysis of financial flows is required to detect reserves that the company uses to generate resources required for the purchase of additional assets with the aim of subsequent development, settlement of debts, and financing company activities.

The financial flows of an agricultural company represent a complex field for analysis. The main tasks of financial flow analysis in companies of the agricultural sector are as follows: define tendencies and trends in the development of company financial flows; analyze ways of formation of financial flows, evaluate the degree of their reasonable use, as well as to detect and prevent possible problems with company financial solvency and prerequisites for company bankruptcy and find reserves and ways of enhancing money turnover.

Many Russian and foreign academicians have conducted researches on the development of fundamental theories and methods of financial flows management.

Conceptual framework of the essence, emergence, and movement of financial flows, as well as methods of efficient financial flow management have been widely described in books on economics. Russian scientists [4, 5, 13] and foreign economists [2, 3, 7–9, 12, 14, 15, 17–19] have analyzed financial flows.

However, analysis of available scientific publications shows that applied specification of agricultural entities is not taken into account and there are practically no researches on the interaction of a material flow and assistant financial flow in companies of the agricultural sector in a part of materials related to theoretical and practical aspects.

Thus, works related to the development of conceptual and methodological frameworks specifying cooperation and interrelation of these flows are quite topical.

The purpose of the research is to develop models and methods of financial flow management taking into account analysis of financial resources immobilized into the stock of material resources to increase efficiency and competitiveness of an economic agent in the agricultural sector of economics.

According to the purpose of the research, the following tasks are defined:

- analyze theoretical aspects of financial flows produced by cash flows;
- clarify the definition of a financial flow for agricultural entities;
- analyze the possibility of financial flow management applying models and methods used in stock management under the influence of various factors, particularly, seasonal patterns in manufacturing and sales of products;
- evaluate offered models and methods.

2   Materials and Method

The economic efficiency of an agricultural company is based on complex analysis that must be carried out using the system of comprehensive factors covering the state and development of the evaluated company.

Drawn generalizations of practical experience show that currently, most agricultural economic entities do not pay proper attention to financial flow management, which leads to constant unbalanced incomings and expenditures of financial resources. Due to this fact, issues related to the development of the system of factors required to evaluate the efficiency of management of financial flows generated by economic entities of the agri-industrial complex become important within the framework of current economic formation. The research devoted to major disadvantages of financial flow management helped to form the most frequent financial flows:

- surplus of commodities in warehouses that helps to increase sales volumes not covered by financial resources;
- passive policy concerning relations with debtors, i.e., active measures aimed at reducing average debt recovery period are not taken and modern forms of payment are not used;
- temporarily available cash is underutilized.

The major purpose of activities aimed at increasing efficiency of agricultural company management is synchronous incomings and expenditures of financial resources as well as enhanced cash flow [11].

In the following research, a financial flow means the direct flow of financial resources inside/outside the company economic system generated with the aim of constant movement of other flows as well as accumulation of stock required by an economic agent.

Visual information given in Fig. 1 helps to acknowledge that financial flows management shall be regarded as a consecutive process that includes setting and accomplishing the following tasks:

- find resources and define the number of financial incomings, as well as the possibility to increase them;
- specify amount and priority of payments to be made within the certain period;
- adjust expenses according to resources required to cover them for the due fulfillment of financial obligations;
- specify the amount of "temporarily available cash" and its efficient application.

Efficient and reasonable management of company working capital serves as a major source of positive financial flows and guarantees of company financial solvency.

Efficient formation and application of working capital are based on certain specific aspects of management of its basic elements: stock, A/R, and financial resources.

Optimization of warehouse stock means a reduction of the amount of warehouse stock owing to proper purchase planning. As far as agricultural commercial activities

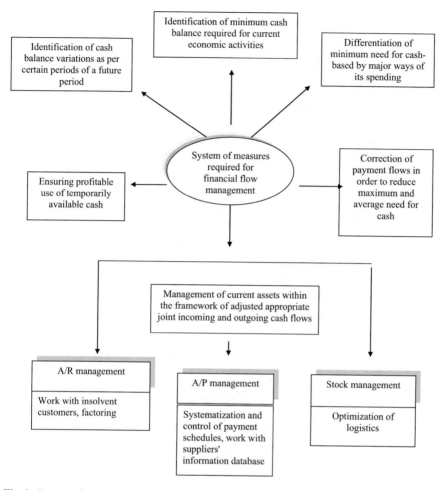

Fig. 1 System of measures required for efficient financial flow management. *Source* Developed and compiled by the authors

are concerned, the amount of stored commodities is inversely related to the amount of company income. Basically, commodities stored in a warehouse form frozen working capital and occupied space. There is a risk of failure to use a surplus of materials that shall lead to an inevitable overstock of a warehouse with the surplus.

SEC Collective Farm Pobeda (an agricultural company located in Stavropol Territory) serves as an example for this research as ABC and XYZ methods of stock management are adapted within the framework of financial flows produced by flows of commodities. Thus, financial flows are adjusted due to the proper system of stock management.

As known, ABC is the simplest method of stock management and involves dividing stock into three categories [6]:

- category A: the most important stock that includes 20% of the stock and 80% of sales;
- category B: important stock that includes 30% of the stock and 15% of sales;
- category C: less important stock that includes 50% of the stock and 5% of sales.

The specific aspect of this method lies in the fact that a relatively small amount of stock can help to acquire the largest amount of financial resources.

Several methods can be applied to define borderlines of item groups A, B, and C [6, 10, 16], i.e., empirical, differential, graphical, and analytical.

The analytical method was used in the following research.

When analytical methods are applied in calculations, the sequence is as follows:

1. Amount of stock N is rated 0–1 and x argument is introduced.
2. Functional relation $y = f(x, a_p)$ is introduced, where a_p are coefficients.
3. The least-squares method is applied to define a_p coefficients.
4. When defining a_p coefficients the following terms shall be observed: 1. if $x = 0$, $y = 0$; 2. if $x = 1$, $y = 1$. Thus, the number of equations that are used to define a_p coefficients can be reduced.
5. In order to define the location of point O, Lagrange's theorem (mean value theorem) is used. According to the theorem,

$$f(x) = \frac{f(b) - f(a)}{x_b - x_a}, \tag{1}$$

where f'(x) means derivative of f(x) at the tangent point;
f(b), f(a) mean values of f(x) function at start and endpoints.
Then the following item nomenclature is introduced:

$$N_A = x_A N, \tag{2}$$

which divides items into two groups.

6. Then a new coordinate system is introduced; x_A-axis and y-axis (x_A) are reference points.

Results of XYZ analysis help to divide material resources into classes following steady demand and possible consumption forecasts.

- category X: resources are characterized by a steady consumption rate and high accuracy of forecasts. The coefficient of variation is from 0 to 10%;
- category Y: resources are characterized by a rather high consumption rate due to certain factors, e.g., seasonal fluctuations, offsets, excess or deficiency of products, and medium accuracy of forecasts. The coefficient of variation is from 10 to 25%;
- category Z: resources are characterized by low consumption rate and low accuracy of forecasts. The coefficient of variation is more than 25%.

The specific aspect of the XYZ method is the coefficient of variation for the demand that helps to define the degree of integrity for the aggregate and similarities of its values. This coefficient apart from, say, standard deviation (σ) is a relative value, i.e., it does not depend on absolute values and average value (x):

$$V = \frac{\sigma}{x} \cdot 100. \tag{3}$$

Information about forecasts related to the amount of consumed stock for one supply shall be used to calculate the coefficient of variation. Thus, the amount of consumed stock for one period ahead shall be forecast. Then variability of the amount of remaining surplus shall be defined based on the calculated coefficient of variation.

The coefficient of variation that outlines the boundaries of classes shall be calculated based on ABC method of grouping material resources.

Integration of the results of both methods in the ABC-XYZ matrix is a reasonable solution since comparison of observed results shall help to develop useful planning and control tools required both for the supply system in general and for the management of the financial flow in particular.

3 Results

Calculations made in this research with the use of the ABC method helped to collect final analytical data related to the distribution of commodities by SEC Collective Farm Pobeda as given in Table 1.

Following Table 1, 80% (6,520 thousand rubles) of sales account for sales of 34.7% of commodity items; 15% of sales account for sales of 16.1% of commodity items. The rest 49.2% of commodity items cover only 5% of total company sales.

Thus, only half of the product range offered by SEC Collective Farm Pobeda produces positive cash flow. With limited financial resources stock of 479 (55.5%) commodity items are not advisable as a result of their small contribution to total sales volume and, consequently, to incoming cash flows.

Table 1 Results of ABC analysis of SEC Collective Farm Pobeda for Q4 2019

Description	UOM	Groups			Total
		A	B	C	
Sales of stored commodities	thousand rubles	6,520	1,222.5	407.5	8,150
Sales pattern of stored commodities	%	80.0	15.0	5.0	100.00
Quantity of stored commodity items sold	pcs	338	157	479	974
Structure of stored commodity items sold	%	34.7	16.1	49.2	100.00

Source Developed and compiled by the authors

Table 2 Results of the XYZ analysis of SEC Collective Farm Pobeda for Q4 2019

Description	UOM	Groups			Total
		X	Y	Z	
Sales of stored commodities	th. rubles	5395.3	2241.3	513.5	8150
Sales pattern of stored commodities	%	66.2	27.5	6.3	100
Quantity of stored commodity items sold	pcs	343	216	415	974
Structure of stored commodity items sold	%	35.2	22.2	42.6	100

Source Developed and compiled by the authors

Integrated results of the XYZ analysis given in Table 2 allow us to come to the following conclusion: 343 commodity items (35.2%) out of 974 commodity items are in steady demand and their sales volume is 66.2% (5,395.3 thousand rubles); 216 commodity items (35.2%) are characterized by medium accuracy of forecasts and account for 27.5% of sales (2,241.3 thousand rubles); 415 commodity items (42.6%) are in occasional demand and account for 6.3% of company sales (513.4 thousand rubles).

Thus, a surplus of 415 commodity items shall be minimized or sold out. The surplus of 343 commodity items shall be limited.

Moreover, the company has to revise its product portfolio policies or find new sales markets for commodities that currently go unsold.

Following Table 3, commodities belonging to groups AX, BX, AY are shipped in the amount equal to 5,599.1 thousand rubles (68.7%) for 379 commodity items (38.9%). Seamless sales of such commodities require a sufficient amount of their stock.

Commodities belonging to groups AZ, BY, CX include 159 commodity items (16.3%) shipped at the amount equal to 1,255.1 thousand rubles (15.4%). Shipment in the amount equal to 1,295.9 thousand rubles (15.9%) is covered by sales of 436 commodity items (44.8%). The stock of these commodities in warehouses is not advisable as the demand is occasional and their rate in total shipment is really small.

Table 3 Result of the ABC–XYZ analysis of SEC Collective Farm Pobeda for Q4 2019

Description	UOM	Groups			Total
		AX, AY, BX	AZ, BY, CX	BZ, CY, CZ	
Sales of stored commodities	th. rubles	5,599.1	1,255.1	1,295.9	8,150
Sales pattern of stored commodities	%	68.7	15.4	15.9	100
Quantity of stored commodity items sold	pcs	379	159	436	974
Structure of stored commodity items sold	%	38.9	16.3	44.8	100

Source Developed and compiled by the authors

Prevailing sales of commodity items with occasional demand that have the least sales rates are mainly based on the renovation of commodities in the agricultural company as well as on the search for new sales markets. Most similar commodities are sold upon orders and as a result amount of commodity items with high sales rates is quite small.

Calculations, made by the authors, show that the use of the ABC and XYZ stock management methods in SEC Collective Farm Pobeda for two months helped to reduce the amount of stock by 11% in terms of money and increase the sales rate by 8%.

4 Conclusions

Analysis of specific aspects related to the functioning of companies in the agricultural sector of economics of the Russian Federation shows that the formation, use, and management of financial flows generated by these companies are influenced by certain factors. This adapted system of efficient financial flows management is offered. This system is based on the management of company working capital and its optimization, budget forecasts made using both standard and rolling planning methods.

Developed methodical approach to analysis and optimization of financial flows generated by economic entities helps to carry out clear control over sustainability of company income and expenses, to increase company financial and production flexibility, to reasonably operate company resources, to carry out control over company debentures as well as to increase company liquidity and financial solvency.

Thus, according to the results of this research aimed at increasing efficiency of agricultural company financial flows management, it is advisable to:

- implement analysis and management at production facilities as an integral part of their activities, calculate system of values for financial flows as factors required to measure financial stability and financial solvency;
- analyze industry patterns of financial flows and use them in agricultural company analysis and management activities;
- define current, including latest information required to provide timely and full information support for analysis of company financial flows.

References

1. Altukhov AI (2003) Russia's accession to the WTO and problems of agricultural development. Agri-food policy and Russia's accession to the WTO. Russian Academy of Agricultural Sciences, Nikon readings, Moscow, pp30–31
2. Begleiter D (2001) A Free Cash Flow (chemical companies and free cash flow). http://www.findarticles

3. Berndt R (2020) Risk Management im Marketing [Handwnrterbuch des Marketing]. pp. 2247–2254.
4. Blank I (2004) Basics of financial management, 2(2), as am. and app. Elga, Nika-Center, Kyiv, p 624
5. Bocharov VV (2008) Financial analysis: guide, 4th edn., as am. and app. Piter, Saint Petersburg, p 218
6. Brodetsky GL, Gusev DA (2012) Economic and mathematic methods and models in logistics. Optimization procedures. Akademiya Publishing Center, Moscow, p 143
7. Bromwich M (1992) Financial reporting, information and capital markets. Pitman Publishing,
8. Domar E (1957) Essays in the theory of economic growth. New York
9. Hachmaister D (1998) The discounted cash flow as the increase in enterprise value [Der Discounted Cash Flow als der Unternehmenswertsteigerung]. Francfurt am Main, p 308
10. Lychkina NN (2011) Simulation modeling of economic processes. INFRA-M, Moscow, p 98
11. Melnikov EN (2011) Contrastive analysis of current models of financial flows management. Audit Financ Anal 4. www.auditfin.com/fin/2011/4/2011_IV_03_11.pdf
12. Mills JR, Yamamura JH (2003) The power of cash flow ratios Electronic resource. Electronic. http://www.findarticles
13. MoiseevaEG (2010) Financial flows management: planning, balancing, synchronization. In: Guidebook for the economist, vol 5, pp 54–59
14. Ross SA, Westerfield RW, Jordan BD (1996) Essentials of corporate finance. Boston, p 528
15. Rushti L (1942) The importanve of depreciation for operations [Die Bedeutung der Abschreibung fuer die Betrieb]. Berlin
16. Sergeev VI (2014) Management of supply chains. Textbook for bachelors. Urait Publishing House, Moscow, p 412
17. Sharp AM, Register CA, Grimes PW (1998) Economics of social issues. Boston
18. Siener F (1991) Cash flow as an instrument of balance sheet analysis: practical significance [Der Cash-Flow als Instrument der Bilanzanalyse: praktische Bedeutung]. Stuttgart, p 350
19. Stevenson HH, Roberts MJ, Grouusbeck HI (1994) New business ventures and the entrepreneur, 4 edn. Boston

Improving the Labor Remuneration System in Agriculture as the Basis for the Food Security of the Country

Aktalina B. Torogeldieva, Kseniya A. Melekhova⊙, and Elena A. Yaitskaya⊙

Abstract One of the most important problems of the agricultural economy is the achievement of food security and increasing economic growth. It is vital to improve the standard of living of the Russian population. The long period of economic reforms had an extremely negative impact on all aspects of the socioeconomic policy of the state. Income inequality, low quality of social infrastructure, and insufficient funding for agriculture lead to the fact that qualified and promising young people often prefer living in urban areas. The paper aims to develop a new labor remuneration system for agricultural workers based on the use of grading. The growth of production can be achieved by using technical and technological innovations in the production process. In turn, the degree of mastering new technologies depends on those employees who will apply modern methods and techniques for mastering innovations in production. The improved system of human capital management in the organization is built in such a way as to stimulate the desire of employees to get an education, improve their skills, work for the company for a long time with a good performance, strive to move up the career ladder, and engage in self-improvement.

Keywords Agriculture · Economic growth · Quality of life · Income · Expenses · Grade · Remuneration

JEL codes Q5 · Q52 · Q59

A. B. Torogeldieva (✉)
K. Sh, Toktomamatova International University, Jalal-Abad, Kyrgyzstan
e-mail: torogeldieva.aktalina@mail.ru

K. A. Melekhova
Altai State University, Barnaul, Russia
e-mail: kschut@mail.ru

E. A. Yaitskaya
Kabardino-Balkarian State Agrarian University Named After V.M. Kokov, Nalchik, Russia
e-mail: ElenaY-1978@yandex.ru

© The Author(s), under exclusive license to Springer Nature Singapore Pte Ltd. 2022   439
E. G. Popkova and B. S. Sergi (eds.), *Sustainable Agriculture*,
Environmental Footprints and Eco-design of Products and Processes,
https://doi.org/10.1007/978-981-16-8731-0_43

1 Introduction

Agriculture is a strategic sector of the economy, which performs one of the most important functions of the country—providing the population with quality food. However, the long period of economic reforms has resulted in the agricultural industry having residual financing and experiencing significant difficulties related to the quality of life of employees. In turn, this threatens the achievement of the main indicators of the Food Security Doctrine and, as a consequence, leads to a decrease in the life expectancy of the country's population. In current conditions, solving the problem of food security is a paramount task. In Russia, it can be achieved through the intensification of available production resources.

2 Materials and Methods

The research object is the quality of life in rural areas.

The research subject includes economic factors and relations affecting the reproduction of labor resources in rural areas.

The information base of the research consists of normative-legal acts, establishing the most important principles of regulation and management of quality of life, statistical data, and proceedings of international and Russian conferences.

The research uses sociological, monographic, abstract-logical, statistical, comparative, and other methods of economic research.

3 Results

Human capital is the main productive force of society [11]. In the economic literature, this term was separated from the concept of labor resources in the mid-twentieth century [1, 15]. The works of foreign and Russian researchers suggest that effective economic growth can be achieved primarily through the use of intensive production factors in post-industrial conditions [4, 5]. Currently, the basic factors of production contribute to economic growth through the use of human capital. This is evidenced by the data presented in Fig. 1.

An important problem of the domestic agro-industrial complex (AIC) is that the growth of production can be achieved by using technical and technological innovations in the production process. However, the degree of mastering new technologies also depends on those employees who will apply modern methods and techniques for mastering innovations in production [7, 8, 14, 17].

Figure 2 shows data on the labor productivity index for the main sectors of the economy for the period from 2008 to 2019, determined according to data provided by state statistics agencies (Fig. 3).

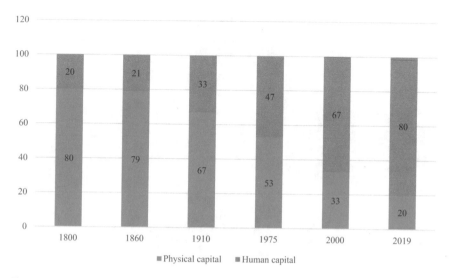

Fig. 1 Changes in the structure of the use of physical and human capital, %. *Source* Compiled by the authors based on [9]

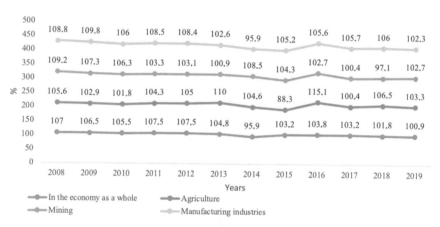

Fig. 2 Dynamics of the labor productivity index by sectors of the economy for 2008–2019. *Source* Compiled by the authors based on [3]

There are different interpretations characterizing human capital and the effective performance of the workforce (Fig. 2) [10–12].

An important factor influencing the efficiency of human capital is the level of wages. Therefore, we consider it advisable to review the most remunerative types of economic activity in Russia (Fig. 4).

Figure 4 shows a significant difference in income in different sectors of the economy. Financial activities were the most paid (80,286 rubles). Agriculture (22,724 rubles) and restaurant business (22,041) were the least paid activities in 2019.

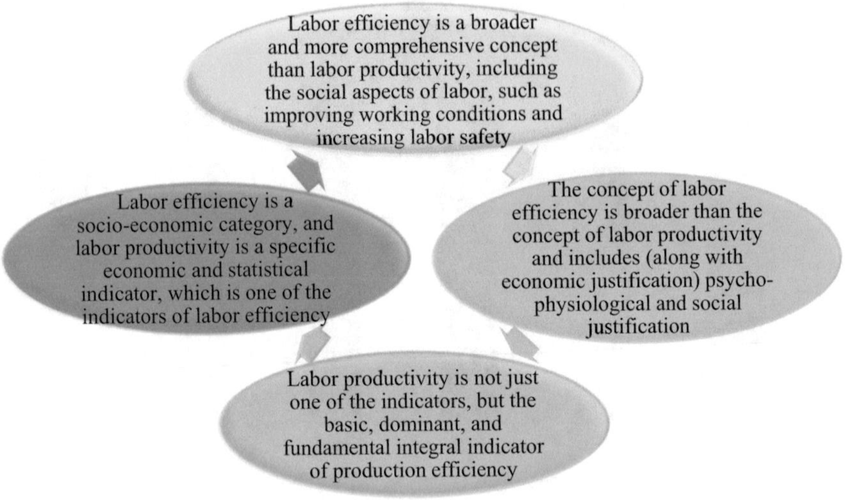

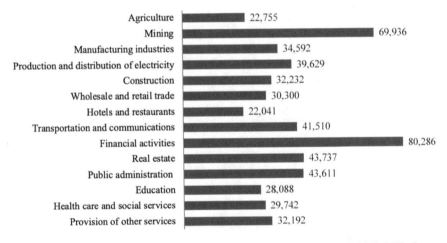

Fig. 3 Correlation of the economic categories of labor efficiency and labor productivity. *Source* Compiled by the authors based on [10, 12]

Agriculture — 22,755
Mining — 69,936
Manufacturing industries — 34,592
Production and distribution of electricity — 39,629
Construction — 32,232
Wholesale and retail trade — 30,300
Hotels and restaurants — 22,041
Transportation and communications — 41,510
Financial activities — 80,286
Real estate — 43,737
Public administration — 43,611
Education — 28,088
Health care and social services — 29,742
Provision of other services — 32,192

Fig. 4 Average wage in the Russian Federation by type of economic activity, in 2019, RUB. *Source* Compiled by the authors based on [3]

The analysis has shown that it is necessary to improve the methodological apparatus of the wage calculation system to equalize the income of employees in the agricultural industry. This necessity is because the future of the AIC depends on the effective and coordinated functioning of agricultural production. Therefore, to solve the above problem, the authors propose to use a fundamentally new system of remuneration, which is based on a grading system. The system includes six grades. The proposed system is used to encourage employees to improve their skills and expand their professional profiles. The bonuses for qualification range from 2,070 rubles for the second category in the first grade to 20,500 rubles for the sixth category in the sixth grade. The first category is assigned to students hired for an internship period. The salary of the employees of the first category includes the following:

$$S_{1c} = S_{min} + P_{wc} + P_{ns}, \tag{1}$$

where:

S_{1c}—salary of the employees of the first category per month, RUB;

S_{min}—minimum salary for the range of the grade in which the corresponding profession is included, RUB. The distribution of professions by grades is presented in the regulation "On the tariff part of employees' wages";

P_{wc}—additional payment for the prevailing working conditions, RUB;

P_{ns}—extra pay for night shifts, RUB.
    The second category is assigned to employees when they are hired. The salary of the employees of the second category includes the following:

$$S_{2c} = S_{min} + B_{2c} + P_{wc} + P_{ns}, \tag{2}$$

where:

S_{2c}—salary of the employees of the second category per month, RUB;

S_{min}—minimum salary for the range of the grade in which the corresponding profession is included, RUB;

B_{2c}—qualification bonus for the 2nd category of the grade, which includes the corresponding profession, RUB;

P_{wc}—additional payment for the prevailing working conditions, RUB;

P_{ns}—extra pay for night shifts, RUB.
    The third category is assigned to employees under the following conditions:

- At least 12 months of uninterrupted experience in the second category;
- At least four operations in the current workplace.

    The salary of the employees of the third category includes the following:

$$S_{3c} = S_{min} + B_{3c} + P_{wc} + P_{ns}, \tag{3}$$

where:

S_{3c}—salary of the employees of the third category per month, RUB;

S_{min}—minimum salary for the range of the grade in which the corresponding profession is included, RUB;

B_{3c}—qualification bonus for the 3rd category of the grade, which includes the corresponding profession, RUB;

P_{wc}—additional payment for the prevailing working conditions, RUB;

P_{ns}—extra pay for night shifts, RUB.

The fourth category is assigned to employees under the following conditions:

- At least 18 months of uninterrupted experience in the third category;
- At least six operations in the current workplace.

The salary of the employees of the fourth category includes the following:

$$S_{4c} = S_{min} + B_{4c} + P_{wc} + P_{ns}, \qquad (4)$$

where:

S_{4c}—salary of the employees of the fourth category per month, RUB;

S_{min}—minimum salary for the range of the grade in which the corresponding profession is included, RUB;

B_{4c}—qualification bonus for the 4th category of the grade, which includes the corresponding profession, RUB;

P_{wc}—additional payment for the prevailing working conditions, RUB;

P_{ns}—extra pay for night shifts, RUB.

The fifth category is assigned to employees under the following conditions:

- At least 24 months of uninterrupted experience in the fourth category;
- At least eight operations in the current workplace;
- Additional conditions for brigade leaders: at least three employees must be certified for each operation of the team, each of whom must know at least three operations.

The salary of the employees of the fifth category includes the following:

$$S_{5c} = S_{min} + B_{5c} + P_{wc} + P_{ns}, \qquad (5)$$

where:

S_{5c}—salary of the employees of the fifth category per month, RUB;

S_{min}—minimum salary for the range of the grade in which the corresponding profession is included, RUB;

B_{5c}—qualification bonus for the 5th category of the grade, which includes the corresponding profession, RUB;

P_{wc}—additional payment for the prevailing working conditions, RUB;

P_{ns}—extra pay for night shifts, RUB.
    The sixth category is assigned to employees under the following conditions:

- At least 36 months of uninterrupted experience in the fifth category;
- At least 11 operations in the current workplace;
- At least 3 for employees,
- At least 15 for brigade leaders.

    The salary of the employees of the sixth category includes the following:

$$S_{6c} = S_{min} + B_{6c} + P_{wc} + P_{ns},\qquad(6)$$

where:

S_{6c}—salary of the employees of the sixth category per month, RUB;

S_{min}—minimum salary for the range of the grade in which the corresponding profession is included, RUB;

B_{6c}—qualification bonus for the 6th category of the grade, which includes the corresponding profession, RUB;

P_{wc}—additional payment for the prevailing working conditions, RUB;

P_{ns}—extra pay for night shifts, RUB.
    The employee loses all eligibility for the bonuses, or the bonuses may be partially reduced to the level of any lower category under the following conditions:

- Decline in qualifications: loss of skills to perform the work, systematic failure to perform shift assignments;
- Improper work performance;
- Violation of the internal labor regulations.

    The evaluation of qualifications and summarizing the results is made in the "Qualification evaluation card of the main or assistant employees." The bonus is paid monthly in proportion to the time actually worked until new information on the employee's promotion is received [2, 6].
    If an employee loses the right to receive a bonus in full or in part, the head of the structural subdivision provides the Department of Labor Regulation, Compensation, and Benefits with a "List for the removal or reduction of the qualification bonus." An employee who has lost the right to receive an allowance in whole or in part shall be familiarized with the order and sign it.
    Employees who receive the title of "Best in Profession" based on the results of professional skill competitions are set the next level of qualification allowance ahead of schedule.

When an employee is transferred to another profession or another subdivision, the qualification bonus is set by the decision of the head of the structural subdivision to which the employee is transferred. In this case, the qualification bonus should not exceed the level previously set for the previous profession [13, 16].

The improved system of human capital management in the organization is built in such a way as to stimulate employees to get an education, improve their skills, work for the company for a long time with a good performance, strive to move up the career ladder, and engage in self-improvement.

4 Conclusion

To summarize, we would like to draw attention to the fact that the agricultural industry is a strategic sector of the economy. The outcome of production activities largely depends on the qualifications and motivation of employees. Therefore, when designing programs to support low-skilled workers, the focus should be made on increasing the incomes of employees, which would ultimately benefit not the material well-being of employees and their families but also boost macroeconomic indicators.

References

1. Becker GS (1994) Human capital: a theoretical and empirical analysis, 3rd edn. University of Chicago Press, Chicago, IL. (Original work published 1964)
2. Buranshina N, Ivanova N (2011) Human capital of a municipal entity: notion and structure. Sci Bull Ural Acad Public Serv Polit Sci Econ Sociol Law 4(17):160–169. Accessed from http://vestnik.uapa.ru/en/issue/2011/04/24/
3. Federal State Statistics Service. (n.d.). Russia in numbers. Accessed from https://rosstat.gov.ru/folder/210/document/12993
4. Glotko AV, Okagbue HI, Utyuzh AS, Shichiyakh RA, Ponomarev EE, Kuznetsova EV (2020) Structural changes in the agricultural microbusiness sector. Entrep Sustain Issues 8(1):398–412. https://doi.org/10.9770/jesi.2020.8.1(28)
5. Glotko AV, Polyakova AG, Kuznetsova MY, Kovalenko KE, Shichiyakh RA, Melnik MV (2020) Main trends of government regulation of sectoral digitalization. Entrep Sustain Issues 7(3):2181–2195. https://doi.org/10.9770/jesi.2020.7.3(48)
6. Gurban IA, Krutikova MA (2011) The state of educational capital as a system-forming factor in the formation of human capital. In: Bulletin of Ural Federal University. Series: economics and management, vol 4, pp 136–148
7. Kuznetsova IG, Polyakova AG, Petrova LI, Artemova EI, Andreeva TV (2020) The impact of Human capital on Engineering Innovations. Int J Emerg Trends Eng Res 8(2):333–338. https://doi.org/10.30534/ijeter/2020/15822020
8. Kuznetsova I, Okagbue H, Plisova A, Noeva E, Mikhailova M, Meshkova G (2020) The latest transition of manufacturing agricultural production as a result of a unique generation of human capital in new economic conditions. Entrep Sustain Issues 8(1):929–944. https://doi.org/10.9770/jesi.2020.8.1(62)

 9. Loshkareva E, Luksha P, Ninenko I, Smagin I, Sudakov D (2018) Skills of the future: How to thrive in the complex new world. Global Education Futures; WorldSkills Russia, Moscow, Russia. Accessed from https://worldskills.ru/assets/docs/media/WSdoklad_12_okt_eng.pdf
10. Mensch G (2000) Technologie- und Innovationsmanagement in diversifizierten Unternehmen. In: Hinterhuber HH (ed) Die Zukunft der Diversifizierten Unternehmen. Munchen, Deutschland, pp 185–200
11. Mezger CL (2004) Humancapital-der Schlussel fur Wirtschaftliches Wachstum? Ibidem-Verlag, Stuttgart, Deutschland
12. Nonaka I, Takeuchi H (1995) The knowledge-creating company: How Japanese companies create the dynamics of innovation. Oxford University Press, New York, NY; Oxford, UK.
13. Nureyev R (2000) Development theories: new models of economic growth (contribution of human capital). Econ Issues 9:126–145
14. Rudoy EV, Shelkovnikov SA, Matveev DM, Sycheva IN, Glotko AV (2015) "Green box" and innovative development of agriculture in the Altai territory of Russia. J Adv Res Law Econ 6(3):632–639. https://doi.org/10.14505/jarle.v6.3(13).17
15. Shultz T (1968) Human capital in the International Encyclopedia of the social sciences, vol 6. Macmillan, New York, NY
16. Shvakov EE (2021) How to train highly qualified personnel? The challenge for state policy in the Altai Krai. In: Maximova SG, Raikin RI, Chibilev AA, Silantyeva MM (eds) Advances in natural, human-made, and coupled human-natural systems research, vol 3. Springer, Cham, Switzerland
17. Stadnik AT, Shelkovnikov SA, Rudoy YV, Matveev DM, Maniehovich GM (2015) Increasing efficiency of breeding dairy cattle in agricultural organizations of the Russian Federation. Asian Soc Sci 11(8):201–206

Innovative Calculating of Products in Industry Enterprises

Igor E. Mizikovsky⬚, Elena P. Polikarpova, Victor P. Kuznetsov⬚,
Ekaterina P. Garina⬚, and Elena V. Romanovskaya⬚

Abstract In this work, the main aim is to improve the quality of the information field for making economic decisions. This is done by developing and introducing into accounting process an innovative methodology for direct costs, as well as calculating, on their basis, the cost of production of enterprises in the animal husbandry industry. An instrumental space of the research methodology is structured on the basis of the complex use of a set of theoretical and empirical methods, including: observation, objectification of resources in the accounting and calculation system and their semantic analysis; identification of cost accounting objects, their decomposition and structural classification, verification and validation; systematization and subsequent recording of direct costs in analytical accounting registers; graphic visualization of results. Along with them, we used: structural–functional approach, which allows formalization of information-tool space of cost management and possibilities of adjusting their normative values and functional modeling. Theoretical and methodological bases of accounting and calculating work are considered. The result of the research was developed by the authors' innovative method of accounting for direct costs and calculating the cost of products of the industry in question on the basis of accounting (management) "Directcosting" system, which allows to significantly improve the quality of accounting and calculating process of investigated enterprises, improve the efficiency of management decision-making and efficiency

I. E. Mizikovsky (✉)
Lobachevsky State University of Nizhny Novgorod, Nizhny Novgorod, Russia

E. P. Polikarpova
Ryazan State Agrotechnological University Named After P.A. Kostychev, Ryazan, Russia
e-mail: Dikusar85@mail.ru

V. P. Kuznetsov · E. P. Garina · E. V. Romanovskaya
Minin Nizhny Novgorod State Pedagogical University, Nizhny Novgorod, Russia
e-mail: kuzneczov-vp@mail.ru

E. P. Garina
e-mail: e.p.garina@mail.ru

E. V. Romanovskaya
e-mail: alenarom@list.ru

of business activity of economic entities in the sphere of livestock husbandry. An introduction of innovative accounting and calculation methods allows to correlate the costs and benefits of livestock enterprises, which largely contributes to the inflow of investments into the sector and its financial stability.

Keywords Cost price · Direct costs · Cost accounting method · Calculation · "Directcosting" · Decision-making

JEL Codes Q10 · Q11

1 Introduction

Development of economic relations in the context of increasing competition in all sectors of agricultural production implies the objective necessity of implementing a balanced enterprise cost management policy for the production and final results, including providing a flexible variation of the sales prices [1]. A solution to the problem of maintaining "low prices" largely depends on "cost leadership," which implies a reasonable and well-planned reduction in the level of the latter. The need for enterprise's efforts to permanently reduce costs through a consistent decrease in the volume of resource consumption of an economic entity at all stages of value creation is emphasized in works of [2–4] which involves a debugging cost management mechanism that allows, among other things, to structure a sustainable implicative dependence Q of cost reduction (CR), flexible pricing policy (FPP), and improve competitiveness (IC) (Formula 1):

$$Q: \ CR \ \Rightarrow FPP \Rightarrow IC \tag{1}$$

An important element of this mechanism is the accounting and calculating function of accounting activities of an enterprise. Where information on costs of production and sale of products play a key role in enterprise management [5, 6].

Current industry is one of the leading in the agro-industrial sector. The research concluded that the problem of complete and qualitative formation of information on the actual cost of production and, therefore, on the received financial remains unsolved due to: (a) duration of production cycles, assuming the costs of future periods; (b) flow of value creation, which involves the realization of biological processes that do not coincide with calendar reporting periods.

The use of innovative methods of cost calculation and financial results of the industry due to a number of technological features. They are: the presence of perishable goods and, therefore, the need for additional expenditure on the maintenance of the equipment for its storage and prompt delivery to the point of sale; the fact that cows give milk only after calving and milking should be carried out on a daily basis without interruption, at least twice a day; litter production process has a long-term nature of time and suffers from uneven yields.

Fig. 1 The ratio of
approaches for attributing
direct costs to the types of
dairy products of studied
enterprises. *Source* compiled
by the authors

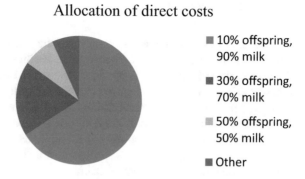

It is quite obvious that technological limitations largely determined the application in the accounting practices of the studied enterprises of an accounting and calculation model based on the conventional ratio of direct costs of milk production and offspring. In most accounting practices of the studied economic entities, it is assumed that 10% of the costs incurred are attributed to the cost of offspring; 90%—for the cost of milk. Figure 1 shows the proportions of approaches for solving this important calculation problem.

An existing accounting and calculation system does not take into account the irreversible, permanent, and objective natural nature of biological processes. It includes the cost of producing offspring, which is lasting about 9 months, and can be attributed to two annual calendar periods following each other. In the accounting of studied enterprises, the costs for the past year are not taken into account. Instead, the calculation includes the costs of a larger number of months of the current year that are carried out after calving, which leads to a distortion of the calculation result in the direction of a significant overestimation of the level of actual production costs. Dropping out from the calculation are the costs of keeping the heifer in the process of forming the cost of the offspring at first pregnancy when she is included in the rearing animals, and not in the main herd. Corresponding costs are attributed entirely to the cost of the heifer gain without further dividing their amount and attributing part of it to a received offspring.

All these facts, obtained by the authors in the course of the study, confirm the need to use innovative calculation approaches in order to improve the quality of the generated production cost and the financial result of the business activity of dairy cattle breeding. Our objectives in this study include substantiating a set of techniques and methods of economic calculations that make it possible to achieve a designated goal.

2   Materials and Method

The research methodology is based on scientific approaches that imply long-term observation of the dynamics of the state of production resources at enterprises and their semantic analysis. Results of which are used to objectify and identify specific types of costs in the accounting and calculation system; their decomposition and structural classification, systematization and subsequent fixation of direct costs in analytical accounting registers; graphic visualization of the results [6–8]. Along with them, we used: structural–functional approach, which allows formalization of information-tool space of cost management and possibilities of adjusting their normative values and functional modeling.

Theoretical and methodological approaches. According to the authors, in order to solve this problem, it is necessary to apply a set of techniques and methods of the highest quality reflection of direct costs, which is feasible by using the method of accounting for costs and calculating the cost of production "Directcosting" [9–12]. Scientists emphasize that the implementation of this method is based on "the classification of costs for fixed (periodic) (indirect–by authors) and variables (for a product) (direct–by authors)." The authors of many studies in this subject area base their professional judgments on these positions. Thus, the presence of analytical accounting registers, which are separately identified direct and indirect costs, allows us to refer to the cost of production. The authors note, "Only direct production costs and all indirect costs are considered costs of the current period and are written off as the cost of sales." This statement should be supplemented, as when using the method "Directcosting" in the cost of sales is written off not only the indirect costs related to the class of product but also on marketing costs (non-manufacturing costs).

3   Results

According to the authors, the following principles should be used as the basis for innovative calculation of dairy products:

- Costs must be charged during the period that begins with the month insemination of cows (heifers) and ends next month insemination;
- Distribution by type of product costs—milk and offspring, should be implemented monthly;
- Costs of maintaining pregnant cows and heifers in terms of offspring formation at the end of the year must be taken into account during work process;
- Part of the costs of maintaining a pregnant heifer related to the formation of offspring should be taken into account separately;
- To organize analytical accounting of costs by groups of cows, formed according to the month of insemination or calving;
- Milk costing should be done on a monthly basis and the cost of litter—at the end of the month of calving.

Fig. 2 Context diagram of an innovative cost accounting and costing model for dairy products. *Source* compiled by the authors

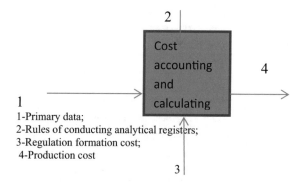

1-Primary data;
2-Rules of conducting analytical registers;
3-Regulation formation cost;
4-Production cost

Based on these principles, it is proposed to present an innovative model of cost accounting and calculating the cost of dairy products in the form of the following context diagram (Fig. 2).

The basis for making entries in the accounting registers of costs for dairy products of the studied enterprises is a set of primary documents approved by an economic entity. They are confirming the basis and volume of costs incurred, the movement of which is regulated by the approved accounting document flow scheme. It is quite obvious that it is necessary to incorporate information verification procedures into the complex of accounting and calculation, information, and instrumental space, both received for processing and generated at the output of the system. Verification is an iterative process, the completion criterion of which is the compliance of the information results with the established parameters, i.e., a given level of validity.

As stated by authors, a grouping of expenses expediently carried out in the context of the following products: dairy cows (calf before pregnancy); dry cows; calf heifers.

Objectification in the information field of cost accounting and calculating the cost of dairy products involves the following calculations (2, 3):

$$S_J = \sum_{i=1}^{N} a_{ij(2)}, \text{ where } S_J \text{ is the cost of the } J\text{-th object}$$
$$a_i j\text{-th cost item of } i\text{-th object} \tag{2}$$

$$S_K = \sum_{i=1}^{N} S_{J(3)}, \text{ where } S_K \text{ is the cost of production} \tag{3}$$

Costing accounting object "cash pregnant cows" that is generated by "Direct-costing," can be represented as follows (Table 1).

Table 1 Calculation of the cost of production for April 2020 of the enterprise LLC "XXXXX"
Product name: Milk cows, code: 8254, date: 06.04.2020 (fragment)

No. p/n	Cost items	Amount, RUB
A	1	2
Direct costs		
1	Material resources, including:	
1.1	Means of protection and animals	40,000
1.2	Animal feed, including:	
1.2.1	– acquired and own production of previous years	20,000
1.2.2	– own production of the current year	30,000
1.3	Petroleum products	9,000
1.4	Fuel and energy for technological purposes	11,000
1.5	Work and outsourced services	9,000
Total material resources		110,000
2	Wages, including:	
2.1	Basic	60,000
2.2	Additional	20,000
2.3	Natural	–
2.4	Other payments	10,000
Total remuneration		90,000
3	Social contributions	30,000
4	Other direct costs	10,000
Total direct costs		240,000

Source compiled by the authors

Consolidated statement of accounting for direct costs incorporates the content of the calculations, which is formed by the above method and has the following form (Table 2).

It should be noted that according to the consumption of feed produced in the current year is estimated at the planned cost, and at the end of the year, according to the calculation of the cost price, their planned cost is brought to the actual cost. A purchased feed is written off at the purchase price, including the cost of delivery to the farm. Costs under the item of fuels and lubricants spent on the performance of technological and transport work for the maintenance of production, etc., (N) are calculated according to the following formula (4):

$$N = C + P \qquad (4)$$

where C—the purchase price; P—the cost of delivery to the farm.

The item "Fuel and energy for technological purposes" is formed in accordance with the cost of purchased fuel from all types spent on maintaining technological

Table 2 Statement of account for direct costs on production in April 2020 of company LLC "XXXXX" rub. Date: 06.04.2020 (fragment)

No. p/n	Name of of product	The code of product	Material resources	Wages	Deductions for social needs	Other	Total C.7 = c.3 + c.4 + c.5 + c.6
A	1	2	3	4	5	6	7
1	Milk cows	8,254	110,000	90,000	30,000	10,000	240,000
2	Dry cows	8,250	80,000	30,000	10,000	5,000	125,000
3	Pregnant heifers	8,251	60,000	18,000	6,000	2,000	86,000
Total			250,000	138,000	46,000	17,000	451,000

Source compiled by the authors

processes in the production and acquisition of all types of energy consumed for technological and other industrial, economic, and administrative needs. Also, the cost of third-party work and services is reflected in the above calculation only insofar as that can be directly attributed to the cost of certain products. It should be noted that the rest of these costs are usually charged to general production costs.

The item "Remuneration" is formed from "cash and in-kind payments, which have the nature of remuneration and are included in the cost of products (works, services), to workers of various categories directly involved in the technological process of a corresponding production." As practice shows accounting calculations on payment of dairy farming workers, that a significant proportion of payments are related to overtime work, work on weekends and holidays.

The use of "Directcosting" makes it possible to calculate the margin of the financial result of CF (5) and the norm of the marginal profit HMB (6):

$$MB = B - S_K \tag{5}$$

$$HMB = \frac{MB}{B} * 100\% \tag{6}$$

where B—Revenue from sales of products.

HMB index reflects the degree of influence of the proceeds from sales to the amount of the margin of a financial result. So, upon receipt of proceeds in the amount of 6,290,000 rubles at the enterprise LLC XXXXX, the marginal financial result (margin profit) will be 178,000 rubles, the margin profit rate is 28.3%.

4 Discussion

An innovative model for the formation of the cost of each type and, in general, the gross output of dairy cattle breeding at the studied enterprises should be carried out on the basis of uniform principles and rules enshrined in local acts in strict accordance with the list of costs established by the current regulatory documents. At the researched enterprises it is advisable to keep records at direct costs since this method allows them to obtain the most reliable set of information about expended resources and receive margin financial results.

A company independently chooses a method for determining the cost of production and marginal profit, taking into account the specifics of a particular production, and secures it in accounting policies. During this procedure, it is possible to use the method "Directcosting," which allows not only significantly improve the accuracy and reliability of economic calculations but also to effectively fill the content of the information field of decision management solutions. Also, it forms employees' objective view of what resources are used directly in the creation of use-value of the product (direct costs) and which only allow the implementation of this process (indirect costs). As an important direction for future research, the author considers the study of the possibilities of using artificial intelligence and processing big data, as well as the development of a methodology for assessing the cost of stocks in work progress for dairy products.

5 Conclusion

Formation of conditions for improving the competitiveness of agro-industrial enterprises requires the introduction of innovative managing methods of costs and their results at all stages of value creation. An important sector of the agricultural sector is the breeding of dairy cattle and the production of raw milk. The authors of the study showed that enterprises of two large areas of Russia still have an unsolved problem. It is the problem of complete and high-quality formation of information about the actual cost of production and obtained financial results. Structuring the information space for cost accounting and calculating the cost of dairy products involves clarification of the cost concept. Concerning this area, its interpretation is an integrated indicator that quantitatively characterizes the process of resource consumption in the value stream used for internal monitoring of costs, search for ways to reduce them, and determine economic benefits.

Innovations in the area of accounting and the calculation of dairy cattle suggest the choice of accounting method to organize the direct costs in order to create high-quality information base management decisions and generate the most complete and reliable on-farm reporting. Studies have shown the dominance of conventional approaches in the accounting and calculation process, which do not allow to solve this problem [13, 14]. Under these conditions, according to the authors, the best solution is to use "Directcosting" method. The cost of products formed through its application literally "highlights" information about the resources that directly form value. It is most relevant for the development of cost reduction strategies. Also, it allows us to find the most advantageous combinations of price and volume, pursue an effective price policy, simplify rationing, planning, accounting, and control. It should be noted that the cost of production, formed by the "Directcosting" method, allows us to significantly expand the information field of the costs and results produced by indicators of the marginal financial result and the rate of marginal profit.

References

1. Tetushkin VA (2016) Estimation of competitiveness of Russian business under adverse conditions. Financ Credit 48:46–64
2. Garina EP, Kozlova EP, Sevryukova AA (2016) Study of alternative strategies and methodological tools for the development of complex systems in the context of the product being created. Azimuth Sci Res Econ Adm 5, 2(15):58–62
3. Goryanskaya IV (2019) Development of competitive advantages in the system of strategic management. Econ Railw 10:15–21
4. Oganesyan AS, Ignatieva TA, Kuznetsov AO (2017) Actual problems of ways and methods of reducing the cost of production. Mod Econ Success 6:192–195
5. Atkinson EA, Banker RD, Kaplan RS, Jung MS (2016) Management accounting. Translated from English. I.D. Williams, Moscow
6. Serebryakova TYu, Biryukova OA, Kondrashova OR (2018) Institutional approaches to the classification of management accounting. Int Acc 2(440), 21: 204–212
7. Nikolaeva SA (2000) Incomes of the organization: the practice, the theory of perspective. Research Press, Mocow 224, p 69
8. Morozova NS, Merkulova EYu (2016) Analysis of the cost of production. Socio-economic Process 8(11):66–71
9. Mizikovsky IE (2005) Technology and organization of management accounting at the enterprise. Monograph. Nizhny Novgorod: Publishing house of NNSU named after N.I. Lobachevsky, p 150
10. Krasikova ET (2018) Problems of attribution of indirect costs in the cost of production in the dairy industry. Journal of Kolomna Institute (branch) of the Moscow Polytechnic University. Series: Social and Human Sciences, vol 13, pp 144–153
11. Mizikovsky IE (2015). An innovative model for the distribution of indirect costs of an industrial enterprise. Bulletin of the Nizhny Novgorod University named after N.I. Lobachevsky. Series: Social Sciences Publishing House of Nizhny Novgorod State University named after N.I. Lobachevsky, 2, vol 1, pp 39–45
12. Mukhina ER (2016) Study of the issue of distribution of indirect costs in electrical enterprises. Bull Mod Sci 1:92–94
13. Romanovskaya EV, Kuznetsov VP, Andryashina NS et al (2020) Development of the system of operational and production planning in the conditions of complex industrial production. In: Popkova E, Sergi B (eds) Digital economy: complexity and variety vs. rationality. ISC 2019. Lecture notes in networks and systems, vol 87. Springer, Cham
14. Yashin SN, Kuznetsov VP, Okhezina GM (2016) Assessment of the prospects and feasibility of process innovations at an industrial enterprise: monograph. Minin University, Nizhny Novgorod, p 152

A Critical Look at Circular Agriculture from a Perspective of Sustainable Development (Conclusion)

The results of the first volume of the book are controversial. On the one hand, scientific and methodological foundations of agricultural sustainability were formed, and the global trend of increasing compliance with the criteria of sustainability of agricultural economies was revealed. On the other hand, the SDGs in agriculture are implemented in an unsystematic manner, which is the reason for the impossibility of achieving full sustainability of agriculture (meeting all criteria at once).

The first volume provides a critical perspective on circular agriculture from the perspective of sustainability. It shows that the minor concessions to the environment applied in circular agricultural practices are insufficient to ensure the sustainability of the agricultural economy. The new philosophy of sustainable development imposes greater demands on agriculture, which must conserve natural resources and act as a source of their improvement.

It turns the perception of agriculture and its role in the economy upside down. It is no longer natural resources that are being used for agriculture, but rather the opposite—agriculture is being rebuilt to improve natural resources. This change marks the beginning of the transition from circular agricultural practices to restorative land use. It changes the adaptation to climate change to the support of reverse climate change (restoration) and replaces soil conservation with increased soil fertility. The first volume of the book sufficiently elaborates and systematizes circular practices. However, regenerative agriculture is a fundamentally new scientific concept.

The vagueness of this concept, its essence, technologies, and prospects for its achievement is a research gap, which is filled in the second volume of this book. The second volume of this book scientifically elaborates and studies the international practical experience of regenerative agriculture. Moreover, it develops scientific and methodological recommendations and applied solutions for the transition to regenerative agriculture.

E. G. Popkova and B. S. Sergi (eds.), *Sustainable Agriculture*,
Environmental Footprints and Eco-design of Products and Processes,
https://doi.org/10.1007/978-981-16-8731-0

Printed in the United States
by Baker & Taylor Publisher Services